Spatial Analysis Techniques Using MyGeoffice®

João Garrott Marques Negreiros
University of Saint Joseph, China & Macau Polytechnic Institute, China

A volume in the Advances in Geospatial Technologies (AGT) Book Series

IGI Global
DISSEMINATOR OF KNOWLEDGE

Published in the United States of America by
IGI Global
Engineering Science Reference (an imprint of IGI Global)
701 E. Chocolate Avenue
Hershey PA, USA 17033
Tel: 717-533-8845
Fax: 717-533-8661
E-mail: cust@igi-global.com
Web site: http://www.igi-global.com

Library of Congress Cataloging-in-Publication Data

Names: Negreiros, Joao Garrott Marques, author.
Title: Spatial analysis techniques using MyGeoffice? / by Joao Garrott Marques Negreiros.
Description: Hershey PA : Engineering Science Reference, [2018]
Identifiers: LCCN 2017015198| ISBN 9781522532705 (hardcover) | ISBN 9781522532712 (ebook)
Subjects: LCSH: Geographic information systems. | Geographic information systems--Mathematics. | Spatial analysis (Statistics) | Geospatial data. | Image processing--Digital techniques
Classification: LCC G70.212 .N44 2018 | DDC 910.285/53--dc23 LC record available at https://lccn.loc.gov/2017015198

This book is published in the IGI Global book series Advances in Geospatial Technologies (AGT) (ISSN: 2327-5715; eISSN: 2327-5723)

British Cataloguing in Publication Data
A Cataloguing in Publication record for this book is available from the British Library.

Advances in Geospatial Technologies (AGT) Book Series

ISSN:2327-5715
EISSN:2327-5723

Editor-in-Chief: Nilanjan Dey, Techno India College of Technology, India

MISSION

The geospatial technology field is a fast-paced, high growth industry that is involved in a variety of fields, including military planning, public health, land use, environmental protection, and Google Earth mapping. With such a diverse body of applications, the research in geospatial technologies is always evolving and new theories, methodologies, tools, and applications are being developed. **Advances in Geospatial Technologies (AGT) Book Series** is a reference source and outlet for research that discusses all aspects of geographic information, including areas such as geomatics, geodesy, GIS, cartography, remote sensing, and other areas. Because geospatial technologies are so pervasive in such a variety of areas, AGT also includes books that address interdisciplinary applications of the technologies.

COVERAGE

- Public Sector Use of Geospatial Technologies
- Environmental Monitoring
- Geographical information systems
- Military Security and Geospatial Technologies
- Disaster Informatics
- Geomatics
- Public Health Applications of Geospatial Technologies
- Geospatial Analysis
- Land Surveying
- Remote Systems

IGI Global is currently accepting manuscripts for publication within this series. To submit a proposal for a volume in this series, please contact our Acquisition Editors at Acquisitions@igi-global.com or visit: http://www.igi-global.com/publish/.

The Advances in Geospatial Technologies (AGT) Book Series (ISSN 2327-5715) is published by IGI Global, 701 E. Chocolate Avenue, Hershey, PA 17033-1240, USA, www.igi-global.com. This series is composed of titles available for purchase individually; each title is edited to be contextually exclusive from any other title within the series. For pricing and ordering information please visit http://www.igi-global.com/book-series/advances-geospatial-technologies/73686. Postmaster: Send all address changes to above address.

Titles in this Series

For a list of additional titles in this series, please visit:
https://www.igi-global.com/book-series/advances-geospatial-technologies/73686

Emerging GIS Applications for Emergency and Disaster Management
Shruti Mantri (Kirti College, India) and Seema Purohit (Kirti College, India)
Information Science Reference • ©2017 • 117pp • H/C (ISBN: 9781683180098) • US $165.00

Volunteered Geographic Information and the Future of Geospatial Data
Cláudio Elízio Calazans Campelo (Federal University of Campina Grande, Brazil) Michela Bertolotto (University College Dublin, Ireland) and Padraig Corcoran (Cardiff University, UK)
Information Science Reference • ©2017 • 357pp • H/C (ISBN: 9781522524465) • US $205.00

Remote Sensing Techniques and GIS Applications in Earth and Environmental Studies
Abhisek Santra (Haldia Institute of Technology, India) and Shreyashi Santra Mitra (Haldia Institute of Technology, India)
Information Science Reference • ©2017 • 335pp • H/C (ISBN: 9781522518143) • US $185.00

Handbook of Research on Geographic Information Systems Applications and Advancements
Sami Faiz (LTSIRS Laboratory, University of Tunis El Manar, Tunisia) and Khaoula Mahmoudi (LTSIRS Laboratory, University of Tunis El Manar, Tunisia)
Information Science Reference • ©2017 • 650pp • H/C (ISBN: 9781522509370) • US $315.00

Geo-Intelligence and Visualization through Big Data Trends
Burçin Bozkaya (Sabanci University School of Management, Turkey) and Vivek Kumar Singh (Rutgers, The State University of New Jersey, USA & Massachusetts Institute of Technology, USA)
Information Science Reference • ©2015 • 348pp • H/C (ISBN: 9781466684652) • US $225.00

Handbook of Research on Emerging Digital Tools for Architectural Surveying, Modeling, ...
Stefano Brusaporci (University of L'Aquila, Italy)
Engineering Science Reference • ©2015 • 829pp • H/C (ISBN: 9781466683792) • US $525.00

For an enitre list of titles in this series, please visit:
https://www.igi-global.com/book-series/advances-geospatial-technologies/73686

IGI Global
DISSEMINATOR OF KNOWLEDGE

701 East Chocolate Avenue, Hershey, PA 17033, USA
Tel: 717-533-8845 x100 • Fax: 717-533-8661
E-Mail: cust@igi-global.com • www.igi-global.com

This book is dedicated to my father João Garrott Negreiros and to my mother Ana Rosa Marques Cabaço, who taught me that the best kind of knowledge to have is that which is learned for its own sake.

Success consists of going from failure to failure without loss of enthusiasm.
-Winston Churchill (1874-1965)

Table of Contents

Preface

This book introduces myGeoffice.org, an innovative WWW platform for spatial analysis, where its architecture, interface, innovations and technical details are presented. It is designed primarily for teaching and learning purposes, particularly at the high-school and university level on a wide range of introductory spatial subjects such as geography, geology, natural sciences, agriculture, geomatics, meteorology and hydrology. It is commonly accepted that the field of Geographical Information Systems (GIS) is an integrated technology and the result of the fusion of various systems, moving from a cartography view to one of virtual reality modeling. At first, spatial analysis shifted from a classical exploratory and descriptive statistical framework to a spatial analysis methodology embracing only non-spatial data. Overlaying, networking, measuring, raster and vector buffering, spatial searching, raster spreading or contouring operations became the next phase in a spatial common sense approach (analytical cartographic analysis). In its present second phase, advanced spatial analysis is needed for interpolation, extrapolation, stochastic simulation and space-time analysis, where the use of both topology and non-spatial information is mandatory. Without it, spatial analysis loses its identity. After all, spatial data are special (Penn State, 2014).

As well, conventional statistics must be reformulated to properly account for spatial autocorrelation (Chun & Griffith, 2013) and spatial heterogeneity within georeferenced data. For instance, if autoregressive residuals reveal a medium-strong spatial autocorrelation then any omitted variable, within the initial regression model, can be significant.

Statistical packages are a possible software implementation to relate GIS with existing statistical data. SAS-GIS®, Glim®, Systat® or Minitab® are some of those packages (Negreiros, Painho, Aguilar, & Aguilar, 2010). The difficulty involved in this solution is that statistical packages do not provide the GIS required functionality because they were not developed to handle spatial data and the specific structure of spatial data switching back and forth. The requirement of learning multiple packages can also be problematic.

A second implementation strategy is to add statistical functionality into commercial GIS by modular integration such as Geostatistical Analyst of ESRI®. A major advantage of ESRI® software is the ready availability of spatial analysis functionality. Curiously, users prefer to buy complete systems. According to Negreiros (2004), ninety-two people were surveyed in some prestigious Portuguese institutions and 80% of them felt more comfortable working with a known commercial product than unknown independent software. The common user blindly trusts the software company trademark attached (probably) because of the brand marketing that lies behind the software.

For simple descriptive analysis, the potential of spatial analysis within GIS is of little interest. Yet, regarding nearest neighbor methods, connectivity measures directly derived from the topology matrix can be extremely important. After all, road network distances may be more relevant than straight line distances. The W weighted matrix specification for autoregressive functions is another example of this greatest linkage potential. Since topological relationships are already stored within GIS, this goal can be achieved with little effort. On the other hand, major GIS module deficiencies include their financial costs and the availability of new tools depends on software developers and not on the research community.

With independent geosoftware, the major issue is the effort required to rewrite a complete package from scratch, reflecting a well-known dilemma: customization versus time/money savings of existing software modules. GRASS® and SAGA® are considered two of the finest pieces of GIS independent software today. Another highlight should be given to Geostokos® (created mainly for teaching) and SPRING®, a non-cost software developed by the National Institute for Space Research, Brazil, including a number of remarkable innovations such as an object-oriented data model for spatial data and an object-oriented data query and manipulation language.

A careful review of all pros and cons of these three software implementations (GIS, independent software and spatial extensions of statistical products) does not give much of a clear trend for the near future and, in the end, the future is partially controlled by the past. Generally, software solutions start with end-users needs. Because of the object embedding capability of Windows®, Linux® and Macintosh®, the potential of computer languages overlaps quite often, leading to the possibility, for instance, of assessing regression estimators or visualizing the same plots/maps with different computer codes. Although certain tools fit best for a specific purpose, programmers commonly use their background and field of expertise in their creative process, which is reflected in several types of implementation. In addition, as new academic methods for spatial analysis emerge, more programming is needed.

In terms of speed and performance, executable languages are certainly the programmers' best choice, especially when integrated with the standards of .NET and Java platforms. From the user's viewpoint, this fuzzy trend becomes clear. The aim for non-expert and expert users is different. The former is looking for simple descriptive statistical tools and data property summarization while the latter requires spatial auto-regression and hypothesis testing. According to Worboys (1996), GIS for utility applications requires network processing operations while geological ones demand 3D views. Thus, the same GUI should offer various possibilities for a wide range of users.

Software should encourage scientific computing visualization where statistical views are interpreted, validated and explored in a dynamic and linked graphical mapping. Still, these graph operations on the Internet are still quite difficult to achieve (Negreiros, Costa, Painho, Santos & Lopes, 2007). Another Web complexity is access to topological data (myGeoffice© suffers from these constraints, for instance).

Furthermore, the computer GIS environment should not embrace a completely automatic procedure. This means users must be aware of decisions at each stage with an explicit interface. As a result, the stages have to respect a linear sequence, where researchers begin at the beginning and go on till the user come to the end, progressing from left to right across menu choices. Therefore, what the users appreciate and interact with is the graphical user interface (GUI) because users do not care about the technical structure as long as the results are trustworthy, prompt and compatible with their operating system and hardware. Users want intuitive and easy-to-use software that give immediate results without having to read pages of documentation.

The standard Web browser, regardless of the background computer code adopted, fulfills this strategy quite well (Negreiros & Painho, 2005). It is cost free and included with any operating system. Does the user need to see a Word document? It can be viewed in a browser. Does the user need to work on an Excel spreadsheet? It can be opened in a browser. Does the user need to find a local or a network file? The browser can search for him/her. Ultimately, the user works with all available software in the same way, regardless of the location of the data and its purpose. In the beginning, the trend was to find the data with the browser (Web 1.0); at present, just run the programs/services from the Internet (Web 2.0).

This represents a significant advance in interface design because users will no longer have to worry about the software location and the technical knowledge required to connect to their data. Even today, the keyboards available on the market hold special keys for the browser such as Home, Search, Back, Mail and Refresh. The standard Web browser is the interface for computers, in general, and apps for mobiles, in particular. The capability to cover the technical implementation to the final user via WWW is essential. After struggling for years with digital information, spatial analysis needs to concentrate on what the information means and on sharing it through the World Wide Web. The shift from mainframes to Apache® intranets and wireless transmissions via PDAs, laptops or smartphones is already a reality. In effect, the wireless Web is developing rapidly nowadays enabling millions of people to access the Internet while on-the-go. Therefore, improving spatial exploration by providing the available tools of spatial analysis within the Internet becomes critical.

As an evolution of SAKWeb© (Negreiros & Painho, 2006), myGeoffice© is an Internet software that provides access to a wider GIS audience by undertaking the Web solution platform, which includes the Moran scatterplot, Kriging with measurement error and several nugget-effect solutions, declustering based on the nearest neighborhood analysis, geographical weighted regression (GWR), Dijkstra shortest path, raster image processing, index of Knot and Mantel, Kernel Gaussian density, sequential simulation and deterministic interpolators, to name a few. myGeoffice© is not a comprehensive statistical package in the traditional sense. Written for an Internet Information Server® (IIS) environment, it was developed with the philosophy of being a learning tool for individuals with limited geostatistical knowledge. myGeoffice© deals with interpolation and simulation in conjunction with spatial association measures in a Web continuum process (second menu) and many local spatial functions (first menu), as well. From this viewpoint, this element of originality and innovation can thus be appreciated.

For space visualization purposes, the near future concerns the creation of virtual reality worlds and 3D Web pages, offering additional functionally compared to 2D representations. The use of virtual reality in major fields such as medicine, commercial movies, military flight simulators and industrial applications has increased significantly.

By definition, X3D is the integration of a 3D Application Program Interface (API) in a Web environment using XML concepts (Negreiros, Painho, Aguilar & Aguilar, 2009). Having VRML 97 as its predecessor and developed by the WebW3 consortium (web3d.org/x3d) for integration of 3D graphics and multimedia content, its main goal is to allow the creation of Web based virtual worlds with interactivity. Lower bandwidth, navigation and exploration are some features of this API. More specifically, X3D supports the inclusion of graphics (polygonal geometry, lighting, materials, textures, hierarchical transformations), animations (timers, interpolators to drive continuous simulations), audio and video, user interactions (mouse picking and dragging, keyboard input), navigation (cameras, collisions, visibility detection), scripting for programming (scene access interface) and the ability to compose a single X3D scene with other assets located in the World Wide Web. Certainly, georeferenced information may benefit from these X3D concepts, which may increase users' engagement and motivation to gather spatial information. Following this trend, the plausible future of myGeoffice© as a Software as a Service (SaaS) is to generate 3D space travel based on the presently generated 2D maps.

A central component of GIS is technology management. "There is a world market for maybe five computers" was a famous statement by the IBM® chairman in 1943. He moved on a few years later and IBM® was saved. Ken Olson, from DEC® Digital Corporation, the second largest computer manufacturer in the seventies did not survive ("There is no reason why anyone would want a computer in their home"). When he woke up, Open VMS® was already dead and DEC® had been bought by Compaq® Corporation. Bill Gates recovered from the "640K is enough for everyone" statement on time.

It is expected that myGeoffice© will not follow this trend. Certainly, this software (and this book) covers an important piece of information for people who dare to explore spatial data and its undercover meaning. To make this project come to life, several different software packages were used such as Active Server Pages (ASPs 2.0), IIS, PHP®, Fireworks®, WebChart®, Java Applets®, JavaScript and HTML, Fortran and C languages (Negreiros, Painho, Aguilar, & Aguilar, 2009). Two international domain names, mygeoffice.org and mygeooffice.org, were the other requirements to be fulfilled for the full implementation of this Web 2.0 service around the globe. Finally, if the Internet is part of our lives, in the end, the common user will inherit all WWW intrinsic features such as the absence of site management obligations and installation procedures. After all, GIS expects the results to be displayed in appealing maps while the arduous work should be hidden in a W3 wireless environment. myGeoffice© follows this belief.

REFERENCES

Chun, Y., & Griffith, D. (2013). *Spatial Statistics and Geostatistics: Theory and Applications for Geographic Information Science and Technology*. SAGE Advances in Geographic Information Science and Technology Series.

Negreiros, J. (2004). *SAKWeb - Spatial autocorrelation and Kriging web.* Ph.D. Thesis.

Negreiros, J., Costa, A., Painho, M., Santos, J., & Lopes, I. (2007). Geostatistical analysis: software flashpoint. Paper presented at Geocomputation 27, Dublin, Ireland.

Negreiros, J., & Painho, M. (2005). *The web platform for spatial statistical analysis.* The Portuguese Conference of Information Systems 5 (APSI), Bragança, Portugal.

Negreiros, J., & Painho, M. (2006). *SAKWeb©–Spatial autocorrelation and Kriging web service. In Geo-Environment & Landscape Evolution II* (pp. 79–89). WIT Press.

Negreiros, J., Painho, M., Aguilar, F., & Aguilar, M. (2010). Geographical information systems principles of ordinary Kriging interpolator. *Journal of Applied Sciences, 10*, 852-867.

Negreiros, J., Painho, M., Aguilar, M., & Aguilar, F. (2009). *Technical outline of a W3 spatial decision support prototype. In Handbook of Research on Social Dimensions of Semantic Technologies and Web Services* (Vol. 2). Hershey, PA: Information Science Reference.

Penn State. (2014). *Lesson 1: Why spatial data are special - problems and opportunities.* Retrieved 2014, from Geography 586: https://www.e-education.psu.edu/geog586/l2.html

Worboys, M. (1996). *GIS: A computing perspective*. Taylor & Francis.

Acknowledgment

Over the past ten years, I have received support and encouragement from a great number of official institutions towards the building of this project: Fundação Ciência e Tecnologia (Praxis XXI/BD/11233/97 and SFRH/BPD/35541/07), Nova Information Management School (former ISEGI-UNL) from Portugal, Almeria University from Spain, Saint Joseph University, Macao Polytechnic Institute and The Science and Technology Development Fund from Macao, China. Certainly, myGeoffice© and this book would never have become a reality without their valuable contribution.

My special thanks to Fernando Aguilar and Manuel Aguilar for the share of new spatial statistical insights. Paulo Pereira helped me through this Internet software inspiration for Geographical Information Systems (GIS) users. I could not forget Mafalda Matias for her myGeoffice© logo, Gabriel Soares, Carlos Nascimento, Herbert Lee and Ana Neves for their English proof reading.

In addition, a thank you to a wide range of researchers and institutions that introduced me to GIS in general and geostatistics (science originally developed in geology and mining) in particular, through their research papers, conferences, books, manuals, courses and software. Without them, as a computer geek, I would be lost. Among others, these include: Pierre Goovaerts, Amilcar Soares, Jorge de Sousa and Maria Pereira, Daniel Griffith, Georges Matheron and Margaret Armstrong, Luc Anselin, Isobel Clark and Bill Harper, Michael Goodchild, Noel Cressie, Deutsch and Journel, Isaaks and Srivastava, David Ebdon, Wikipedia©, CrimeStat©, JHlabs©, LightBox© and ESRI©, Fotheringham, Brunsdon and Charlton, David Wong and Jay Lee, Christopher Lloyd, Peter Burrough and Rachael McDonnell, Cynthia Leong, Carlos Felgueiras, Cristina Costa and Marco Painho.

Finally, my unique appreciation to Claudia Chin and Daniel Negreiros for providing me the crucial peace, love and stability needed during my Andalusia years.

Section 1
GIS Functions

Chapter 1
Help

ABSTRACT

An overview of the aids given by myGeoffice© is described in this chapter. As a rule of thumb, the philosophy of myGeoffice© towards help consists of a set of definitions, notes, links, examples and suggestions to help the regular researcher fulfill his/her duty on each function.

COPYRIGHT AND DISCLAIMER

Copyright

myGeoffice© was setup by João Garrott Marques Negreiros and registered at name.com in 2008. With the exception of the Fortran 77/90 code of GSLIB® (Deutsch & Journel, 1998), the Bootstrap framework and several free Java Applets available on the Internet (check the Help/Credits section), most of the remaining C, classic ASP, PHP, JavaScript, HTML, Jquery and CSSs code were written by João Garrott Marques Negreiros. myGeoffice© was also built with an official license of Component One Web ActiveX WebChart® 8 (www.componentone.com) for mapping generation and Fotolia® (www.fotolia.com) for images. The present owner grants anyone worldwide a non-exclusive, royalty-free license to view/use myGeoffice© on a computer device via a Web browser for their own (personal, teaching, research and non-commercial) use. No registration fee or software installation is required. In fact, the single aim of this project was and remains to provide a spatial analysis (Web 2.0) tool for the learning and teaching of key GIS concepts in an Internet environment.

DOI: 10.4018/978-1-5225-3270-5.ch001

Disclaimer

This Website is provided "as it is" without any representations or warranties, expressed or implied. João Garrott Marques Negreiros will not be liable to you in relation to the content of this site. By using myGeoffice©, you agree that the exclusions and limitations of liability set out in this Website disclaimer are reasonable. If you do not think they are reasonable, you must not use this site.

Privacy

myGeoffice© will take technical and organizational precautions to prevent the loss, misuse or alteration of your personal data. Every Sunday at 8 AM of Macao, China, time (+8 hours UTC/GMT), myGeoffice© stops this service for 30 minutes maintenance (defragment and reboot operations). During this scheduled maintenance, it will automatically delete all personal data that users provided to this service. Finally, this Website contains links to other Websites. João Garrott Marques Negreiros is not responsible for the privacy policies or practices of any third party.

Cookies

myGeoffice© uses cookies. Further, this Web 2.0 tool uses Google Analytics® to evaluate the use of this Website to generate statistical data on its users.

Anti-Spam

In the context of electronic messaging, spam refers to unsolicited, bulk or indiscriminate messages, typically sent for a commercial purpose. myGeoffice© has a zero-tolerance spam policy.

Linking Policy

myGeoffice© welcomes links to this Website. Links pointing to this site should not be misleading. In addition, myGeoffice© includes links to other websites owned and operated by third parties. These links are not endorsements.

Legal-Disclaimer

João Garrott Marques Negreiros does not warrant that (A) the legal information on this website will be constantly available at all times and (B) the legal information on myGeoffice© is complete, true, accurate and non-misleading.

Keywords

See Figure 1.

OTHER FUNCTIONALITIES

Several capabilities are presented under this option: (A) Blogger; (B) Calculator; (C) Glossary; (D) Internet GIS software/university degree resources (including myGeoffice© YouTube channel and PPTs); (E) Free Java download link; (F) Key instructions on how to enable Java in your web browser; (G) Google maps; (H) Exploratory exercises supplied by myGeoffice© researchers; (I) Security requirements for running Java Applets in your web browser (avoid the use of Opera®, Edge® and Chrome® to run Applets). While the first eight options are self-explanatory, the present author stresses that some Java Applet pre-considerations are needed to run the Raster Image Processing options (chapter 3) and Voronoi/Thiessen (section 9.8) interpolation. For instance, Java Runtime Environment® (JRE 1.7, at least) must be installed previously to run any Java Applet. As demonstrated in Figure 2, a digital signature will

Figure 1. The key main word cloud of terms related to myGeoffice©
(Created with wordle.net)

Figure 2. Due to security concerns, under the Java Control Panel (Windows OS => Check Control Panel/Programs and click on the Java icon; Mac OS X => Check System Preferences and click on the Java icon), the user must add www.mygeoffice.org, mygeoffice.org, www.mygeooffice.org and mygeoffice.org to the Exception Site List in the Security tab in order to visualize myGeoffice© output

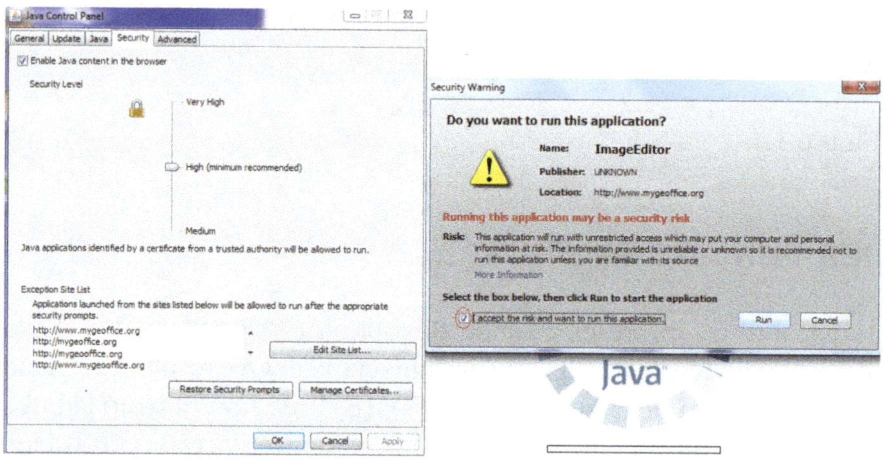

be required, which the user should enable using the available option box to visualize the final output.

By way of clarification, any Java Applet requires a specific plugin to control the launch and execution of the programming code. By creating a thread for every Java Applet instance within the Java Runtime Environment (JRE), all Applets run in the same instance of the JRE. As a client-side technology, security does become an important issue. Java does provide enhanced security features to make the users system less vulnerable to external abuse. Starting with Java 7 Update 51, Java does not allow users to run unsigned applications. To overcome this limitation, the user may use the Exception Site list feature to run these applications blocked under the security settings by adding the URL of the blocked application to the list. This step is shown in Figure 2. Concerning IE®, one should uncheck the ActiveX Filtering property in the Safety option.

Another clarification is regarding the Google Maps JavaScript API (Application Programming Interface), which lets any specialist to customize maps within Web pages (W3schools-Google Maps, 2016). According to the code presented in Table 1, five basic steps should be taken into account for the use of this wonderful spatial API: (A) Load of the Google API with the <script src="http://maps.googleapis.com/maps/api/js"></script> command;

Help

Table 1. Example of a java script program using google maps API

```
<!DOCTYPE html>
<html><head>
<script src="http://maps.googleapis.com/maps/api/js"></script>
<script>
function initialize() {
var mapProp = {
center:new google.maps.LatLng(48.209003,16.37289),
zoom:11,
mapTypeId:google.maps.MapTypeId.ROADMAP };
var map=new google.maps.Map(document.getElementById("googleMap"), mapProp);
}
google.maps.event.addDomListener(window, 'load', initialize);
</script>
</head>
<body><div id="googleMap" style="width:500px;height:380px;"></div></body>
</html>
```

(B) Set the map properties with the initialize function, where the mapTypeId property specifies the map type to display, i.e., ROADMAP (normal, default 2D map), SATELLITE (photographic map), HYBRID (photographic map + roads and city names) and TERRAIN (map with mountains, rivers…); (C) Setup a map HTML container by utilizing the <div> element; (D) Create of a map object with the new operator; (E) Add an event listener to load the map by using the google.maps.event. addDomListener() method.

Figure 3. Output of the previous Java Script code

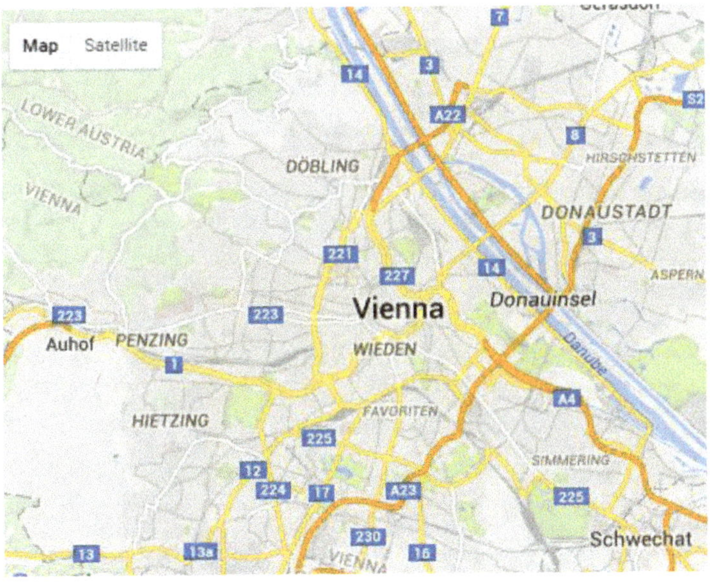

This particular Help section highlights eight possibilities. The Session Variable Status allows the control of internal variables whenever the user starts the login process into myGeoffice©. The Web Analytics section displays the graphical results of Google Analytics® data, which generates several detailed statistics about website traffic. By definition, Web analytics is a generic jargon meaning the study of the impact of a website on its users. E-commerce and other website publishers often use Web analytics software to measure such concrete details as how many clients visited their site; how many of those visitors were unique visitors; how they came to the site (if they followed a link to get to the site or connected there directly); what keywords they searched for within the site's search engine; how long they stayed on a given page or on the entire site; and what links they clicked on and when they left the site (Webopedia-Web analytics, 2016). This type of software can also be used to monitor whether or not a site's pages are working properly. Thus, Web site administrators can determine which areas of the site are popular and which areas of the site do not get traffic.

Under the exploratory exercises option, myGeoffice© welcomes the PDF submissions by any researcher regarding relevant theoretical and practical cases for sharing purposes and for the maintenance of this academic project. As for the Web Server Information, nine items are presented for general knowledge

Figure 4. Partial view of the fourteen items of the User Guide Reminders regarding the use of myGeoffice©

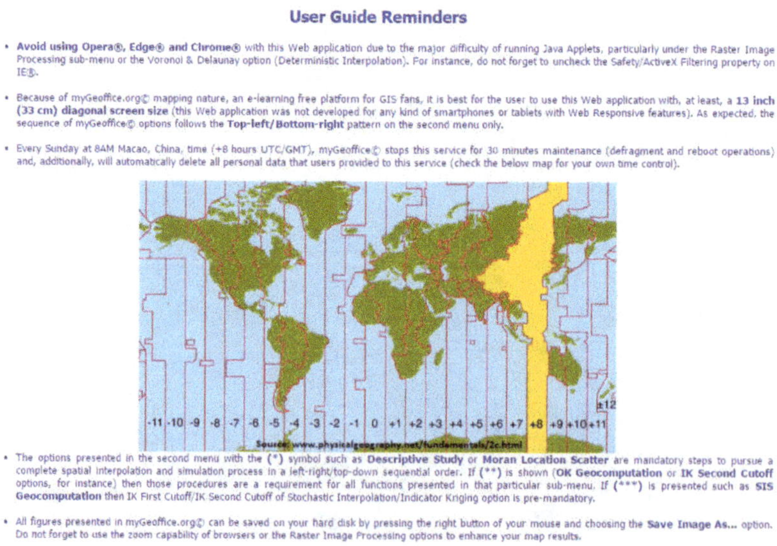

User Guide Reminders

- **Avoid using Opera®, Edge® and Chrome®** with this Web application due to the major difficulty of running Java Applets, particularly under the Raster Image Processing sub-menu or the Voronoi & Delaunay option (Deterministic Interpolation). For instance, do not forget to uncheck the Safety/ActiveX Filtering property on IE®.
- Because of myGeoffice.org© mapping nature, an e-learning free platform for GIS fans, it is best for the user to use this Web application with, at least, a **13 inch (33 cm) diagonal screen size** (this Web application was not developed for any kind of smartphones or tablets with Web Responsive features). As expected, the sequence of myGeoffice© options follows the **Top-left/Bottom-right** pattern on the second menu only.
- Every Sunday at 8AM Macao, China, time (+8 hours UTC/GMT), myGeoffice© stops this service for 30 minutes maintenance (defragment and reboot operations) and, additionally, will automatically delete all personal data that users provided to this service (check the below map for your own time control).

- The options presented in the second menu with the **(*)** symbol such as **Descriptive Study** or **Moran Location Scatter** are mandatory steps to pursue a complete spatial interpolation and simulation process in a left-right/top-down sequential order. If **(**)** is shown (**OK Geocomputation** or **IK Second Cutoff** options, for instance) then those procedures are a requirement for all functions presented in that particular sub-menu. If **(***)** is presented such as **SIS Geocomputation** then IK First Cutoff/IK Second Cutoff of Stochastic Interpolation/Indicator Kriging option is pre-mandatory.
- All figures presented in myGeoffice.org© can be saved on your hard disk by pressing the right button of your mouse and choosing the **Save Image As...** option. Do not forget to use the zoom capability of browsers or the Raster Image Processing options to enhance your map results.

Figure 5. Overall structure of myGeoffice©

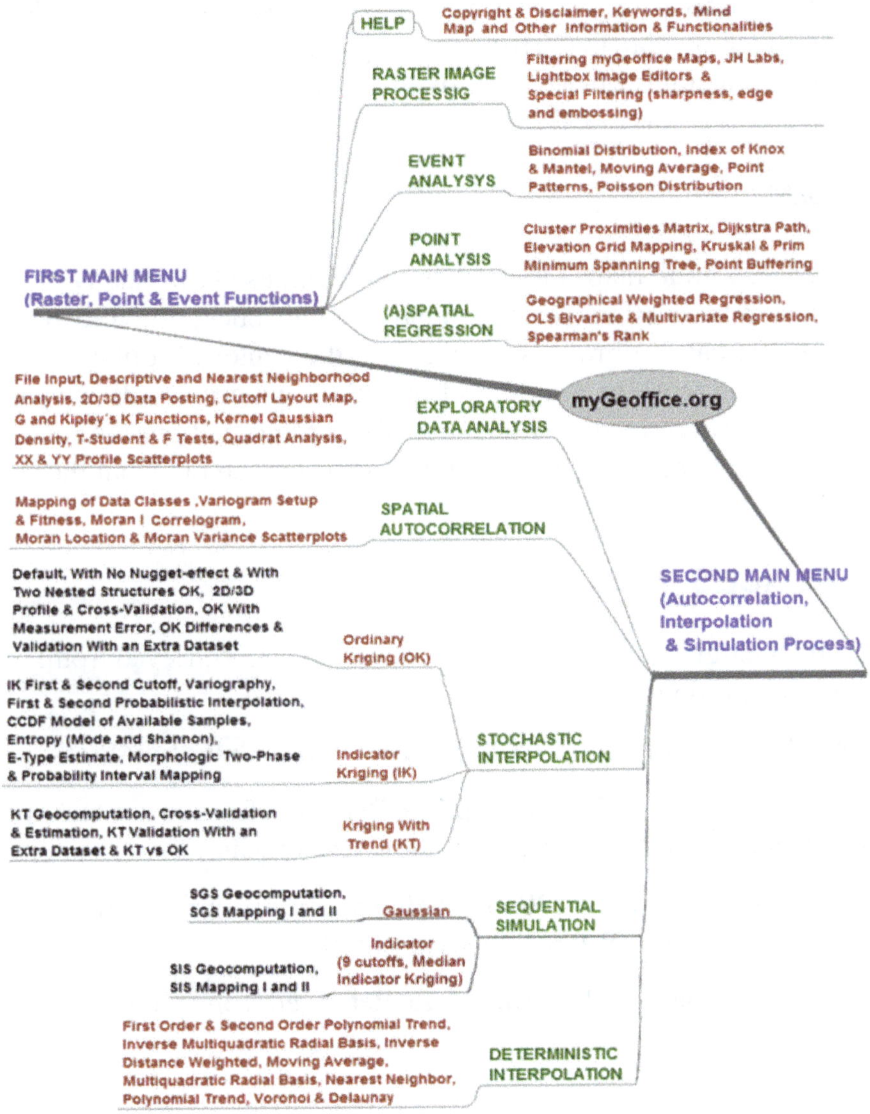

purposes such as Web server local time and date, location, IIS® version and Web port used by myGeoffice©. As a reference, IIS® (Internet Information Server) is a group of Internet servers (including a Web or Hypertext Transfer Protocol and a File Transfer Protocol server) with additional capabilities for

Microsoft's Windows Server® operating system. IIS® competes in the Internet server market that is also addressed by Apache®, Sun Microsystems®, O'Reilly® and others (Techopedia-IIS, 2016). In a complementary form, the Web Server Geolocation illustrates Macao position in the world map (see Figure 4).

MIND MAP

For reference, a mind map (Figure 5) is a diagram used to visually organize information and it is often created around a single concept (myGeoffice.org in this specific case), drawn as an image in the center of a blank page, to which associated representations of ideas such as images, words and parts of words are added (Wikipedia-Mind Map, 2016). Major notions are connected directly to the central concept while other ideas branch out from those.

REFERENCES

W3schools-Google Maps. (2016). *Google maps.* W3. Retrieved from: www.w3schools.com/googleapi/google_maps_ basic.asp

Deutsch, C., & Journel, A. (1998). GSLIB. Oxford University Press.

Techopedia-IIS. (2016). *IIS.* Technopedia. Retrieved from: www.techopedia.com/definition/24953/internet-information-services-iis

Webopedia-Web Analytics. (2016). *Web analytics.* Retrieved from: www.webopedia.com/TERM/W/Web_analytics.html

Wikipedia-Mind map. (2016). *Mind map.* Retrieved from: en.wikipedia.org/wiki/Mind_map

Chapter 2
Digital Image Processing

ABSTRACT

The first two sections of this chapter introduce the motivation and fundamental theoretical perceptions towards DIP and encoding formats. The RGB system color and the visual quality improvement (by increasing the contrast with histograms enhancement techniques) are emphasized in section "Digital Image Enhancement" and "Elementary Operations With Digital Images", respectively. Spatial filtering and masks are analyzed in section "Filtering", which some Java code has been included for illustrative purposes. The next section explains how to save a myGeoffice© generated image to your hard disk and provides examples of image processing. Section "JHLABS® Image Editor" and "LightBox® Image Editor" depicts JHLabs®, a Web editor built on Swing technology, and Lightbox® of myGeoffice©. Its filtering capabilities are presented in the last section.

INTRODUCTION

By definition, digital image processing (DIP) consists of the capability to perform operations on raster images to produce enhanced illustrations or extract some useful information from them. This includes analyzing and manipulating the matrix image, including data compression and image enhancement such as sharpening and restoration. In this chapter, a brief description of these issues related to DIP is provided.

DOI: 10.4018/978-1-5225-3270-5.ch002

The primary goal of DIP is to simulate environmental color and shading. Alternatively, DIP aims to apply changes to images to better retrieve information from the scene in question. This concept is mainly used on images taken by microscopes, digital cameras and remote sensors. Thus, the input process is always an image while the output is a new and transformed one. Among its foremost fields of application are medicine, meteorology, agriculture, geology, meteorology and cartography. A third fundamental concept is closely related with pattern recognition to identify its features. This last technique is often applied in spatial analysis in general and remote sensing in particular.

CONCEPTS OF DIGITAL IMAGE PROCESSING

Digital images can be obtained through scanners, digital cameras and remote sensors. A scanner is an optical reading device that converts images, photos, illustrations and texts into a digital format that can be manipulated by a computer. The most common scanner is the flatbed, which is similar to a photocopier. Other types are barcode readers and hand and drum scanners. In an operational context, all scanners are based on the principle of light reactance, which consists of positioning the image so that the intensive light (generated by fluorescent or cathode-cold type lamp) hits it. A sensitive sensor captures this reflected radiance thereby forming a digital image. Currently, the most commonly used sensors are as follows (Felgueiras & Negreiros, 2008):

- **Photo Multiplier Tube (PMT):** The standard hardware for drum scanners in the current printing industry for high-quality prints. This scanning device is placed in a glass cylinder that rotates at high speed around the PMT sensor, dividing the reflected light into three beams (which passes through filters) and generating the scanned image.
- **Charge Coupled Device (CCD):** Used in almost all domestic scanners (fax, digital cameras), it stands out for its good quality and low price. This type of sensor converts the reflected light into electrical signals that in turn are converted into bits by an analogue-to-digital converter (ADC) circuit.
- **Contact Image Sensor (CIS):** Uses a series of red, blue and green LEDs to produce white light. This allows greater energy savings. However, the image quality is not as good as compared to CCDs.

The scanning process enables analog signals to be converted into quantized levels of digital information. Naturally, the higher the number of digital levels, the better the representation of the analog signal becomes. Yet, one need a larger number of bits to represent the image and, hereafter, more (virtual and real) RAM memory to store that image is required.

With regard to digital cameras resolution, Table 1 relates the standard sizes of a digital image (columns versus rows) to the quantity of megapixels incorporated in the conventional digital camera.

For computation purposes, a digital image is represented by a bi-dimensional array. It is said then that a digital image has a matrix representation. The entire image has a certain number of lines (nlin) and a certain number of columns (ncol). Naturally, the number of rows and columns can be different. It has become a convention that the original row of an image is the top line (0). The last row is labeled (nlin-1). The initial zero column of the image corresponds to the left column. The right most column is labeled (ncol-1). Obviously, the total size of an image is obtained by multiplying the total number of lines with the total number of columns (size=nlin×ncol).

By default, CCD scanners generally have an optical resolution of 600x1200 dpi (dots per inch). However, there are scanners on the current market working with resolutions of 2400, 4800 or 9600 dpi. This is only possible due to a technique named interpolation, which increases the capability of the scanner software resolution from a lower one. Technically, this interpolation software creates new dots between pixels obtained at the expense of the original optical resolution (Felgueiras & Negreiros, 2008).

At a more detailed level, this interpolation process is based on mathematical calculations implemented by algorithms that generate new pixels from the original image. This is usually accomplished by using the nearest pixels. On

Table 1. For reference, remote sensors are Earth mapping devices placed on airplanes and satellites that enable imaging over large distances using three types of technologies: Thematic mapping, infrared and CCD sensors

Size	Megapixels	Recommended Use
640x480	0.3	Only for Internet
1024x768	0.78	Printings up to 9x13
1289x960	1.2	Printings up to 10x15
1600x1200	2.0	Printings up to 13x18
2048x1536	3.1	Printings up to 20x30

Figure 1. Each element of this matrix is known as pixel (picture element). In 3D images, they are called voxel. Again, the (0,0) pixel becomes the upper-left pixel while the (nlin-1,ncol-1) position represents the lower-right pixel

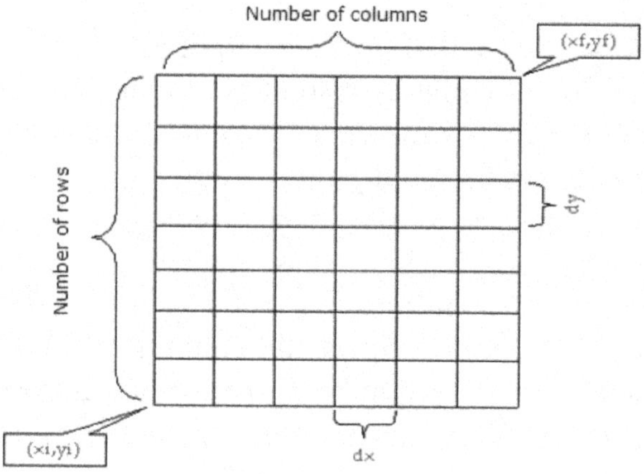

the other hand, interpolation by local means is based on the determination of the digital pixels levels of the output image according to the following criteria: The value of the digital level of each pixel is assessed based on the distance between the new image pixel and its neighbors. That is, the new digital value is computed as an average (weighted by the inverse distance factor) of the values of the nearest neighbors. Again, the neighborhood total number for this computation equals 4, 8 or 16. It should be stressed that this neighborhood approach preserves the digital values set of the original illustration. On the other hand, interpolation by local means generates digital levels that were not presented in the input image.

Sampling is the reverse process, which consists of the selection from a digital image (and in a regular manner) of a few rows and columns for the generation of a new image. It is possible, for example, to subtract all odd rows and columns of the original image. One can also use the criterion of withdrawing m rows (or columns) from each n rows (or columns), in which n>>m. The result of this processing is designated reduction of the overall image size with a quality detriment side-effect (as expected).

DIGITAL SIGNAL CODING AND STORAGE FORMAT

The codification of an image is intimately related to the number of bits required to represent the quantization of any analog information. Thus, images with 2 bits can represent $2^2=4$ different levels, $2^3=8$ levels with 3 bits and $2^5=32$ levels with 5 bits. Theoretically, n bits would represent 2^n levels. Digital levels are also recognized as color and image gray levels. These modes describe how pixels are organized and represented within the computer's memory. In general, the most commonly used encodings are as follows:

- **Black and White (1 Bit):** Known as bitmap because each pixel is represented by a single bit (where each pixel can only assume the 0 value, black, or 1, white). However, several design manipulation programs usually offer four conversion modes: Art line (50% threshold), ordered, halftone and error diffusion. As expected, this bitmap format results in an image of a smaller size.
- **Gray Scale (8 Bits):** Internal representation system that uses 256 gray levels per pixel (byte per pixel). The 0 value corresponds to black and 255 to white. In Photoshop®, for example, a gray scale is represented as a percentage of black, where 0% corresponds to white and 100% is equivalent to black. This mode is highly recommended to store black and white images to maintain uninterrupted tones.
- **RGB (24-Bit or True Color):** Uses three colors per pixel, allowing a reproduction of up to 16.7 million (256x256x256) colors. Each color is represented with 8 bits (1 byte), enabling 256 values or levels per color. Analogous to the gray scale, the (0,0,0) value is equivalent to black while the (255,255,255) value equals white. Known as the primary colors, the final color representation is based on the combination of light (additive system) emitted by the three sources of light: Red, Green and Blue.
- **Indexed Colors (1 to 8 Bits):** Known as the system of 256 colors. In this mode, each pixel takes on an existing value of the 256-color palette. This mode is extremely useful for multimedia applications and image publication on the Web.

Digital images are stored in binary files with specific formats. In general, all pictures contain a header or information fields. Under the file header, one finds the following information: Number of rows, number of columns, number of bits used to represent the image radiometry, geographical references (when the image is georeferenced), horizontal resolution (dx), vertical resolution (dy), number of image bands, type of compression used to save the data, date and time of purchase, type of sensor that captured the image, parametric data from sensors and relevant supplementary information. Among the most used storage formats of digital images, the following are some of the most commonly used:

- **BMP (Windows® Bitmap):** It enables 24-bit color storage (used by MS-Paint®).
- **TIFF (Tagged Image File Format):** This is the preferred format accepted by almost all image processing software. It was developed in 1986 by Aldus® and Microsoft® in an attempt to create an image standard and generated by digital equipment. It usually makes use of LZW compression, an algorithm invented by Abraham Lempel, Jacob Ziv and Terry Welch. The reader may check Appendix 7 regarding the RLE (Run Length Encoding) compression format.
- **GIF (Graphics Interchange Format):** Currently, this is probably the most popular graphic file format. It was created by CompuServe® to transmit bitmap images across the Internet. Its first version appeared in 1987 but in 1989, CompuServe® launched the GIF89a specification, which implements the use of transparent color. GIF images are always compressed and encoded by the LZW specification (without loss of information).
- **RAW (or Rustic):** This is an image format in which pixels are stored in binary (with 8, 16 or 24 bits). Its main feature is that only the image data is stored within the file (check Java example code of Tables 2, 3 and 4). Due to the lack of a file header, reading this format requires the user to enter some additional parameters for the image description such as the number of rows, number of columns and number of bits.

Table 2. Using a raw input picture with 500 lines and 270 columns (middle), this tiny Java code allows a non-whiting effect on the same picture (bottom) by dividing each original pixel value (8 bits of length) by two. The free IrfanView® (with all installed plugins) software was the chosen visual image editor to check both input and output outcomes

```
import java.io.*;
public class Copy_div2 {
public static void main (String[] args) throws IOException {
String nome_arq_ent = "nemo500-270orig.raw";
String nome_arq_sai = "nemo500-270div2.raw";
InputStream arq_ent;
OutputStream arq_sai;
arq_ent = new FileInputStream(nome_arq_ent);
arq_sai = new FileOutputStream(nome_arq_sai);
int data, cont=0;
while ((data = arq_ent.read())>=0){
arq_sai.write(data/2);}
arq_ent.close();
arq_sai.close();}
}
```

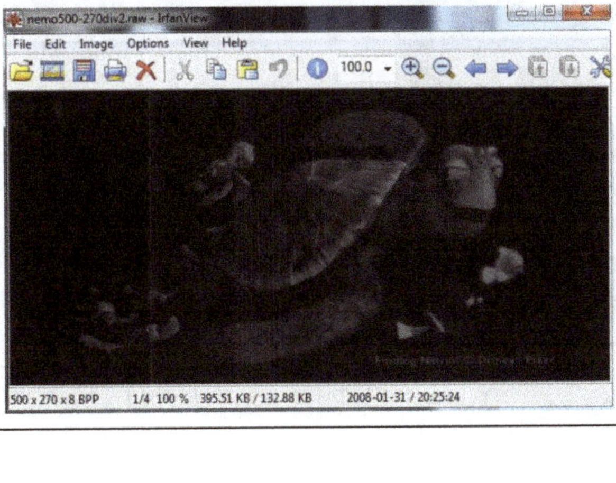

Table 3. This Java program reads a RAW color image (24 bits) and turns it into a gray one (8 bits). As expected, the file size of the color image (top) is close to three times larger than the bottom one

```
import java.io.*;
public class Cor2nc {
public static void main (String[] args) throws IOException {
String nome_arq_ent = "nemo500-270interleaved.raw";
String nome_arq_sai = "nemo500-270nc.raw";
int data1,data2,data3;
InputStream arq_ent;
OutputStream arq_sai;
arq_ent = new FileInputStream(nome_arq_ent);
arq_sai = new FileOutputStream(nome_arq_sai);
while ((data1 = arq_ent.read())>=0){
data2 = arq_ent.read();
data3 = arq_ent.read();
arq_sai.write((data1+data2+data3)/3);}
arq_ent.close();
arq_sai.close();}
}
```

Table 4. This Java code allows the conversion of a gray 8 bit image into a black and white one (all pixel values above 127 are assigned to 255; otherwise, they will be converted to zero)

```
import java.io.*;
public class Nc2Pb{
public static void main (String[] args) throws IOException {
String nome_arq_ent= "nemo500-270nc.raw";
String nome_arq_sai= "nemo500-270pb.raw";
FileInputStream arq_ent;
FileOutputStream arq_sai;
int nlin=270, ncol =500;
char buf[]= new char[1000];
arq_ent = new FileInputStream(nome_arq_ent);
arq_sai = new FileOutputStream(nome_arq_sai);
for(int l=0; l<nlin; l++){
for(int c=0; c<ncol; c++){
buf[c]= (char)arq_ent.read();
if(buf[c]>127)buf[c]=255; else buf[c]=0; }
for(int c=0; c<ncol; c++) arq_sai.write(buf[c]); }
arq_ent.close();
arq_sai.close();}
}
```

DIGITAL IMAGE ENHANCEMENT

As any reader may have noticed, the human eye is more sensitive to color variations than gray scale changes. In fact, the human eye functions as an optical instrument. Light passes through the pupil eye and gets focused on the retina by the lens. It turns out that the retina contains a large number of photo light sensitive receivers: Cones (daytime) and rods (night). These receivers transform the electromagnetic energy (light) into neural energy thus creating the image in our brain. In turn, color perception is related to the composition of the light wavelength that reaches the eye.

The RGB system allows color creations derived from the primary ones in an additive composition. For example, pure black color is generated by the absence of these three (R,G,B) components, as is pure white by mixing each maximum component value. In general, this intensity ranges from 0 to 1. As expected, the 0 value represents complete absence of color component while 1 denotes maximum intensity. Each color representation is then modeled as a linear combination of the intensity levels of the three elements. Thus, we have: $\text{Final color} = r \times R + g \times G + b \times B$, where r, g and b represent the proportions (between 0 and 1) of the primary (R, G, B) components (Felgueiras & Negreiros, 2008).

As previously stated, the digital color images are generated by converting analog information to digital signals, for each pixel, through the scanning process of the analog signal. Analog information needs to be quantized in a specific number of digital levels. The higher this number becomes, the better the representation of analog data is achieved. In parallel, the table color is comprised of the index color number and the (R,G,B) intensity assigned to each color. Thus, any color image requires the respective color table to be recorded in the image file header, leading to larger size files.

In the particular case of remote sensing, the mapping system usually acquires various spectral bands that together form a multispectral image. Generically, this multispectral procedure is composed of several bands that correspond to multiple photos captured simultaneously in the same region of the Earth's surface.

Figure 2. Representation of the primary and secondary colors under the RGB (left) and CYM (right) system. For reference, printer devices are based on the Cyan, Yellow and Magenta subtractive scheme

(Source: forums.createspace.com/en/community/message/191630)

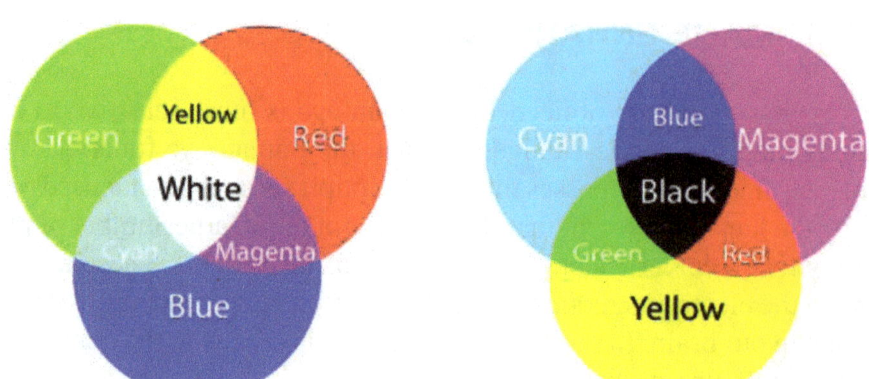

The histogram of a digital image is a table that lists each gray level value to their frequency of appearance of the digital image. As a rule, a histogram is displayed by a graph showing the relationship between the gray levels (x-axis) and the number of pixels associated with each one (y-axis). Based on this statistical concept, this technique involves modifying the radiometric response of an original image using a linear transformation that depends on the image histogram (check the example illustrated on Figure 3, where the number of blue pixels exceed the total number of red and green ones). The ultimate goal is to obtain an image whose histogram is completely scattered in the field of values determined for the image encoding (Felgueiras & Negreiros, 2008).

When a figure presents a low pixel range (low variance), certainly a diffused image will be displayed and a contrast effect should be applied to enhance it to the human eye. This contrast can be made through a transfer function that lays out the original images values to a better distributed radiometric range. Frequently, a straight-line function, Y=a×X+b, may be applied. This algorithm is described as follows (check developed Java code of Appendix 6):

- Scroll the image, in terms of its internal values, and calculate the minimum and maximum digital values, Xmin and Xmax. Optionally, one can set the Xmin and Xmax values based on the image histogram.
- Compute the slope of the linear equation: a=255/(Xmax-Xmin).
- Assess the Y intercept of the same linear equation: b=-a×Xmin.
- Apply the linear relation (Y=a×X+b) to each pixel of the input image (X), yielding the digital level of the output image (Y).

Let us consider the following hypothetical 3×3 pixel image (encoded with 8 bits) of Table 5 (left). The result of its histogram analysis becomes as follows: Xmin=4; Xmax=55; a=255/(Xmax-Xmin)=255/51=5; b=-a×Xmin=-5×4=-20. Thus, the linear relation becomes Y=20-5×X. If this linear transformation is applied to each element of the input image, one obtains the right image of Table 5, containing a wider variance of its output. However, this automatic enhancement method may not be effective in some cases. For this to happen, it is enough that there is, at least, one pixel with a minimum digital level and a pixel with a maximum digital one. Can the reader explain why one cannot get any enhancement when this situation occurs?

Figure 3. Histogram of a standard digital image

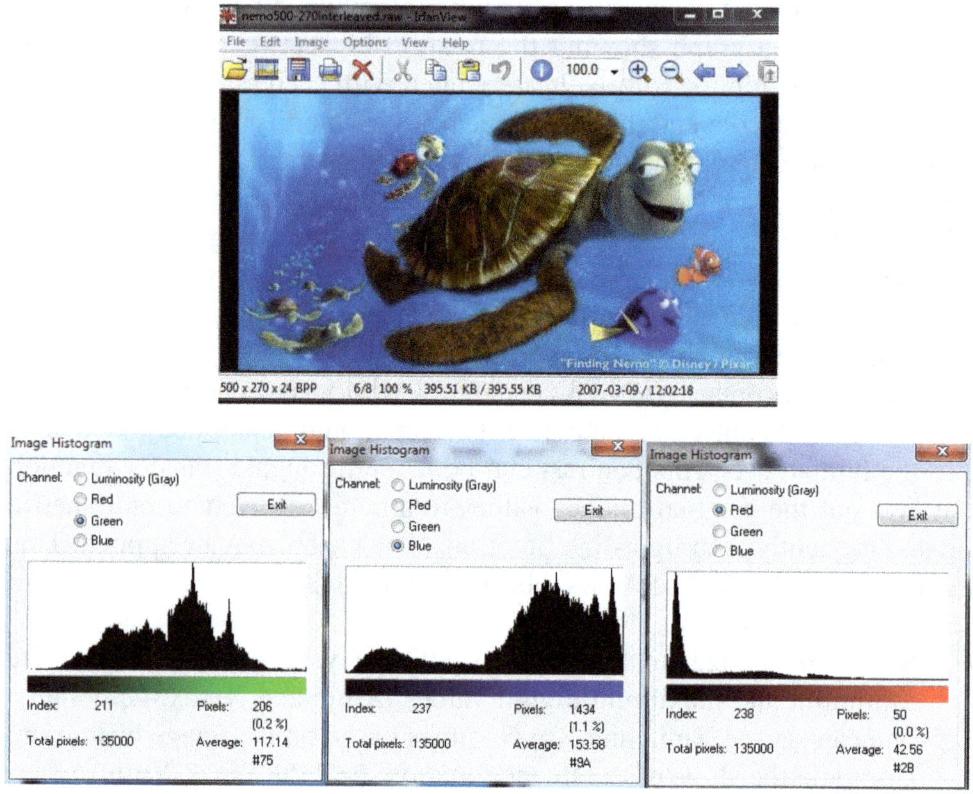

Table 5. As an exceptional rule, all digital final values below 0 and above 255 are mapped to 0 and 255, respectively

12 20 30 4 14 10 55 8 13		40 80 130 0 50 40 255 20 45

There are other possibilities of transfer functions such as square root, $y=f(x)=a^{1/2}$ (whose characteristic is to enhance darker values) or quadratic, $y=f(x)=a^2$ (whose characteristic is to enhance lighter values). It is also possible to modify the contract of individual bands of a multispectral image. Imagine a database of different image frames. Stressing each spectral composition with the appropriate mathematical relations of transformation, Figure 4 shows a colorful composition enhanced by three different spectral bands.

Figure 4. Example of a spectral composition
(Felgueiras & Negreiros, 2008)

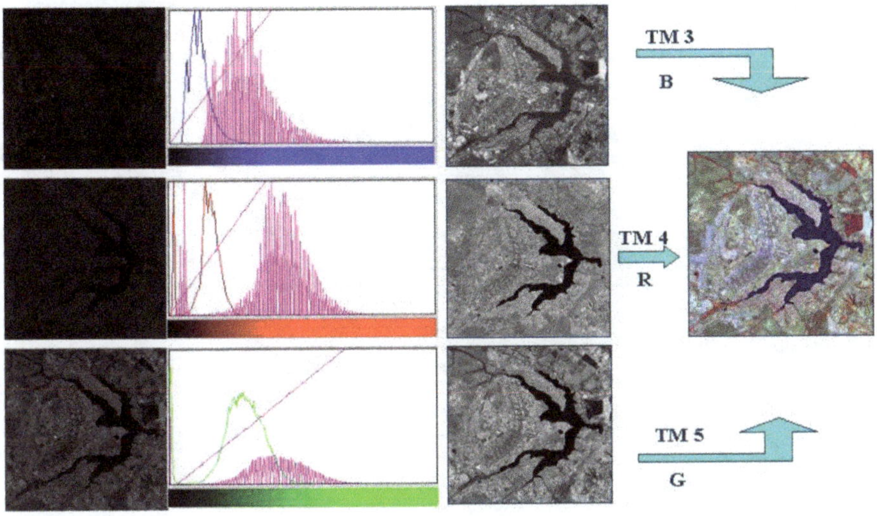

ELEMENTARY OPERATIONS WITH DIGITAL IMAGES

Let us consider matrix operations between digital images such as the application of mathematical, arithmetical and logical calculations between digital levels of two pictures. These operations are one-pixel operation in which the outcome is always a new digital image. Among the main objectives of these procedures, the following ones stands out: Elimination of systematic noise, improvement of the overall visual quality and the ability to highlight specific targets or objects in a digital image.

The basic principle of any mathematical operator can be applied at any digital level of images. Arithmetic operators such as addition, subtraction, division and multiplication can be used on a pixel-by-pixel basis. Similarly, one can use logical operators. For instance, arithmetic operations for noise suppression are usually applied for elimination of systematic noise. Let us suppose that X and Y images are contaminated by the same additive R noise (imagine that this multiplicative error consists of shadows stemming from solar ground lighting). It can be said that $X=X'+R$, $Y=Y'+R$ and $Z=X-Y=X'-Y'$. Obviously, the resulting Z image is proportional to X′ and Y′ images but untainted by R error.

Figure 5. Illustration of an operation between two input images, X and Y, generating the Z output

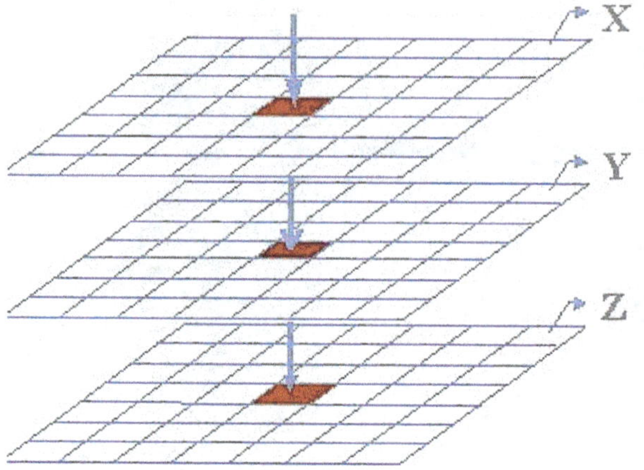

Image division can also be applied when the noise factor is multiplied by each digital image value. Imagine now that both X and Y images are contaminated by the same multiplicative R noise. It is evident that X=X´×R; Y=Y´×R and Z=X/Y=X´/Y´. Thus, the resulting Z image is proportional to images X´ and Y´ but untainted by R error.

Rhetorically, consider two inputs, X and Y, and a Z output image. Table 6 shows the effects summary of each operation including the typical application for each type.

Table 6. Effects and applications of arithmetic operations on images

Operation	Effect on the Image	Most Common Usage
Addition	Z is the sum of X and Y intensity values. If Y is a positive scalar, Z will become a lighter version of X.	Normalization of image brightness. Noise removal.
Subtraction	Z is the result of the difference between X and Y. If Y is a positive scalar, Z will become a darker version of X.	Detection of differences between two images of the same scene.
Multiplication	Z is the scalar product of the intensity values of X and Y. If Y is a positive scalar, Z intensity are directly proportional to X by a factor of Y.	Brightness adjustment (adjustment to different lightning values on the same scene).
Division	Z is the ratio of the intensity values of X with the corresponding values on Y. If Y is a positive scalar, Z intensity values are inversely proportional to X by a factor of Y.	Brightness normalization.

(Filho & Neto, 1999)

In addition to the usages listed, one can state with some certainty that subtraction is capable of enhancing spectral differences between images, addition can be used for obtaining the average of images and division increases the contrast between objects in a scene. In fact, the division between bands of a multispectral image is also widely used to obtain the Vegetation Difference Index (VDI). The idea is to use a band with high vegetation sensitivity response and divide this band by one with low vegetation sensitivity.

Mathematically, the result of arithmetic operations between images cannot be less than the minimum value or exceed the maximum allowed by its digital level encoding. For example, an encoded image with 8 bits cannot contain negative resulting values or rates greater than 255. Therefore, mapping these results to the present coding domain becomes mandatory. Imagine two X and Y images encoded with 8 bits and defined as below, where the Z image equals the result of the subtraction of X from Y.

The Z enhancement for the minimum and maximum values of the present coding (0...255) can be performed by the following mathematical operation: Since Zmin=-20 and Zmax=31 then a=(28-1)/(Zmax-Zmin)=255/51=5 and b=-a×Zmin=-5×(-20)=100. Under this linear transformation, the multiplier value of "a" is referred to as the gain value while the additive "b" parameter is called the offset.

Table 7. A linear transformation of Z image is mandatory to respect the coding domain due to the presence of negative outcomes on Z image

132	120	138	-	152	92	107	=	-20	28	31
84	110	200		80	124	210		4	-14	-10
255	92	177		230	100	164		25	-8	13
X			-	**Y**			=	**Z**		

Table 8. When applying the linear relationship S=a×Z+b=5×Z+100, each picture element of S is obtained from the Z image. One should note, however, that the new values range follows between 0 and 255

0	240	255
120	30	50
225	60	165
	S	

In the same way one can apply Boolean operations between different images. Figures 6-9 depict the application of the AND, OR, XOR and NOT logical operators.

In order to recall the logical operators, the Boolean truth table is shown in Table 9.

Let us now consider two 3×3 images (X and Y) encoded at 8 bits. The Z image, resulting from the AND logical operation between X and Y, would be as Table 10 shows.

FILTERING

Bearing in mind the current pixel context, filtering is a local operation that modifies the values of the digital levels of each pixel by using a neighborhood operation, which takes into account the digital levels of its adjacent pixels. Figure 10 illustrates this operator concept of proximity.

The diffuse definition of neighborhood begins by considering a generic pixel of the digital image and the four closest pixels around it. Naturally, one could also consider as neighborhood, among other possibilities, the eight closest neighbors pixels. Figure 11 shows both concepts.

The primary use of digital image filters is image homogenization, removal of noise or edge detection of shapes. The process that allows sorting out these issues is called kernel convolution. Generally, this is tantamount to a procedure that walks over the entire original image by modifying its original values and weights associated with a pre-defined mask. Confirmed by Figure 12, the mask effect will assign new digital values for the central pixel whose weights mask can be positive, negative or zero. As expected, a masking element with zero weight stands for a non-consideration condition of that particular digital level in the original image. Computationally, the new digital pixel kernel is assessed by summing the digital values of the neighboring pixels and weighting them with the respective weights mask.

In mathematical terms, for a 3x3 convolution mask, p[i,j] equals a×p[i-1,j-1]+b×p[i-1,j]+c×p[i-1,j+1]+d×p[i,j-1]+e×p[i,j]+f×p[i,j+1]+g×p[i+1,j-1]+h×p[i+ 1,j]+k×p[i+1,j+1], where a, b, c, d, e, f, g, h and k are the nine weights of the considered mask while i and j corresponds to the line and column indexes of the image matrix, respectively. For example, after calculating p[2,5], this procedure is repeated with the same mask but moved one pixel to the right and so on until the last pixel of line three (keep in mind that the first line of the image matrix corresponds to an index of zero). Subsequently, the

Figure 6. Example of an AND operation between two color images

(a) (b) (c)

Figure 7. Example of an OR operation between two color images

(a) (b) (c)

Figure 8. Example of a XOR operation between two color images

(a) (b) (c)

Figure 9. Example of a NOT operation on (a) color image

(a) (b)

Table 9. The reader should note that these Boolean operators are bitwise operations between the values of each pixel of each image

Input Pixel		Logical Operator Result			
A	**B**	**AND**	**OR**	**XOR**	**NOT A**
0	0	0	0	0	1
0	1	0	1	1	1
1	0	0	1	1	0
1	1	1	1	0	0

Table 10. Unlike arithmetic operations when applying logical operations between digital images, one can never obtain values less than zero or greater than the maximum coding

X			AND	Y			=	Z		
132	120	138		152	92	107		128	88	10
84	110	200		80	124	210		80	108	192
255	92	177		230	100	164		25	68	160
X			AND	Y			=	Z		
10000100	01111000	10001010		10011000	01011100	01101011		10000000	01011000	00001010
01010100	01101110	11001000		01010000	01111100	11010010		01010000	01101100	11000000
11111111	01011100	10110001		11100110	01100100	10100100		11100110	01000100	10100000

mask is moved to the beginning of line four and the whole process is repeated until the end. Meanwhile, the digital values from the edge of the original image are not modified since both i and j indexes will hold negative values (a non-sense situation for matrices computation). A possibility to overcome this issue is to consider both i and j indexes greater than zero and less than (nlines-2) and (ncolumns-2), respectively.

The differentiation among filters derives from several masks being applied to a digital image. Linear low-pass type (also known as average filters) holds the capability of noise extraction and smoothing effect by generating a more uniform image if compared with the original one. Technically, several 3x3 masks with its weights are revealed in Figure 13. To avoid bias issues, the sum of the weights should equal one. If the sum is different from 1, one should divide each weight by the total sum of the mask weights.

Now let us consider the X image of Figure 14. By applying the leftmost filter window, we obtain its respective Z image. As a result, the final image becomes smoother, more similar and includes noise suppression. One loses

Table 11. This C program allows the user to compute an AND bitwise operation between two files, nemo500-270orig.raw and nemo500-270.raw

```
#include <stdio.h>
#include <malloc.h>
int main(){
FILE *fpiX, *fpiY, *fpo;
unsigned char *bufferX, *bufferY, *bufferZ;
int nlin,ncol,l,c,cor,Xmin=10000, Xmax=-1000;
float ganho, offset, result;
fpiX = fopen("nemo500-270orig.raw","rb");
if (fpiX==NULL){
printf("Imagem X does not exist");
getchar(); return 1; }
fpiY = fopen("nemo500-270.raw","rb");
if (fpiY==NULL){
printf("Imagem Y does not exist");
getchar(); return 1; }
fpo = fopen("resultadoAND.raw","wb");
if(fpo==NULL){
printf("File resultadoAND.raw does not exist");
getchar(); return 1; }
nlin=500;
ncol=270;
bufferX = (unsigned char *) malloc(ncol*sizeof(unsigned char));
if (bufferX == NULL){
printf("No allocation for buffer X");
getchar(); return 1; }
bufferY = (unsigned char *) malloc(ncol*sizeof(unsigned char));
if (bufferY == NULL){
printf("No allocation for buffer Y");
getchar(); return 1; }
bufferZ = (unsigned char *) malloc(ncol*sizeof(unsigned char));
if (bufferY == NULL){
printf("No allocation for buffer Z");
getchar(); return 1; }
for(l=0; l<nlin; l++){
fread(bufferX, sizeof(char), ncol, fpiX);
fread(bufferY, sizeof(char), ncol, fpiY);
for(c=0; c<ncol; c++) bufferZ[c] = bufferX[c] & bufferY[c];
fwrite(bufferZ, sizeof(char), ncol, fpo);}
fclose (fpiX);fclose (fpiY);fclose (fpo);
free(bufferX);free(bufferY);
printf("\n Press one key to finish...");
getchar();}
```

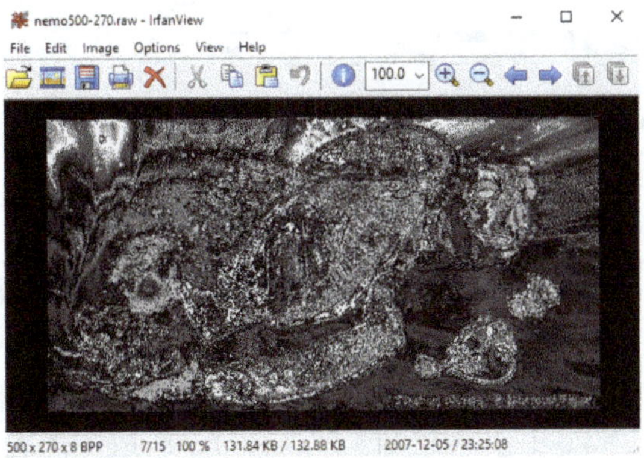

continued on following page

Table 11. Continued

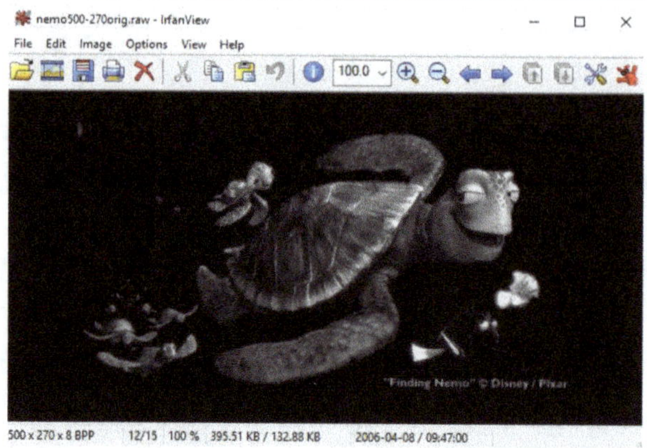

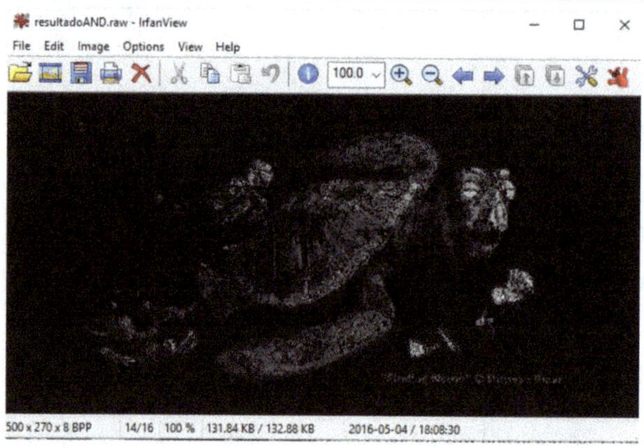

Figure 10. The global concept of neighborhood operations applied to digital images

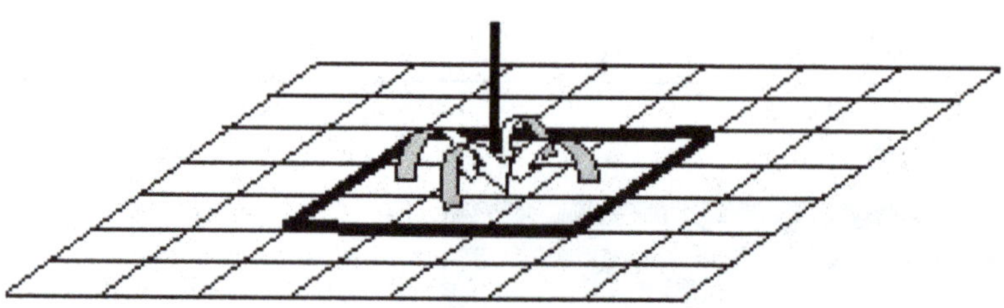

Figure 11. Vicinity concept of four (a) and eight (b) pixels

	X	
X	X	X
	X	

(a)

X	X	X
X	X	X
X	X	X

(b)

Figure 12. Kernel convolution procedure of a weight mask on a generic digital image *(Source: developer.apple.com/library/ios/documentation/Performance/Conceptual/vImage/ConvolutionOperations/ConvolutionOperations.html)*

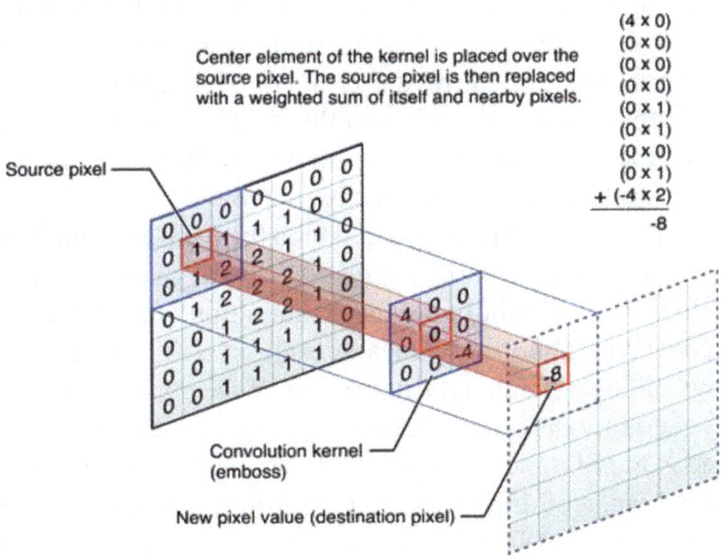

Figure 13. Three generic examples of low-pass masks

1	1	1		1	1	0		1	1	1	
1	1	1	÷9	1	0	1	÷4	1	2	1	÷10
1	1	1		0	1	0		1	1	1	

Figure 14. Specific example of a low-pass mask applied to X figure

1/9	1/9	1/9
1/9	1/9	1/9
1/9	1/9	1/9

120	120	120	120	120
120	210	120	30	120
120	120	120	120	120

X

»

120	120	120	120	120
120	130	120	110	120
120	120	120	120	120

Z

sharpness and contrast but one gains homogeneity of image elements. Obviously, low-pass filters can be applied in cascade on the same digital image. Consequently, the larger the number of iterations, the more uniform the resulting image becomes.

As stressed previously, low-pass filters are used to remove or smooth noise in a digital image. However, they present a major disadvantage, as they strongly affect areas without noise, homogenizing the overall image. Two other alternatives that allow the removal of image noise (without the effects of smoothing) are the mode and the median filters. These filters are non-linear and, therefore, there is a continuous mathematical formula which describes their application process to any image.

The mode filter is a nonlinear one in which the central element of a region is replaced by the most common digital level value in its vicinity, which is the mode value of its neighbors. Its implementation is achieved by building a histogram of the neighboring values of each pixel. The one presenting the most frequent value will match the new value of the considered central pixel. If there are several modes, the user can choose the one whose value is more similar to the value of the central pixel. One should also note that this filter ensures that the range of digital values of the output image is a subset of the domain of the input one. Thus, different digital levels from those presented in the input image are never generated.

The median filter is another nonlinear one. In this case, the central element of the region is replaced by the digital level value corresponding to the median of its neighbors. Statistically, the median calculation of neighbor values requires sorting their digital values. One that divides the neighborhood set into two equal cardinality groups is chosen. An interesting feature is that when considering the vicinity of an odd cardinality, the median filter ensures that the range of the output is a domain subset of the input image.

Figure 15. Applying to X image a mode filter with a neighborhood of eight

X

120	100	100	100	80
95	255	120	0	80
120	110	80	100	80

»

Z

100	100	120	100	80
95	120	100	80	80
110	120	110	100	90

Figure 16. Applying to X image a median filter with neighborhood of nine

X

100	100	120	100	80
95	255	120	0	115
110	120	110	110	80

»

Z

100	100	120	100	80
95	110	110	110	115
110	120	110	100	80

Linear high-pass filters (Figure 17) are characterized by enhancing the image and sharpen the transitions between different regions or objects. They can also serve as detectors of isolated points, lines and edges.

The sum of the weights mask should equal zero. This means that when applied to homogeneous regions, the mark will result in zero or in an extremely tiny value. If applied to a heterogeneous region, it will result in a much higher or lower value than the original one. In addition, if we want to consider a larger neighborhood, it is common to use up to 5x5 or 7x7 size masks. As a rhetorical example, reconsider X image shown in Figure 18. By applying the high-pass filter window, the respective Z output can thus be obtained. Other

Figure 17. Most common three masks of high-pass filters

-1	-1	-1
-1	8	-1
-1	-1	-1

0	-1	0
-1	4	-1
0	-1	0

1	-2	1
-2	4	-2
1	-2	1

Figure 18. Note that the Z values obtained hold a more significant difference than the original image

-1	-1	-1
-1	8	-1
-1	-1	-1

X

10	10	10	10	10
10	0	10	20	10
10	10	10	10	10

»

Z

10	10	10	10	10
10	-80	0	80	10
10	10	10	10	10

masks could be used for the detection of horizontal, vertical and diagonal lines in any image (see Figure 19).

Concerning edge detection, one usually applies linear filters of two sorts: Gradient (first derivatives of the xx-axis and yy-axis directions of the image) and Laplacian (second derivatives of the xx-axis and yy-axis directions of the image). The following masks are referred to as Sobel and Roberts operators (see Figures 20 and 21).

When using high-pass filters, the picture becomes more heterogeneous with a transition enhancement between colors. It increases the clarity of differences but decreases the homogeneity of each image element. The previous (b) image shows that the traditional high-pass filter traditional emphasizes differences in a timely fashion without detecting edge continuities at each

Figure 19. Mask detection regarding horizontal (a), vertical (b) and diagonal (c and d) lines

(a)

-1	-1	-1
2	2	2
-1	-1	-1

(b)

-1	2	-1
-1	2	-1
-1	2	-1

(c)

-1	-1	2
-1	2	-1
2	-1	-1

(d)

2	-1	-1
-1	2	-1
-1	-1	2

Figure 20. Robert's cross and Sobel edge detection filters

Robert operator

0	0	-1		-1	0	0
0	1	0		0	1	0
0	0	0		0	0	0

da = p[i,j]-p[i-1,j+1]
db = p[i,j]-p[i-1,j-1]
$S = ((da)^2+(db])^2)^{1/2}$

Sobel operator

1	0	-1		1	2	1
2	0	-2		0	0	0
1	0	-1		-1	-2	-1

dx = ((p[i-1,j-1]+2.0*p[i,j-1]+p[i+1,j-1])/4.0 -
(p[i-1,j+1]+2.0*p[i,j+1]+p[i+1,j+1])/4.)/2.0
dy = ((p[i-1,j-1]+2.0*p[i-1,j]+p[i-1,j+1])/4.0 -
(p[i+1,j-1]+2.0*p[i+1,j]+p[i+1,j+1]/4.)/2.0
$S = ((dx)^2+(dy)^2)^{1/2}$

Figure 21. Examples of (b) high-pass filter, (c) Robert's cross and (d) Sobel operators to (a) digital image. It is worth underline that there are also nonlinear spatial filters, known as morphological or frequency filters

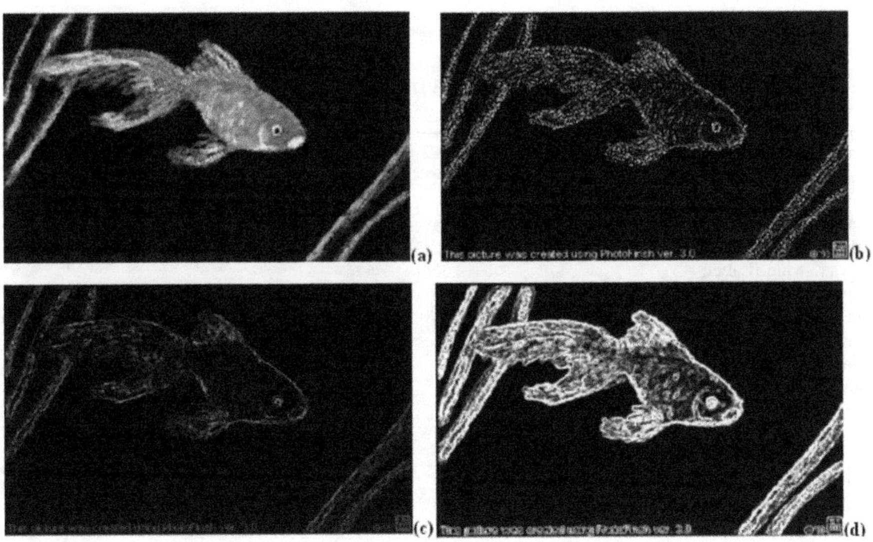

image element. Another conclusion expressed in this example is closely related to the Sobel operator, which stresses brightened edges more than Robert's operator. However, the Sobel operator is more sensitive to noise (by expanding it) as shown in (d) output.

SAVING MYGEOFFICE© MAPS

The foremost goal of this option is to describe the usage of the available filtering capabilities with the layout results produced by myGeoffice©. Do not forget to check your Java Download, Java Enable and Java Security tabs at myGeoffice© Help/Other Functionalities to run Java Applets technology in your browser (again, avoid the use of Opera®, Edge® and Chrome® if you are using Applets technology). As demonstrated in Figures 22 and 23, everything starts by saving the outcome map from any myGeoffice© function onto your local hard disk with the right click of your mouse (in this case, the 2D/3D maps of the quadratic deterministic interpolation method were used).

Table 12. Java code of the average filter with a JPEG file (robert.jpg)

```
import java.io.File;
import java.awt.image.BufferedImage;
import java.awt.image.Raster;
import java.awt.image.WritableRaster;
import javax.swing.JFrame;
import javax.swing.JLabel;
import javax.swing.JScrollPane;
import javax.swing.ImageIcon;
import javax.imageio.ImageIO;
public class FilterAverage extends JFrame {
public static void main(String args[]) {
   JFrame.setDefaultLookAndFeelDecorated(true);
   String ima_name_in = "robert.jpg";
   String ima_name_out = "robertSolution.jpg";
   BufferedImage ima_in, ima_out;
   ima_in = LoadImage(ima_name_in);
   FilterAverage frame1 = new FilterAverage();
   frame1.ShowImage(ima_in, ima_name_in);
   ima_out = ProcessImage(ima_in);
   FilterAverage frame2 = new FilterAverage();
   frame2.ShowImage(ima_out, ima_name_out);
   SaveImage(ima_out, ima_name_out); }
public static BufferedImage LoadImage(String image_name) {
   BufferedImage ima_in = null;
   File file = new File(image_name);
   try {
ima_in= ImageIO.read(file); }
catch(Exception e) {
System.out.println("File Image '" + image_name + "' does not exist.");
System.exit(0);}
   System.out.println("Name of image: "+image_name+" Type of Image:"+ima_in.getType());
   System.out.println("Size of Image: Columns "+ima_in.getWidth()+" Lines "+ima_in.getHeight());
   return ima_in; }
```

continued on following page

Table 12. Continued

```
public static void SaveImage(BufferedImage dest, String image_name) {
try {
ImageIO.write(dest, "jpg", new File(image_name)); }
catch (Exception e) {
System.out.println("Saving image problems... ");
System.exit(0); }
}
public static BufferedImage ProcessImage (BufferedImage ima_in) {
BufferedImage ima_out = new BufferedImage(ima_in.getWidth(),ima_in.getHeight(),
ima_in.getType());
Raster raster = ima_in.getRaster();
WritableRaster wraster = ima_out.getRaster();
int r, g, b;
double valornr, valorng, valornb;
double p[][]=new double[3][3];
for(int i=0;i<3;i++)
for(int j=0; j<3; j++) p[i][j]=1.0/9.0;
for(int y=1; y<ima_in.getHeight()-1; y++)
for(int x=1; x<ima_in.getWidth()-1; x++) {
        r = raster.getSample(x,y,0);
        valornr=p[0][0]*(double)raster.getSample(x-1,y-1,0)+p[0][1]*(double)raster.getSample (x,y-1,0)+p[0]
[2]*(double)raster.getSample(x+1,y-1,0)+p[1][0]*(double)raster. getSample(x-1,y,0)+p[1][1]*(double)raster.
getSample(x,y,0)+p[1][2]*(double)raster. getSample(x+1,y,0)+p[2][0]*(double)raster.getSample(x-1,y+1,0)+p[2]
[1]*(double) raster.getSample(x,y+1,0) + p[2][2]*(double)raster.getSample(x+1,y+1,0);
        valorng=p[0][0]*(double)raster.getSample(x-1,y-1,1)+p[0][1]*(double)raster.getSample (x,y-1,1)+p[0]
[2]*(double)raster.getSample(x+1,y-1,1)+p[1][0]*(double)raster. getSample(x-1,y,1)+p[1][1]*(double)raster.
getSample(x,y,1)+p[1][2]*(double)raster. getSample(x+1,y,1)+p[2][0]*(double)raster.getSample(x-1,y+1,1)+p[2]
[1]*(double) raster.getSample(x,y+1,1) + p[2][2]*(double)raster.getSample(x+1,y+1,1);
        valornb=p[0][0]*(double)raster.getSample(x-1,y-1,2)+p[0][1]*(double)raster. getSample(x,y-1,2)+p[0]
[2]*(double)raster.getSample(x+1,y-1,2)+p[1][0]*(double) raster.getSample(x-1,y,2)+p[1][1]*(double)raster.
getSample(x,y,2)+p[1][2]*(double) raster.getSample(x+1,y,2)+p[2][0]*(double)raster.getSample(x-1,y+1,2)+p[2]
[1]* (double)raster.getSample (x,y+1,2)+p[2][2]*(double)raster.getSample(x+1,y+1,2);
        wraster.setSample(x,y,0,(int)(valornr+.5));
        wraster.setSample(x,y,1,(int)(valorng+.5));
        wraster.setSample(x,y,2,(int)(valornb+.5));}
        return ima_out;
};
```

```
public void ShowImage(BufferedImage ima, String image_name) {JLabel lsrc2 = new JLabel(new ImageIcon(ima));
getContentPane().add(new JScrollPane(lsrc2));
setTitle("Java2DImageDisplay: =>" + image_name);
setSize(ima.getWidth()+40, ima.getHeight()+40);
setVisible(true);
setDefaultCloseOperation(JFrame.EXIT_ON_CLOSE); }
}
```

JHLABS® IMAGE EDITOR

Similar to other image editors, www.jhlabs.com was built on Swing technology for the Internet environment (once more, avoid the use of Opera®, Edge® and Chrome® for this Java Applet option). Other available features can be found here including multiple image layers, brushes (paint, clone, erase, etc.), blending modes (add, screen, darken, etc.), transformations (scale, rotate, shear and perspective), effects and filtering operations. It can also read and write multiple image formats (see Figures 24,25 and 26).

LIGHTBOX® IMAGE EDITOR

This Applet from the Maryland Institute for Technology® is a flexible tool for online GIF/JPG/PNG image processing, in which a few number of images can be moved, juxtaposed, repositioned and resized at will (see Figures 27 and 28). It offers a few simple image processing functions (contrast adjustment, color to greyscale conversion and inversion) with the capability to save the user's session output (.vlb extension file) on a laptop, re-open a previous session in its saved state and add individual images to the original display.

Table 13. Java code of Roberts' cross operator with a JPEG file (robert.jpg)

```
Import java.io.File;
import java.awt.image.BufferedImage;
import java.awt.image.Raster;
import java.awt.image.WritableRaster;
import javax.swing.JFrame;
import javax.swing.JLabel;
import javax.swing.JScrollPane;
import javax.swing.ImageIcon;
import javax.imageio.ImageIO;
public class FiltroRoberts extends JFrame {
public static void main(String args[]) {
JFrame.setDefaultLookAndFeelDecorated(true);
String ima_name_in = “robert.jpg”;
String ima_name_out = “robertSolution.jpg”;
BufferedImage ima_in = LoadImage(ima_name_in);
FiltroRoberts frame1 = new FiltroRoberts();
frame1.ShowImage(ima_in, ima_name_in);
BufferedImage ima_out = ProcessImage(ima_in);
FiltroRoberts frame2 = new FiltroRoberts();
frame2.ShowImage(ima_out, ima_name_out);
SaveImage(ima_out, ima_name_out); }
public static BufferedImage LoadImage(String image_name){
BufferedImage ima_in = null;
File file = new File(image_name);
try {
```

continued on following page

Table 13. Continued

```
ima_in= ImageIO.read(file);
} catch(Exception e) {
System.out.println("Image file '" + image_name + "' does not exist.");
System.exit(0); }
System.out.println("Name of file: "+image_name+" Type of image: "+ima_in.getType());
System.out.println("Size of image: Columns "+ima_in.getWidth()+" Lines
            "+ima_in.getHeight());
return ima_in; }
public static void SaveImage(BufferedImage dest, String image_name) {
try {
ImageIO.write(dest, "jpg", new File(image_name));
} catch (Exception e) {
System.out.println("Not saving file... ");
System.exit(0); } }
public static BufferedImage ProcessImage (BufferedImage ima_in) {
BufferedImage ima_out = new
            BufferedImage(ima_in.getWidth(),ima_in.getHeight(),ima_in.getType());
Raster raster = ima_in.getRaster();
WritableRaster wraster = ima_out.getRaster();
double valornr, valorng, valornb;
int[] v = new int[9];
for(int y=1; y<ima_in.getHeight()-1; y++)
for(int x=1; x<ima_in.getWidth()-1; x++){
            valornr = Math.sqrt(((double)raster.getSample(x+1,y-1,0) - (double)raster.getSample(x,y,0))*((double)
raster.getSample(x+1,y-1,0) - (double)raster.getSample(x,y,0))+((double)raster.getSample(x-1,y-1,0) - (double)
raster.getSample(x,y,0))*((double)raster.getSample(x-1,y-1,0) - (double)raster.getSample(x,y,0)));
            valorng = Math.sqrt(((double)raster.getSample(x+1,y-1,1) - (double)raster.getSample(x,y,1))*((double)
raster.getSample(x+1,y-1,1) - (double)raster.getSample(x,y,1))+((double)raster.getSample(x-1,y-1,1) - (double)
raster.getSample(x,y,1))*((double)raster.getSample(x-1,y-1,1) - (double)raster.getSample(x,y,1)));
            valornb = Math.sqrt(((double)raster.getSample(x+1,y-1,2) - (double)raster.getSample(x,y,2))*((double)
raster.getSample(x+1,y-1,2) - (double)raster.getSample(x,y,2))+((double)raster.getSample(x-1,y-1,2) - (double)
raster.getSample(x,y,2))*((double)raster.getSample(x-1,y-1,2) - (double)raster.getSample(x,y,2)));
            valornr = (valornr+valorng+valornb)/3.;
            valorng = valornr;
            valornb = valornr;
            wraster.setSample(x,y,0,(int)(valornr+.5));
            wraster.setSample(x,y,1,(int)(valorng+.5));
            wraster.setSample(x,y,2,(int)(valornb+.5)); }
            return ima_out;}
```

```
public void ShowImage(BufferedImage ima, String image_name) {JLabel lsrc2 = new JLabel(new ImageIcon(ima));
getContentPane().add(new JScrollPane(lsrc2));
setTitle("Java2DImageDisplay: =>" + image_name);
setSize(ima.getWidth()+40, ima.getHeight()+40);
setVisible(true);
setDefaultCloseOperation(JFrame.EXIT_ON_CLOSE); }
}
```

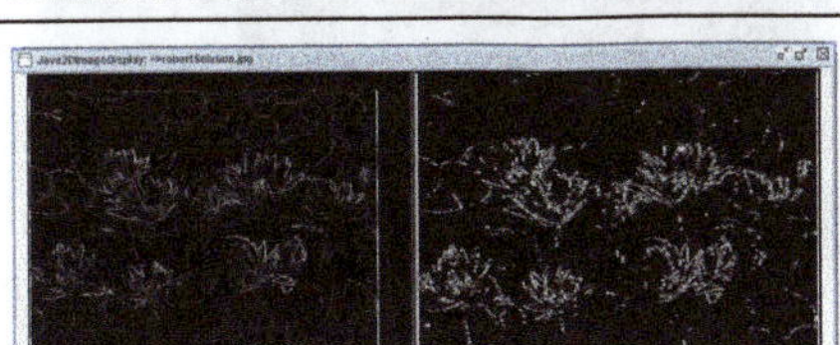

oa

Figure 22. GIF or JPG formats are, quite often, the best option for image processing

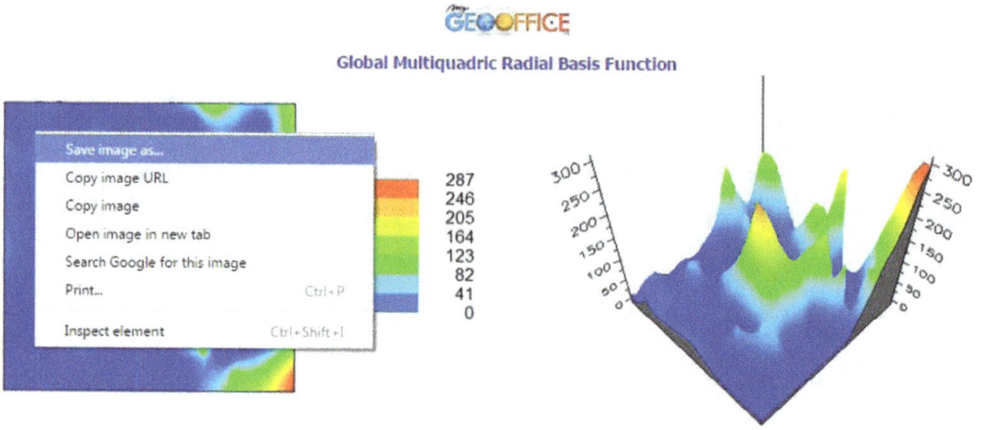

Figure 23. It is then possible to use the saved image (GIF in this particular case) as an input on any myGeoffice© filters for map enhancement

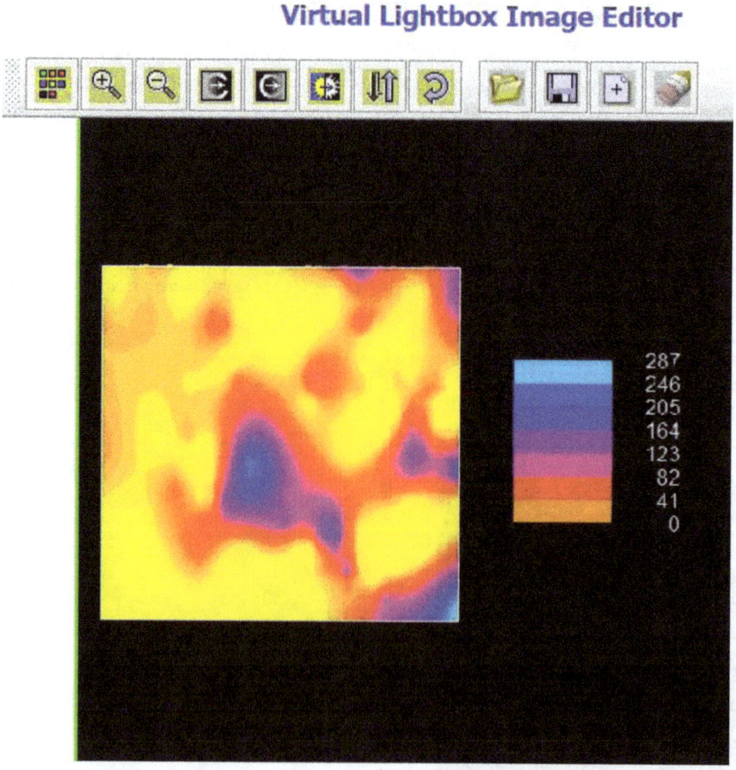

Figure 24. For illustration purposes, JHLabs® was tested with two different filtering capabilities. Lightbox® and Special Filtering® are other possibilities within myGeoffice©

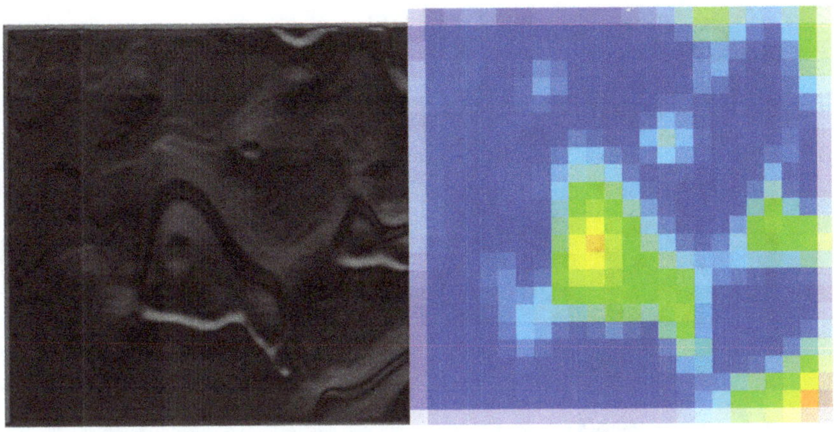

Figure 25. By using one of the three images supplied by myGeoffice©, it is possible to apply filters to produce different versions of the same image, for instance

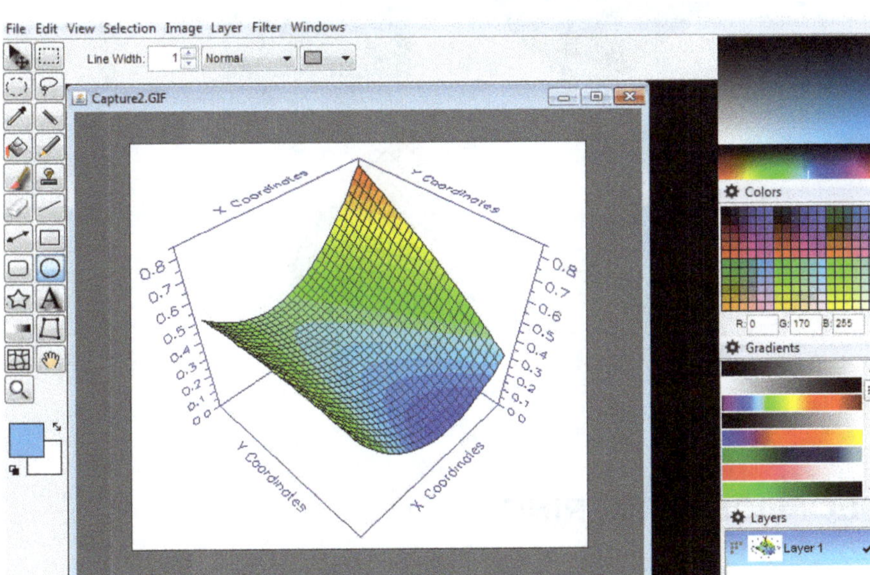

Figure 26. Six examples of JHLabs® special effects of Figure 25: chrome (top-left), sphere (top-right), stamp (middle-left), crop (middle-right), rotate 180⁰ (bottom-left) and kaleidoscope (bottom-right)

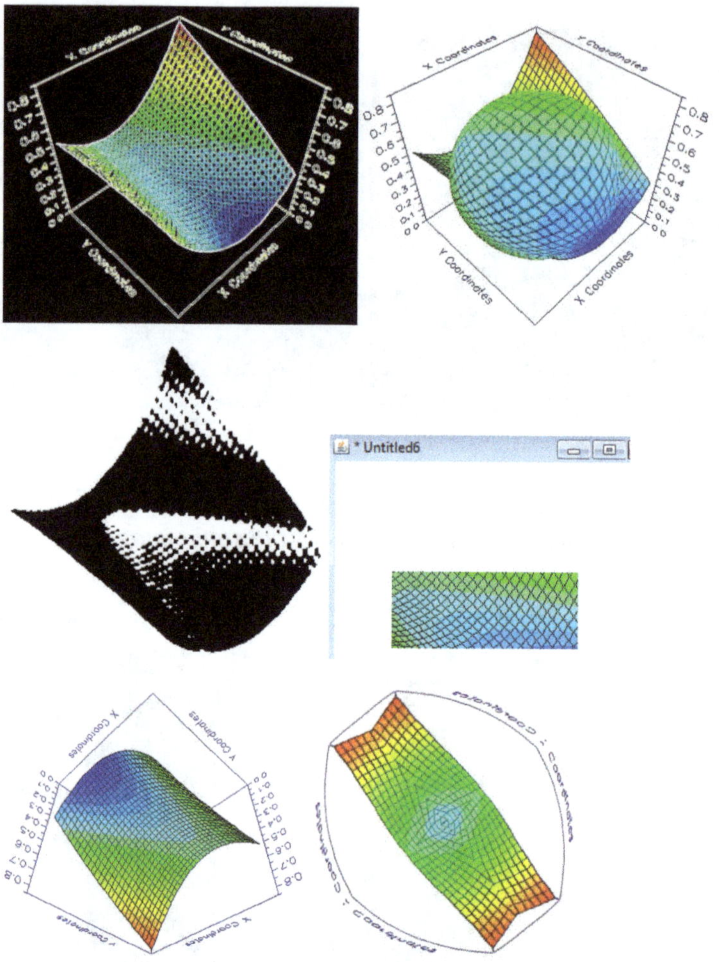

MYGEOFFICE© FILTERING

This preference relies on the concept of a 3×3 convolution matrix that allows the transformation computation of each pixel towards the original image (see Table 14). From the GIS standpoint, this matrix is known as focal operator, i.e., the output values are a function of the neighboring cells whose image processing results will fall into two categories: (1) Smoothing to reduce noise or outliers; (2) Gradient to highlight rapid changes such as borders. Again,

Figure 27. By using two simultaneous images generated by myGeoffice©, Virtual Lightbox© was used to apply a set of limited effects on both pictures

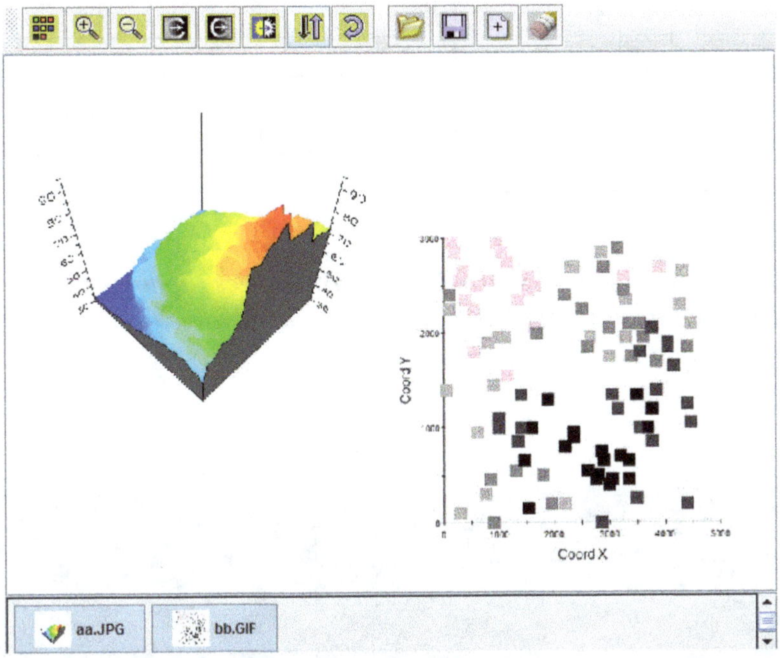

Figure 28. It is also possible to change position or apply dissimilar effects to each picture individually

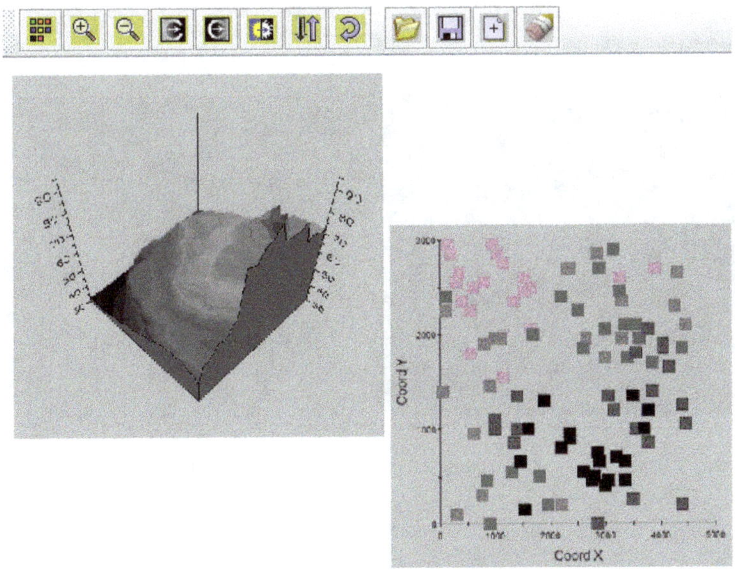

*Table 14. In this particular case, the value of the target pixel becomes (58*0)+(59*1)+(53*0)+(42 *0)+(44*0)+(49*0)+(50*0)+(53*0)+(60*0)=59*

65	55	54	61	62												
63	58	59	53	51			0	1	0							
45	42	44	49	50	×		0	0	0		=			59		
52	50	53	60	52			0	0	0							
60	63	62	67	47												

Figure 29. Demonstration of sharpen (top-right), edges (bottom-left) and emboss (bottom-right) image filtering based on one of the pictures (top-left) supplied by myGeoffice© within Firefox®

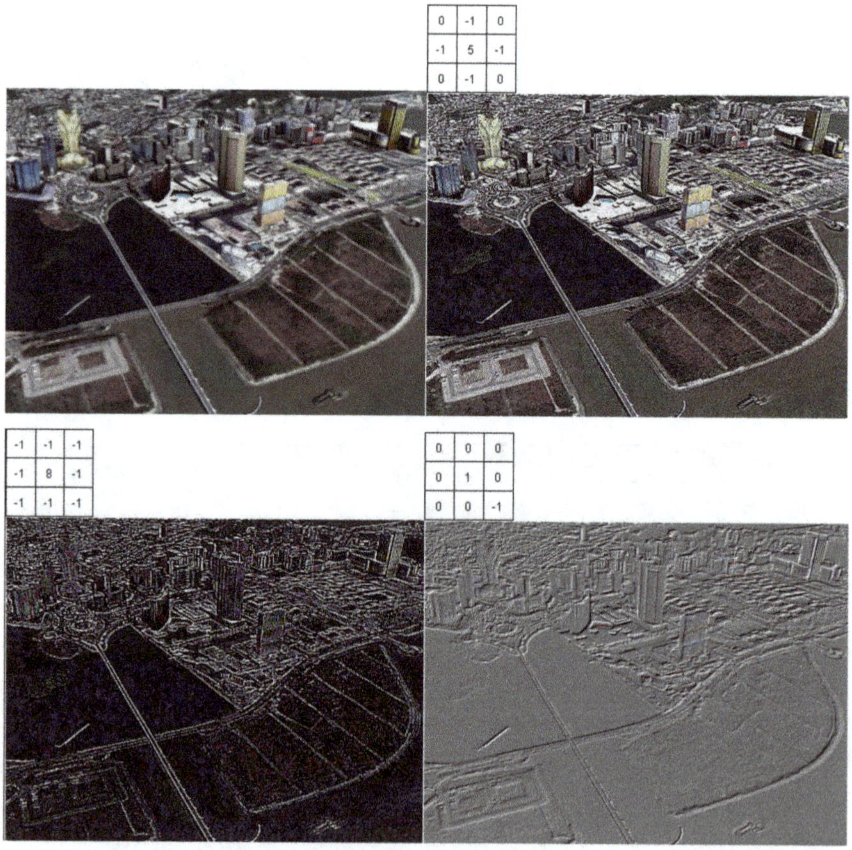

the concept of convolution is the treatment of a matrix (raster image) by the kernel matrix (quite common, 3×3 or 5×5 square matrices). As illustrated in Figure 29 (below-left), the image matrix can be visualized, where each pixel is marked with its value (its range varies between 0 and 255). In the middle lies the kernel. The convolution result is shown on the right. During this continuous image processing, the filter reads successively all pixels into the kernel action area and multiplies/adds each of their values by the corresponding kernel weight.

REFERENCES

Apple. (n.d.). *Developer documentation.* Retrieved from: developer.apple.com/library/ios/documentation/Performance/Conceptual/vImage/ConvolutionOperations/ConvolutionOperations.html

Felgueiras, C., & Negreiros, J. (2008). *Introdução ao processamento digital de imagem – Implementação em Java.* FCA-Lidel.

Filho, O., & Neto, H. (1999). *Processamento digital de imagens.* Brasport.

Chapter 3
Point Analysis

ABSTRACT

Point analysis concerns the study of spatial arrangements of points in space (usually 2D) under a particular time-frame. Its applications include a wide range of areas such as astronomy, ecology, criminology and epidemiology. A case control study that compares point patterns of living organisms to determine if there were significant differences in their arrangements is a good example of this set of techniques. The present chapter comprehends the six main procedures of point analysis of myGeoffice©, including some theoretical background for each approach.

CLUSTER PROXIMITIES MATRIX

Clustering methods are descriptive techniques (not interpretative let alone predictive) for patterns and outliers among available samples. It classifies a set of observations into two or more mutually exclusive unknown groups with the aim of data reduction. The purpose is to discover a system of organizing observations into groups, where members of those groups share common properties. For instance, the researcher may create clusters of customers who have similar buying habits in specific supermarkets for geosegmentation marketing purposes. From the city planning perspective, identifying groups of houses according to their type (financial value and geographical location) is another example of clustering applications.

DOI: 10.4018/978-1-5225-3270-5.ch003

Anything can be clustered but a good methodology should produce a high intra-class similarity and a low inter-class similarity. It is critical to highlight that clustering analysis additionally looks for individual outliers (not only patterns) among the data as well. Different distance (Euclidean, Manhattan, Minkowski or Chebychev) metrics and clustering (partitional K-means or agglomerative hierarchical) algorithms often lead to different outcomes too. Thus, it is up to the researcher to decide which approach and parameters should be applied to his/her problem. A typically hard question raised on statistical cluster analysis regards the input number of clusters to be defined and with a tremendous significance influence on the clustering result. As a principle (and working as a default start value), this parameter should be a number between 3 and 5 but no law should be setup.

From the computational standpoint, partional clustering of K-means works as follows: (1) Set K as the desired number of clusters; (2) Select randomly K representative elements called centroids; (3) Compute the distance of each pattern (points) from all centroids; (4) Assign all data points to the centroid with the minimum distance; (5) Update the centroids as the mean of the element belonging to each cluster and compute a new cluster membership; (6) Check the convergence condition, i.e., if all data points are assigned to the same cluster with respect to the previous iteration (all the centroids remain the same) then stop the process. Otherwise, reapply the assignment process starting from step 3.

Alternatively, the algorithm of the hierarchical clustering (myGeoffice© option) works as follows: (1) Calculate the distance between all data points; (2) Cluster the data points to the initial clusters; (3) Calculate the distance metrics between all clusters; (4) Repeatedly gather the most similar clusters into a higher level of similarity; (5) Repeat steps 3 and 4 for the most high-level clusters. Another extra parameter to be defined from this agglomerative clustering technique is the choice of the distance function between clusters, i.e., single linkage (distance between the closest neighbors), complete linkage (distance between the furthest neighbors), central linkage (distance of centroids) and average linkage (average distance of all patterns in each cluster).

From a GIS perspective, this last methodology has proven to be quite useful in clustering neighborhoods (zip codes) into different sets based on census information (such as population density, income and age) regarding decisions about where to locate future depots. Nonetheless, this procedure is considered a-spatial since the geographical component is not used under this algorithm. Moreover, this numerical approach does not explain the reasons why a specific phenomenon behaves in a particular way. Basically, it only supplies clues or evidence for the scientist to investigate possible explanations or causes for that group's behavior.

Within myGeoffice©, the hierarchical cluster analysis starts with a single data column, where each cell holds the observation value for each region. The key problem consists of how to combine multiple measures into a single number and the similarity between two observations, where the grouping concept is based on value's proximity to each region. myGeoffice© depicts two alternatives for this parameter: (A) Absolute value of the difference between scores (Single Difference); (B) Squared difference between the two measures (Squared Euclidean Difference). After the comparison matrix among data values has been assessed, the last step is to join the present regions into clusters based on the nearest values' distances. This simple linkage computes the distance between two subgroups as the minimum distance between any two members of opposite groups. The following step reshuffles the remaining regions into clusters by re-assigning them based on the similarity among regions. The recalculation of the nearest distances is accomplished and this process continues recursively until no further regions are left out.

Based on Figure 1, Figure 2 and Figure 3, cluster analysis will always produce a group that may or may not prove useful for classifying regions. It starts with each region describing a subgroup and then combines subgroups into more inclusive subgroups until only one group remains. Again, it is up to the investigator to explain this arrangement. Logically, if those groups discriminate certain patterns then cluster analysis can be useful. For example, if someone groups zip code areas into four categories (age, gender, education and income) with the capability to discriminate between wine drinking behaviors, it would be useful information for an entrepreneur who was interested in expanding his/her wine retail business into new areas (Stockburger, 2014).

A good example of this technique is given by Bălan, Toma, Dobre and Soare (2015) to identify viable solutions for organizing organic producers groups and supply chains at a county level. With this purpose in mind, their research goal was to evaluate which was the best networking solution for forty organic farmers in Calarasi County, Romania. Regarding the agricultural pattern of each cluster, their evaluation demonstrated that only one cluster is viable for establishing a global producers' group with the following characteristics: (a) Farmers must be in a proximity range of maximum 40 Km; (b) At least, 75% of farmers should have a similar farming system type; (c) Over 65% of

Figure 1. In this case, the data presented here represents the rainfall level in mm of the six southern districts of Portugal (1-Faro, 2-Santarém, 3-Setubal, 4-Ėvora, 5-Beja, 6-Portalegre) on February 8, 2013. The chosen measure of similarity among the considered regions was the Squared Euclidean Distance

farmers must own big farms; (d) The best place for a joint pooling, storing or selling point is the cluster center (Calarasi). For these authors, the viable solution for this clustering allows networking promotion among farms with similar agricultural patterns to contribute to the creation and establishment of a global cost-effective supply chain.

DIJKSTRA SHORTEST PATH

Before this algorithm is presented, some graph concepts are introduced next. Graphs are one of the most flexible data structures in computing. Formally, graphs are used in problem solving defined by inter-relations between nodes. In a worldwide perspective, web browsing can be modeled by graphs. Another major graph issue, for instance, is the traveling salesman algorithm to know the shortest route for a given tourist itinerary. Floyd and Dijkstra are two

Figure 2. Consistent with the present output, sub-regions 1 and 5 and 3 and 4 form two groups while both region 1 and 6 are considered "runt" because they do not enter any group until near the end of the procedure

Proximities Matrix of Cluster Analysis (Part II)

	Region 1	Region 2	Region 3	Region 4	Region 5	Region 6
Region 1	0	31	10	6	4	16
Region 2	31	0	41	37	27	15
Region 3	10	41	0	4	14	26
Region 4	6	37	4	0	10	22
Region 5	4	27	14	10	0	12
Region 6	16	15	26	22	12	0

0	41	37	27	15
41	0	4	14	26
37	4	0	10	22
27	14	10	0	12
15	26	22	12	0

Merger of (group that includes) region 1 with (group that includes) region 5

0	37	27	15
37	0	10	22
27	10	0	12
15	22	12	0

Merger of (group that includes) region 3 with (group that includes) region 4

0	27	15
27	0	12
15	12	0

Merger of (group that includes) region 4 with (group that includes) region 5

0	15
15	0

Merger of (group that includes) region 5 with (group that includes) region 6

0

Merger of (group that includes) region 2 with (group that includes) region 6

Figure 3. These clustering results are best described using a dendogram or binary tree, a layout not generated by myGeoffice©. In terms of interpretation, it can be said that districts 1 and 5 and 3 and 4 hold a similar rainfall outline for that particular date, a situation not followed by regions 6 and 2. It is the responsibility of the meteorologist, for instance, to explain this climate behavior

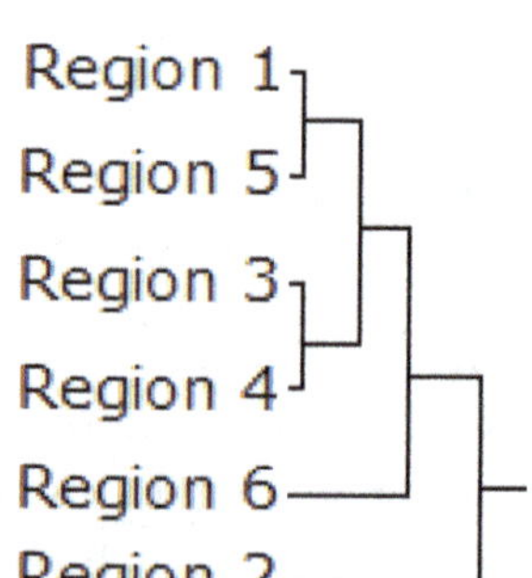

different ways to achieve this aim. If the former is considered a dynamic programming method, the latter is contemplated as a greedy one.

A graph consists of a set of links connecting pairs of vertices. Based on this common concept, the adjacent relationship can be considered symmetric if (a, b) link and (b, a) link between both vertices are the same. If these links are directional (arcs), then they are represented by arrows, resulting into directed graphs. Thus, (a, b) link represents a directional connection from a to b vertices (both vertices are adjacent to each other). Self-loops are vertices whose initial and final vertex is the same (check vertex two in Figure 4, center). Be aware that in non-directional graphs, self-loops are not allowed.

In an undirected graph, the degree of a vertex is the number of links incident on it. For example in Figure 4 (left), vertex zero is considered a degree of two as vertex three is a degree of zero (also called isolated or not connected). In a directed graph, the degree of a vertex is the number of arrows leaving it (out-degree) plus the number of arrows arriving at the vertex (in-degree). Hence, in Figure 4 (center), vertex two has degree four (two arrows leaving it out and two other arriving). On the other hand, the right graph is designated isomorphic because it is vertically and horizontally symmetrical.

By definition, the path is the sequence of vertices that allows connecting two vertices. Technically, one says that both vertices are reachable. Other essential theoretical concepts lie ahead in this field of study:

- An undirected graph can be transformed into a directed graph as illustrated in Figure 5 (left).
- The graph of Figure 5 (right) is a sub-graph of Figure 5 (center-left).
- A weighted graph holds transition weights (costs, distances, times or other impedances) associated with their arrows.

Figure 4. Examples of undirected (left and right) and directed (middle) graphs. Notice that the center graph is strongly connected because almost all vertices are directly reachable from one another while, in the left graph, vertex 3 is unreachable from the remaining ones

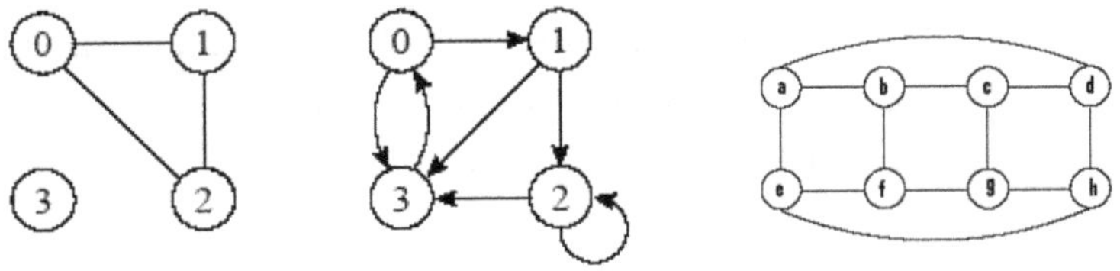

Figure 5. Examples of graphs

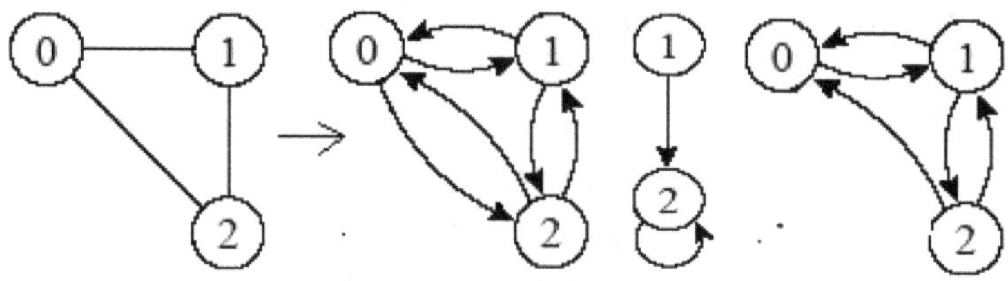

- A graph is said to be free if it is not directed acyclic and connected (commonly called tree).

Another central point of analysis relates to the flow search strategy on a graph. Basically, two techniques can be found:

- **Depth-First Search:** The links are explored from v vertex, the most recently discovered that still has unexplored links (lower level). When all the adjacent links to v have been explored, the search returns to the starting point to explore the vertices left out from the vertex from which v was discovered (upper level). In other words, one starts at the root and explores as far as possible along each branch before backtracking.
- **Breadth-First Search:** Analogous to the movement of a concentric circle generated by a stone dropped into a still water surface, this approach expands the boundary search uniformly between the discovered and the undiscovered vertices. That is, the algorithm discovers all vertices at a distance k from the origin vertex before discovering any vertex at a distance k+1.

Although myGeoffice© only covers Dijkstra regarding the shortest path issue, some highlights will be given next regarding the Floyd approach. Consider L as the initial matrix of l_{ij} elements, which represents the different cost lengths of each (i,j) arc. As a result of an iterative process, everything begins by taking $L^{(0)}=L$ and proceeds to the following $L^{(1)}$ cycle in accordance to the next criteria: $L_{ij}^{(1)}=\min(l_{ij}^{(0)},l_{it}^{(0)}+l_{tj}^{(0)})$, where $L_{ij}^{(1)}$ stands for the minimum path length from i to j while t equals the interaction cycle (for instance, $L_{23}^{(5)}=\min(l_{23}^{(4)},l_{25}^{(4)}+l_{53}^{(4)})$). In general terms, matrix $L^{(k)}$ is determined from $L^{(k-1)}$ matrix and the application of the next expression, i.e., $L_{ij}^{(k)}=\min(l_{ij}^{(k-1)},l_{it}^{(k-1)}+l_{tj}^{(k-1)})$, where $l_{it}^{(k-1)}$ and $l_{tj}^{(k-1)}$ symbolizes the arc weight length of (i,t) and

Figure 6. Example of a graph being analyzed by the Floyd algorithm

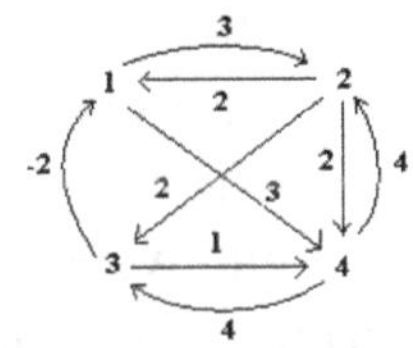

L(1)	1	2	3	4
1	0	3	∞	3
2	2	0	2	2
3	-2	∞	0	1
4	∞	4	4	0

L(2)	1	2	3	4
1	0	3	5	3
2	2	0	2	2
3	-2	∞	0	1
4	6	4	4	0

L(3)	1	2	3	4
1	0	3	5	3
2	0	0	2	2
3	-2	1	0	1
4	2	4	4	0

L(4)	1	2	3	4
1	0	3	5	3
2	0	0	2	2
3	-2	1	0	1
4	2	4	4	0

(t,j), respectively, of the previous matrix. Under this algorithm, the number of iterative cycles equals the number of nodes.

Figure 6 illustrates a graph (including costs) between the four phases of a hypothetical time project. L(1) specifies the initial links and costs between the four states. Naturally, its diagonal is zero. Nodes that hold no connection between them, the ∞ (infinitive) value should be assigned. Considering the L(1) matrix and all links from/to one vertex, the 1-3 and 4-1 links are likely to be minimized through the connections 1-2-3 (five units cost) and 4-2-1 (six units cost), respectively. After these changes are accomplished, the new matrix L(3) results from the next two operations: (A) $L_{2,1}^{(3)}=\min(L_{2,1}^{(2)}, L_{2,3}^{(2)}+L_{3,1}^{(2)})=\min(2,0)=0$; (B) $L_{4,1}^{(3)}=\min(L_{4,1}^{(2)}, L_{4,3}^{(2)}+L_{3,1}^{(2)})=\min(6,2)=2$. The same procedure must be applied to obtain L(4). It turns out that the cost values are already all minimized and, hence, the final result is not subject to change.

Conceived in 1959 by the Dutch scientist Edsger Dijkstra, his algorithm is a graph search procedure that solves the single-source shortest path problem for a graph with non-negative path costs. For a given source vertex in a graph, this procedure finds the path with the lowest cost (the shortest path) between that node and every other node. As an alternative, it can be used for finding costs of shortest connections from a single vertex to another destination vertex by stopping the algorithm once the shortest path to the desired destination vertex has been determined.

If the endpoints of the graph represent cities and the link costs represent driving distances between pairs of cities connected by a direct road, Dijkstra's algorithm can be used to find the shortest route between one city and the remaining ones. As a result, the shortest path first is widely used in network routing protocols among routers, gateways and switches, most notably IS-IS (Intermediate System to Intermediate System) and OSPF (Open Shortest Path First).

Once again, Dijkstra is an algorithm to solve the problem of the shortest path in a graph with weighted edges in which the weights are not negative. Generically, this algorithm maintains three pieces of information of the current v vertex: The shortest way to reach it, the overall length and the predecessors of v. From a practical point of view, consider the graph of Figure 7. The terminology used in this step-by-step analysis is as follows:

- **S:** Vertices set of the graph that has already been incorporated into the current shortest path;
- **$\prod$(j):** Quantity measure that indicates the minimum length from the last vertex i to j;
- **r_j:** Set of vertices that hold direct connections with the last *i* vertex.
- $\bar{S}$: Vertices set of the graph not yet incorporated into the current shortest path; In this particular case, the iterations of this algorithm are as follows:
    a. S={1}, $\prod(1)=0$, $r_1=\{2,3\}$, $\prod(2)=7$, $\prod(3)=1$, $\prod(4)=\prod(5)=\prod(6)=\infty$. The starting point is node one and the nodes connected to it are nodes 2 and 3.

Figure 7. Graph to be analyzed by the Dijkstra's algorithm

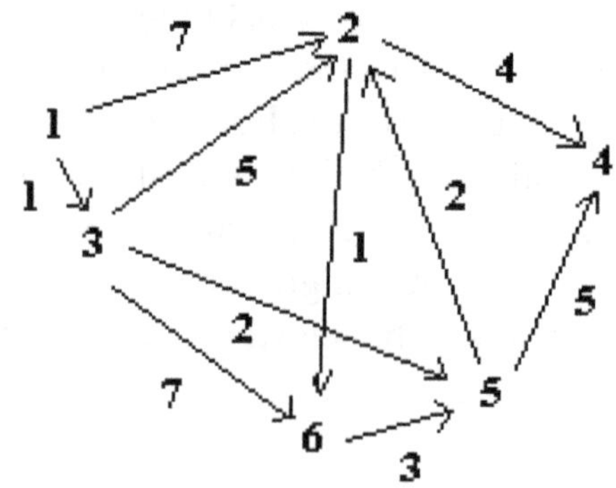

b. As $\prod(3)=1$ presents the minimum distance cost between node 1 and node 2, we obtain j=3, S={1,3} and r_1={2,5,6}.

c. $\prod(2)=\min(7,1+5)=6$, $\prod(5)=\min(\infty,1+2)=3$, $\prod(6)=\min(\infty,1+7)=8$. Again, the initial cost becomes infinitive when no direct connection can be found.

b. As $\prod(5)$ is the minimum, we obtain j=5, S={1,3,5} and r_5={4,2}.

c. $\prod(2)=\min(7,2+1+2)=5$ and $\prod(4)=\min(\infty,1+2+5)=8$.

b. As $\prod(2)$ becomes the minimum, we then obtain j=2, S={1,3,5,2} and r_2={4,6}.

c. $\prod(4)=\min(8,5+4)=8$ and $\prod(6)=\min(8,5+1)=6$.

b. As $\prod(6)$ equals the minimum, we obtain j=6, S={1,3,5,2,6} and r_6={5}.

c. It turns out that node 5 already belongs to S while node 4 still belongs to $\bar{S}$. As expected, the latter should be transferred to S={1,3,5,2,6,4} set through the connection with the lowest available cost path already defined, i.e., from node 5. Therefore, the shortest lengths between vertex 1 and the remainders five nodes becomes $\prod(3)=1$, $\prod(5)=3$, $\prod(2)=5$, $\prod(6)=6$ and $\prod(4)=8$.

In network analysis, the number of distinct links (connections) and vertices (nodes, edges or endpoints) of a planar graph (drawn on the plane in such a way that its connections intersect only at their endpoints) and non-planar

Figure 8. myGeoffice© layout of the first step regarding Dijsktra shortest path, a spatial issue

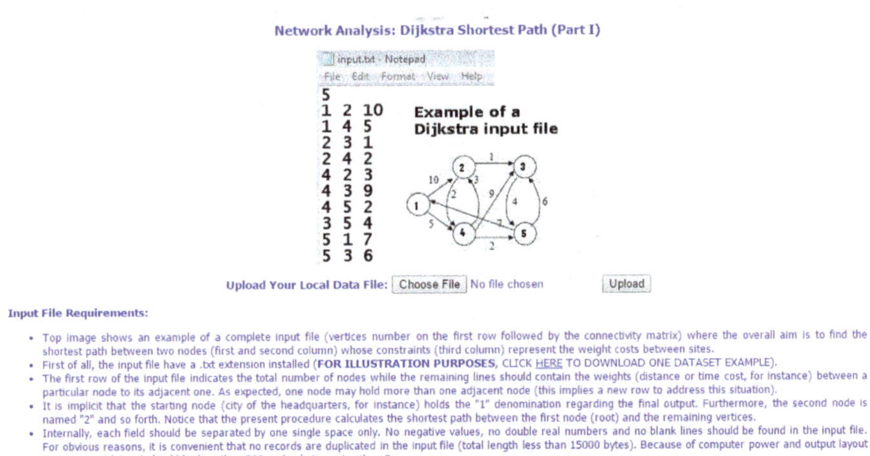

(where links crossing becomes a reality) topology may be used to derive its connectivity level or Gamma index. By definition, this index defines the ratio of the number of connections to the maximum possible number of links. If V denotes the number of vertices (5, for instance) and L denotes the existent distinct number of links between vertices (8, for instance), the minimum number of links (minimal connected network) that is required to connect all these vertices becomes (V-1), i.e., 4. Similarly, the maximum connected network that do not cross or intersect with each other equals 3×(V-2), i.e., 9 in a planar network. So, Gamma equals 0.88=8/9 for this particular case. For a non-planar graph, the maximum number of paths is given by V×(V-1)/2, i.e., 5×4/2=10. Hence, the Gamma Index equals 0.8=8/10. Unsurprisingly, a higher index denotes a great connectivity of the entire network in both situations.

Emergency events such as earthquakes, hurricanes, fires, chemical accidents, nuclear calamities and terror attacks may led to injure or endanger the life and the health of human beings when a large scale crowd has to evacuate from a dangerous area to a safe one. Under this crisis event context, Chen, Shen, Chen and Yang (2014) carried out their research through observing real-time road network and analyzing three different emergency evacuation cases (morning, common and evening peak) including nodes, intersections delay and velocity of vehicles. A dynamic road network model was built for vehicles evacuation based on the Dijkstra algorithm. The obtained outcome provided a predictive method and a theoretical basis for an optimal emergency evacuation path selection and emergency rescue decision in public places, especially for those with high population density.

Figure 9. Based on myGeoffice© dataset example, this layout shows the matrix links outline among the five vertices

Network Analysis: Dijkstra Shortest Path (Part II)

Link Number	From	To	Weight
1	1	2	10
2	1	4	5
3	2	3	1
4	2	4	2
5	4	2	3
6	4	3	9
7	4	5	2
8	3	5	4
9	5	1	7
10	5	3	6

Press here for the Dijkstra geocomputation...

Figure 10. It is then possible to visualize the shortest connections between node one and the remaining four ones (both routes and cost)

Network Analysis: Dijkstra Shortest Path (Part III)

c:\inetpub\wwwroot\executable\dijkstra_final.exe d:\data\499611834dijkstra.txt d:\data\499611834dijkstra.out

The distance from the root (node 1) to vertex 2 is 8 (shortest path: 1, 4, 2)

The distance from the root (node 1) to vertex 3 is 9 (shortest path: 1, 4, 2, 3)

The distance from the root (node 1) to vertex 4 is 5 (shortest path: 1, 4)

The distance from the root (node 1) to vertex 5 is 7 (shortest path: 1, 4, 5)

ELEVATION GRID MAPPING

Digital Terrain Model (DTM) tries to represent height information without any further definition about the surface. Based on a regular grid, myGeoffice© presents this terrain graphical possibility, if the researcher holds a set of elevations in a systematic lattice framework as long as those heights are null or positive. Some restrictions should be highlighted under this selection: (1) The number of rows and columns in the grid must be greater than 9 but less than 91; (2) The ratio between the number of rows and columns must be less or equal to 1.5 in order to avoid major layout distortion; (3) The total number

Figure 11. The input plain file used in this example can be directly downloaded from myGeoffice©. Be aware that the first line of the present input plain file stresses the number of rows (12) and columns (17). As expected, 12×17=204 values should follow next (one per line)

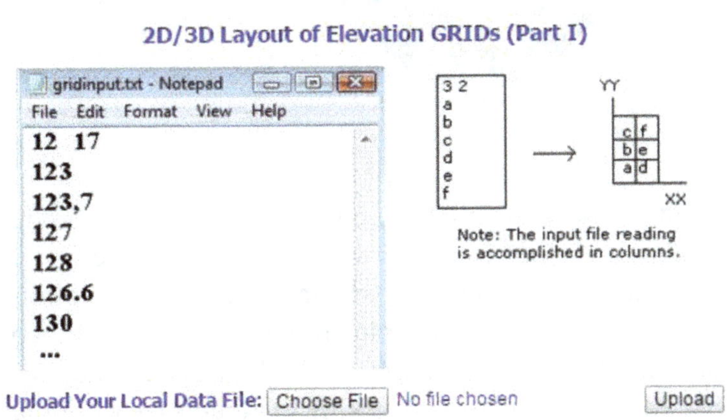

Figure 12. Possible layout of myGeoffice©

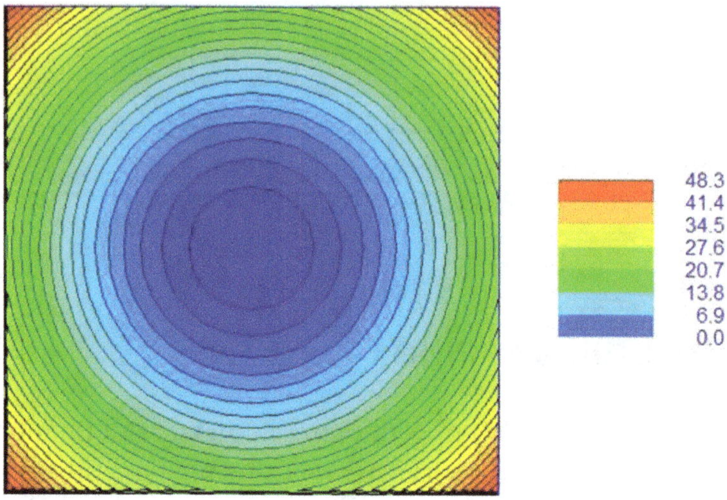

Figure 13. Another possible layout of myGeoffice©

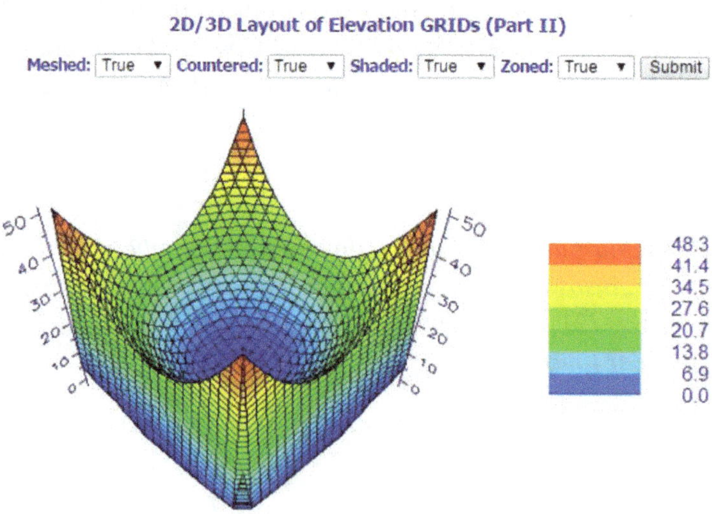

of elevation data should be, at least, equal or greater to the product of rows by columns (if greater, the remaining values are disregarded).

Floodplain maps serve as a basis instrument for determining whether homes or buildings require flood insurance under the National Flood Insurance Program, a $650 billion activity run by the Federal Emergency Management Agency, USA. Meanwhile, FEMA is trying to modernize floodplain maps to better serve this system. Hence, elevation data for floodplain mapping has been used as the main graphical tool to determine whether a building should have flood insurance or not (National Research Council, 2007). Certainly, a set of recommendations can be drawn from a 2D and 3D national digital elevation for policy makers, property insurance professionals, federal, local and state governments and others concerned with natural disaster prevention.

In Santiago, Cape Verde, there is a large population living under a recurrent hydrological risk area, especially along the water streams of the county of Praia. The morphology and configuration of the surface runoff network in the surrounding areas of this county increase that risk mainly due to the confluence of several basins that flow into the urban area. The September and October flash floods do not help, either (Lino, Neves & Negreiros, 2016). Indeed, the areas of the greatest hydrological risks in this district are in the lowland areas (along the flow streams). Despite of the existence of lateral protections, there is water overflow at Vila Nova creek, for instance, because of the large amount of debris carried out from the terrain slopes and the highland areas of Fontes Almeida, St. Martinho and Mitra Barnabé. Moreover, there is a shortage of infrastructure in several ramps of the city, where one can find a high density of soil occupation, especially in the settlements of San Pedro, Safende, Calabaceira, Vila Nova and Fazenda. Using the high precipitation of October 21, 2010, and Macrobuilder of ArcGIS 10.2 of ESRI®, it was possible to calculate the peak accumulated water discharge (almost 1 trillion liters per second) in downtown during that day. As expected, only houses with a minimal altitude of 15m from the margin of the main river were suitable for house insurance.

KRUSKAL MST

Another significant topic on studying graphs concerns the vicinity concept. The neighborhood matrix of a graph with n vertices can be represented as A=[n,n], where A [i,j] equals 1 (or true) if and only if there is an arc from vertex *i* to vertex *j*. Otherwise, the 0 value (or false) should be assumed. Along with the weight cost matrix (if any), all this information may turn out quite useful so that graph algorithms can trigger a set of operations such as:

- Check if there is a linkage between two vertices;
- Get the list of adjacent vertices from any given vertex;
- Add or remove a connection between two nodes;

- Print a graph;
- Remove a given vertex of the graph (including an update of the overall graph);
- Identify the path between two vertices but based on the weight matrix information and according to a specific minimization/maximization criteria (operation widely used within communication routers or within activities planning such as hotels project management at PERT/CPM level).

Under the computation perspective, this adjacent matrix can be saved in two additional ways:

- Through a table of pointers in conjunction with simple linked lists. The pointers table size is given by the number of the available vertices of the graph in conjunction of the associated neighborhood linked list nodes (see Figure 15).

Figure 14. Example of a neighborhood matrix for an undirected graph

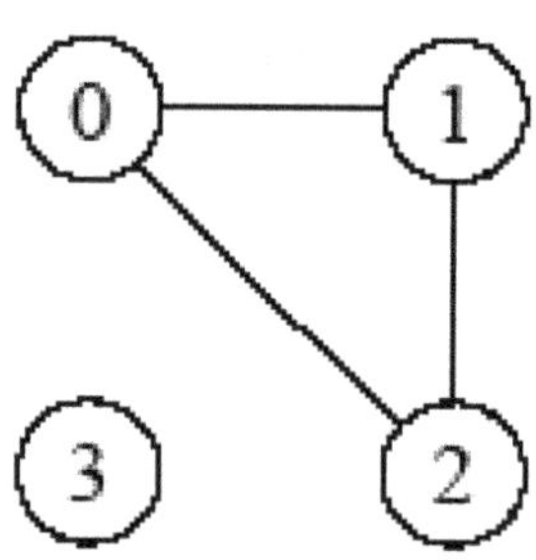

	0	1	2	3
0	0	1	1	0
1	1	0	1	0
2	1	1	0	0
3	0	0	0	0

Figure 15. Example of an adjacent list using dynamic pointers. For instance, node three is isolated and, therefore, there is no linear linked list to it

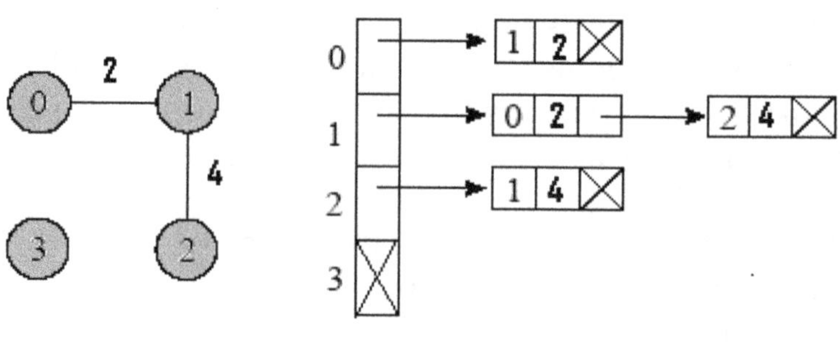

Figure 16. Example of adjacency using two tables

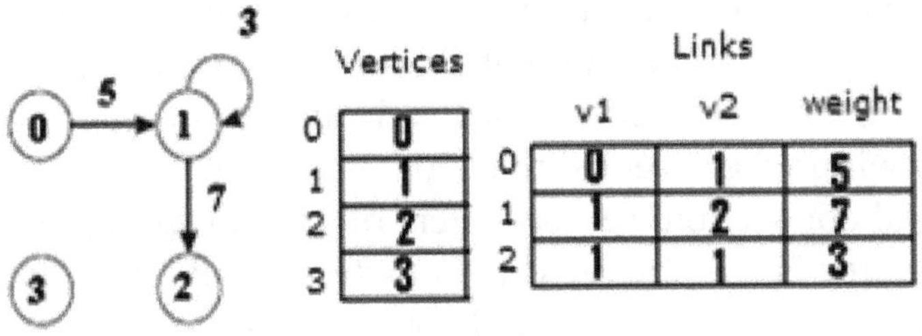

- The array of vertices and links are separated into two different tables, i.e., there is an array of vertices and the associated link matrix, which comprises the following information: Vertex of origin (v1), final vertex (v2) and the impedance weight (see Figure 16).

The minimum spanning tree of a graph defines the connection minimization of all its vertices. Designated by all specialized literature as the minimum spanning tree (MST) issue, telephone, gas and water companies are particularly interested in this problem so as to minimize the telephone cables or the sewage system that connects their customers at various locations. To connect two clients in an *n* client structure, there are *n-1* possible connections. Therefore, MST aim becomes as follows: Within the possible available connections, which configuration uses fewer length cables? Generally, the MST corresponds to one sub-graph containing all vertices connected together so that the cost of the edges is the smallest possible.

Kindly conceive the complete graph shown in Figure 17 with four basic nodes. This graph presents sixteen spanning trees. Suppose now that there are weights or costs associated to each respective link among the four edges.

Figure 17. The sixteen possible configurations for the complete graph with four vertices

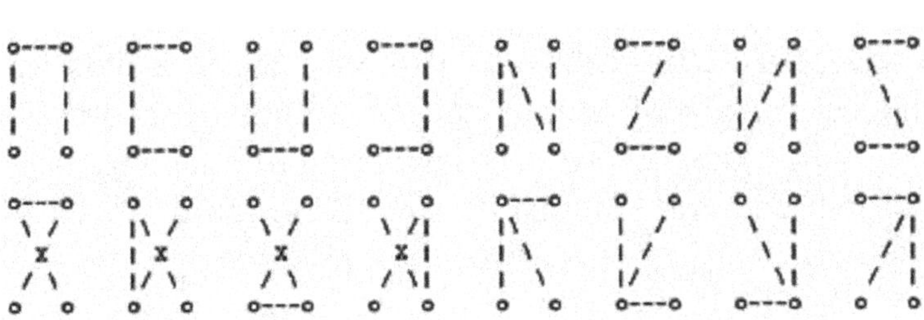

A problem arises: Which spanning tree encloses the minimal cost? Currently, there are two main algorithms for this dilemma: Kruskal and Prim.

Kruskal is a graph algorithm that finds the minimum spanning tree for a connected weighted graph. Specifically, it finds a subset of nodes (towns, for instance) within a connected tree (routes between towns, for instance), where the total connection weights (liters of gas, for example) of all nodes of the present tree is minimized. If the graph is not all connected, it finds a minimum spanning forest, i.e., a MST for each connected component.

Conceptually, this greedy procedure works as follows: (A) Upwardly sort the vertices of G graph according to the weight of the associated edges; (B) Create a S sub-graph (initially empty) of G; (C) For each v vertex belonging to the ordered set of G, ask if it does not belong to S. If this is the case, add v to S. In a more detailed perspective, this procedure works as follows: (1) Creation of a forest F (a set of trees), where each vertex of the graph is a separate tree; (2) Creation of a S set containing all the nodes in the graph; (3) While S is non-empty, (a) Remove a node with minimum weight from S; (b) If that node connects two different trees then add it to the forest by combining two trees into a single one; (c) Otherwise, discard that node; (4) At the end, the forest has only one component and forms a minimum spanning tree of the present graph.

Graphically, Figure 18 illustrates the Kruskal procedure from the graph displayed on the left side, i.e., (A). The reader may check that the selected vertices are ranked according to their ascending weight. Consequently, we have:

- (B) – Connection v1-v4;
- (C) – Connection v5-v6;
- (D) – Connection v5-v3;
- (E) – Connection v5-v4;
- (F) – Connection v1-v2 (bear in mind that although there are other endpoints with lower link weights, both endpoints are already included in the spanning tree);
- (G) – Final configuration.

Under myGeoffice©, the first line of the input file stands for the total number of nodes available of the tree while the following records focus on the available cost connections between each node and the remaining ones (including itself). Consistent with the example illustrated in Figure 19, the second line

Figure 18. A step-by-step visual approach of the Kruskal's algorithm

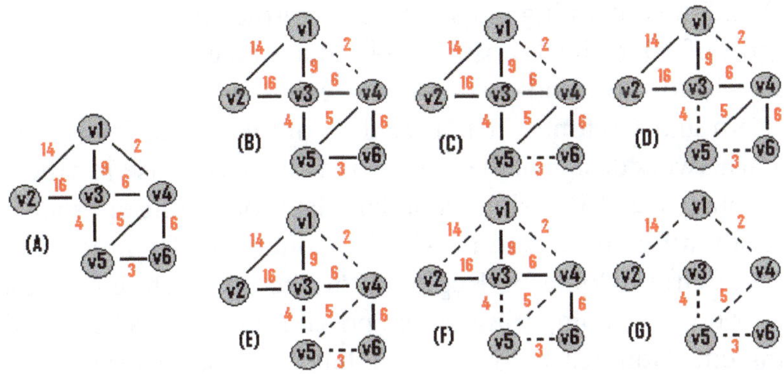

Figure 19. Bear in mind that the weight assigned to each connection symbolizes any particular type of cost such as walking distance, time travel, labor effort or money fee

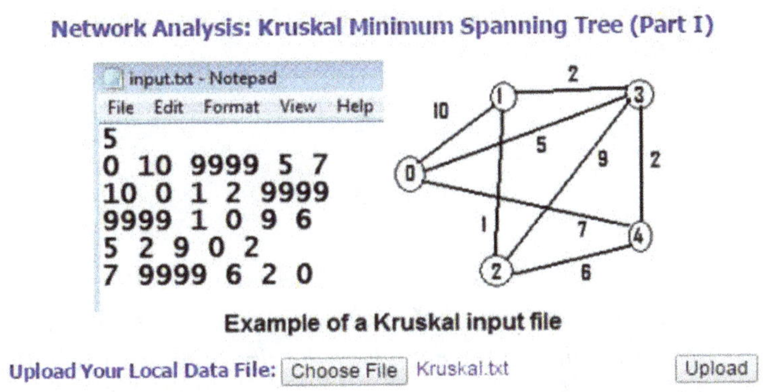

Figure 20. Based on the interpretation of the present output, vertex 1 is connected to vertex 2 and 3 while the latter directly links to nodes 0 and 4. With this overall solution, all nodes can be connected among themselves, with a total cost for running this minimal sub-network of 10 units.

indicates the traveling cost between site 0 (location of the headquarters of the company) and the remaining ones (branches of the company), i.e., cost of 0 to site 0, cost of 10 to site 1, cost of 9999 (due to the lack of connection) to site 2, cost of 5 to site 3 and cost of 7 to site 4.

Lusiantoro and Ciptono (2012) tried to optimize the interrelation and hierarchical network design of the Indonesian's airports (based on the ASEAN Open Sky Policy of 2015). Using secondary data containing distance between airports (both in Indonesia and in ASEAN countries), flight frequency and correlation of Gross Domestic Regional Product for each region, the result analysis was done by comparing others possibilities of MST networks with the existing one. Both researchers found that the existing airport network in Indonesia does not depict the optimal network connections among them when considering the shortest distance and maximization of the regional economic potential correlation of the country. Unsurprisingly, this research identified the airports and regions as hubs and spokes concerning the optimal network and to be considered for a future decision-making reference.

Figure 21. The major outbreaks of Ebola virus epidemic across West Africa in 2014 can be represented with point buffering, where the size of each point buffering circle is proportional to the number of deaths

POINT BUFFERING

Within GIS, buffering refers to the creation of a spatial zone with a specified width around a point, line or polygon (see Figure 21). It also refers to a single zone of a specific distance around spatial objects. These zones or buffers can thus be used in queries to determine which entities occur either within or outside the defined buffer zone.

Since myGeoffice© is not capable of handling shape files (.shp, .dbf), the input file required in this step is a text one (.txt) with the following arrangement: Number of centroids (which cannot exceed 50) and its (x,y) spatial coordinates regarding each point. Therefore, it is mandatory that the number of centroids presented in the first line of the input file matches the next available records. Also, the coordinate's range of those centroids should vary between 0 and 356. As illustrated in Figure 22, the present total buffering region covers an

Figure 22. Due to the considered square grid size of 11.87 pixels, it can happen that, at most, no buffering area (red regions) is displayed. This extreme situation happens if the chosen radius is too small. Further, if the radius is too large, the cells total number may equal the overall available 900 cells and, hence, the above map is considered a whole buffering zone although no red color is shown (dark blue is exhibited, instead)

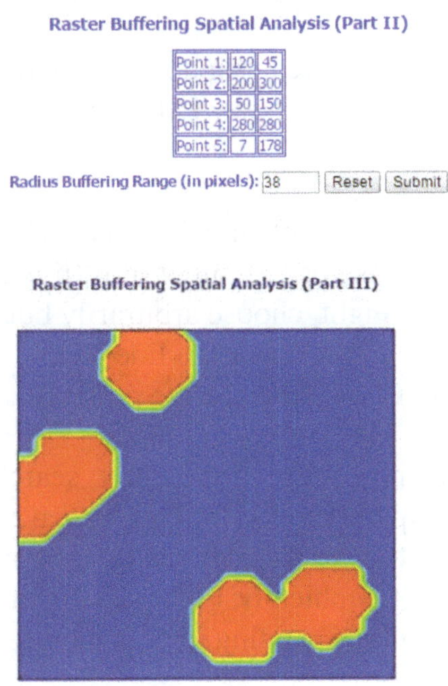

area of 174 cells, i.e., 19,3% of the entire map (this outcome is displayed by myGeoffice© in Part III). Regarding the light blue-green-yellow outlines, two views can be stated: (1) ActiveX® misspecification; (2) Transition borders.

Shoreline along mangrove ecosystem in the East Coast of Surabaya, Indonesia, changed over the years. Prasita (2015) used GIS technology to overlay satellite images between 2002 and 2014 to study shoreline changes, including buffering concepts. According to this researcher, results showed that the shoreline have changed by 5.387m, 5.428m, 5.128m and 7.431m in 2002, 2007, 2011 and 2014, respectively. For the time being, Prasita (2015) reported that the total mangrove conservation area that experienced accretion is close to 287.16ha whereas those with abrasion are 11.02ha. Again, these results are critical for the government in the making of environmental strategies of these coastal areas.

PRIM MST

Analogous to Kruskal, Prim is a graph algorithm that finds the minimum spanning tree (a sub-graph that contains all the available vertices) for a connected weighted graph. Broadly, this procedure continuously increases the size of a tree starting with a single vertex until it spans for the remaining vertices. It works as follows (Wikipedia-Prim, 2015):

- **Input:** A connected weighted graph with V nodes.
- **Initialize:** Enew={} and Vnew={x}, where x is an arbitrary node (starting point) from V.
- Repeat until Vnew = V:
    - Choose connection (u, v) with minimal weight such that u belongs to Vnew while v is not a member of it (if there are multiple edges with the same weight, choose arbitrarily but consistently).
    - Add v to Vnew and add (u,v) to Enew.
- **Output:** Vnew and Enew describe the minimal spanning tree.

Fundamentally, this procedure thoroughly searches for the incident vertices that enable to expand most efficiently the available MST. In a way, this algorithm is similar to breadth-first search (one at a time, it add edges based on the best timing). Graphically, Figure 23 illustrates Prim's procedure that determines the minimum spanning tree. In a more detailed analysis, the

Figure 23. A step-by-step visual approach of the Prim's algorithm

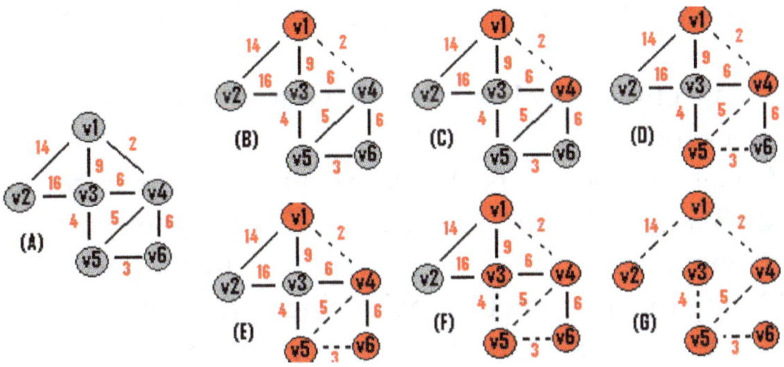

Figure 24. Output of the illustration example supplied by myGeoffice© whose total minimal distance to connect the seven available nodes is 82 units

Network Analysis: Prim Minimum Spanning Tree (Part II)

	Node A	Node B	Node C	Node D	Node E	Node F	Node G
Node A	0	22	0	21	89	123	56
Node B	22	0	38	17	65	2	99
Node C	0	38	0	16	23	99	13
Node D	21	17	16	0	78	100	46
Node E	89	65	23	78	0	61	66
Node F	123	2	99	100	61	0	7
Node G	56	99	13	46	66	7	0

Press here for the Prim geocomputation...

c:\inetpub\wwwroot\executable\prim.exe d:\data\657014042prim.txt d:\data\657014042priml.out

>Adding node A

>Adding edge A-D

>Adding edge D-C

>Adding edge C-G

>Adding edge G-F

>Adding edge F-B

>Adding edge C-E

Total distance: 82

choice of the vertices depends on whether they are adjacent and, in particular, on the lower costs of the corresponding edges. Thus, we have:

- (B) – Connection v1-v4 is setup as connections with v2 and v3 are more expensive.
- (C) – Connection v4-v5 is chosen because the link edges to v2 (from v1) and v3 and v6 (from v2) present higher weights.
- (D) – Connection v5-v6 is associated due to its lower existing cost (considering the nodes not yet belonging to the present MST).
- (E) – Connection v5-v3 for the same previous reason.
- (F) – Connection v1-v2 (note that although there are edges with lower weights, those vertices are already included in the spanning tree).
- (G) – Final configuration.

Under myGeoffice©, the input plain file must follow the next pattern: (1) Number of nodes; (2) Weight costs among nodes in a square symmetric matrix (a connected, non-directed and weighted graph) whose diagonal equals zero. Therefore, the first row of the input file indicates the total number of nodes. The second line should contain the weights (distance or time cost, for instance) between *A* node and the remaining ones (including itself). The lasting records should follow this configuration. Regarding non-connections between nodes, a zero value should be used instead. It is implicit that the starting node (headquarters city, for instance) holds the *A* denomination regarding the final output. Likewise, the second node (city branch) is named *B* and so forth.

For testimonial purposes, Dutta, Patra, Shankar and Verma (2014) have developed a customized ArcGIS® tool using Python script for the MST problem, regarding the Road Transportation Network of Dehradun, India, and considering distance and time as the main impedance factors. For these researchers, this particular implementation can be applied to a nationwide plan by providing the ideal road connectivity to unconnected villages. Additionally, this tool may also be useful for the construction of future railways among cities in an optimal framework.

REFERENCES

Bălan, A., Toma, E., Dobre, C., & Soare, E. (2015). Organic farming patterns analysis based on clustering methods. *Agriculture and Agricultural Science Procedia, 6*, 639-646. doi:.2015.08.11010.1016/j.aaspro

Chen, Y., Shen, S., Chen, T., & Yang, R. (2014). Path optimization study for vehicles evacuation based on Dijkstra algorithm. *Procedia Engineering, 71*, 159-165. doi:10.1016/j.proeng.2014.04.023

Dutta, S., Patra, D., Shankar, H., & Verma, P. A. (2014). Development of GIS tool for the solution of minimum spanning tree problem using prim's algorithm. *International Archives of the Photogrammetry, Remote Sensing & Spatial Information Sciences*, 40–48, 1105.

Lino, J., Neves, A., & Negreiros, J. (2016). *Spatial modulation of the hydrological risk at Praia, Cape Verde.* 8th IGRSM International Conference and Exhibition on Remote Sensing and GIS, Kuala Lumpur, Malaysia.

Lusiantoro, L., & Ciptono, W. S. (2012). An alternative to optimize the Indonesian's airport network design: An application of minimum spanning tree (MST) technique. *Gadjah Mada International Journal of Business, 14*(3), 267–302.

National Research Council. (2007). *Elevation data for floodplain mapping.* Washington, DC: National Academies Press.

Prasita, V. D. (2015). Determination of shoreline changes from 22 to 2014 in the mangrove conservation areas of Pamurbaya using GIS. *Procedia Earth and Planetary Science, 14*, 25–32. doi:10.1016/j.proeps.2015.07.081

Prim. (2015). In *Wikipedia.* Retrieved from: en.wikipedia.org/wiki/prim%27s_algorithm

Stockburger, D. (2014). *Cluster analysis.* Retrieved in 2014 from Multivariate Statistics: www.psychstat.missouristate.edu/multibook/mlt04.htm

Chapter 4
(A)Spatial Regression

ABSTRACT

The global spatial lag model for stationary processes ($Y=\rho WY+X\beta+\varepsilon$, where W represents the neighborhood matrix while ρ is the spatial autoregressive coefficient) and the spatial regimes (individual regression for each sub-region), developed within SpaceStat®, are two possibilities. An extra option is the local continuous variation framework computed by the Geographically Weighted Regression (GWR). The idea here is to adjust a regression model (the βs of the regression model are not the same everywhere) by weighting the neighborhood observations. In this way, the estimation computation will reflect automatic adjustments, according to the distance and values of the available samples. Section "Geographically Weighted Regression" introduces this matter while the next two sections review the conventional bivariate and multivariate OLS regression, respectively. As an extension of the OLS family, non-linear relationships, regression with dummy variables, path analysis and logistic regression are briefly explained under the following section three although myGeoffice© does not cover these practices. Spearman Rank Correlation for ordinal data is presented in the last section.

INTRODUCTION

The goal of linear regression analysis is to uncover which independent variables (IV) contribute in a significant way for a linear relationship of the dependent variable (DV). It is expected, among other factors, that regression errors are independent (within spatial processes, this aspect can be measured by the

DOI: 10.4018/978-1-5225-3270-5.ch004

error term using the Localized Autocorrelation Diagnostic Statistic (LADS), Moran I and Joint Count Statistics, for instance). As a result, Ordinary Least Square (OLS) model usually breaks down due to spatial autocorrelation existence (Negreiros, Painho, Aguilar & Aguilar, 2009).

GEOGRAPHICALLY WEIGHTED REGRESSION

GWR is a specific spatial model, which allows representing non-stationary local phenomena by generating a separate regression equation for every feature analyzed as a means to address spatial variation (Fotheringham, Brunsdon & Charlton, 2002). Hence, GWR allows the modelling of processes that vary over space (non-stationary regions). Since it usually works with aggregated data, this inferential model considers that spatial data may change abruptly (or not) at region boundaries only.

For Lloyd (2007), one key decision emerges from this local inferential approach for spatial non-stationary situations (when the same stimulus provokes a different response in different parts of the study region): The choice of a weighting function (kernel shape VS kernel bandwidth) regarding neighborhood samples (which plays a central role towards each prediction).

In a conventional form, according to Mennis (2006), the GWR equation can be expressed as $\widehat{y_i} = {}^{2}{}_{0}\left(u_i, v_i\right) + \sum_{k=1}^{m} {}^{2}{}_{k}\left(u_{i,}, v_i\right) x_{ik} + \epsilon_i$, where $\widehat{y_i}$ represents the estimate of the dependent study variable for i observation, m equals the number of independent variables, ${}^{2}{}_{0}$ is the intercept value, ${}^{2}{}_{k}$ stands for the

Figure 1. The adaptive (left) and fixed weighted scheme (right) for the weighted matrix. According to Yu & Wei (2011), the fixed scheme might produce large estimate variances where data is sparse. On the other hand, it can mask subtle local variations where data is dense. Indeed, this is the key of GWR in order to measure spatial variation in any study region.

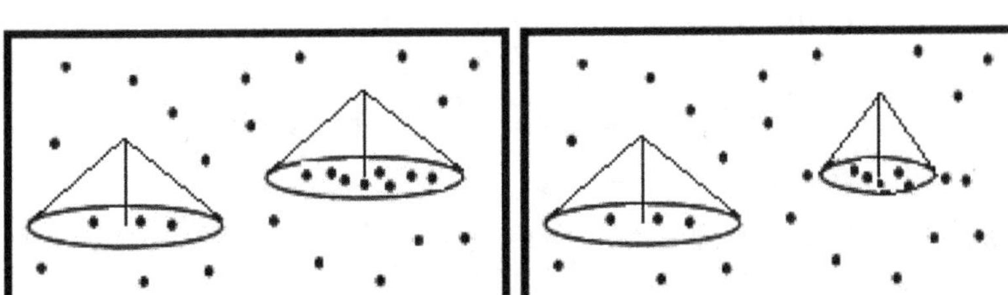

parameter estimate for the independent k variable, x_{ik} denotes the value of the k variable for i observation, (u_i, v_i) regards the 2D coordinate location of i sample while ϵ_i is the residual error. Instead of calibrating a single and global OLS conventional regression equation for the overall area, GWR generates a separate regression equation for each observation. Each equality is then calibrated with different weights of the neighboring observations contained in the dataset, based on the assumption that observations nearby one another have a greater influence on one another's parameter estimates than observations farther apart (Mennis, 2006). In essence, only the nearest samples of each estimated site will be used for the local estimation regression.

Given two independent variables, Y1 and Y2, the local estimation of GWR parameters equals $Z(u_i)=\beta 0(u_i)+\beta 1(u_i)xY1+\beta 2(u_i)xY2+\varepsilon(u_i)$, where (u_i) is the coordinates location at which the estimates of the parameters are being calculated. The GWR solving system becomes $\beta(u_i)=(Y^T\times W(u_i)\times Y)^{-1}\times Y^T \times W(u_i)\times Z(u_i)$, where $W(u_i)$ equals the weight square matrix closely related to u_i position to be estimated towards the neighborhood available samples (the non-diagonal terms equals zero), Y^T stands for the independent variables transverse matrix while $Y^T\times W(u_i)\times Y$ represents the geographically weighted variance-covariance matrix.

Regularly, these weights follow the Gaussian function to incorporate the First Law of Geography or Tobler's Law, i.e., everything is related with everything but closer things tend to be more alike than further away ones: $W(u_i)=\exp(-0.5(d_i/b)^2)$ if $d_i<=b$, where d_i is the Euclidean distance between the location of the neighborhood observations and the u_i location to be estimated, exp() equals e^x function (e is Euler's number, i.e., 2.7182) while b represents the bandwidth of the kernel. As the bandwidth gets larger, the

Figure 2. $W(u_i)$ is an n by n matrix (n is the number of available observations), representing the geographical weights close to i location. This weighting scheme is known as kernel. For instance, W_{i3} represents the weight given to observation 3 for the estimate of the local parameters at i location (Lloyd, 2007).

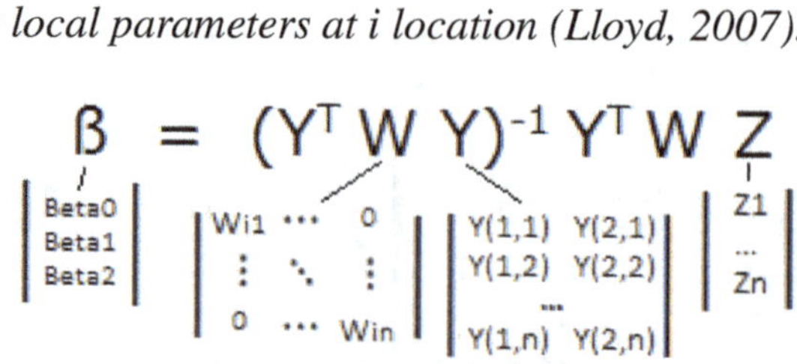

local GWR model approaches the global OLS model. As expected, if $d_i>b$, a zero weight should be produced.

Again, GWR extends the traditional regression framework by allowing local rather than global parameters to be estimated. Regarding the choice of the weighting system W(i), the majority of software follows an adaptive kernel decoded by the following mathematical expression: $W_{ij}=[1-(d_{ij}/b)^2]^2$, where b equals the distance between i and the nth nearest neighbor (quite common, between 8 and 16 neighbors becomes a good start).

A fixed kernel strategy may be another option, where b is a fixed vicinity distance (working as a cutoff boundary). As expected, this kernel function distance varies in space and presents an adaptive bandwidth depending on the density of data points. Thus, b distance becomes relatively small in areas where data points are densely distributed and relatively large where data points are sparsely distributed.

For Fotheringham, Brunsdon and Charlton (2002), the GWR measure of fit goodness is the corrected Akaike Information Criterion. According to these authors, two separate models, which are being compared, are held to be equivalent if the difference between the two AIC values is less than 3. The AIC can also be used to determine the optimal value for the kernel bandwidth (the lowest, the better). As stated by Lloyd (2007), the goodness-of-fit of a GWR model can additionally be assessed by the next geographically weighted coefficient of determination: $R_i^2=(TSS^w-RSS^w)/TSS^w$ (Figure 4).

Figure 3. For a 30 sample dataset with two independent variables, for instance, the number of rows and columns for each matrix will be as follows: Y=[30,2], Yt=[2,30], Z=[30,1] and W(Xi)=[30,30]

$$W(ui)=\begin{vmatrix} Wi1 & 0 & 0 & \cdots & 0 \\ 0 & Wi2 & 0 & \cdots & 0 \\ 0 & 0 & Wi3 & \cdots & 0 \\ \cdots & \cdots & \cdots & \cdots & \cdots \\ 0 & 0 & 0 & \cdots & Win \end{vmatrix}$$

Figure 4. TSS^w denotes the geographically weights total sum of squares while RSS^w is the geographically weighted residual sum of squares

$$TSS^w = \sum_{j=1}^{n} w_{ij}(y_j - \bar{y})^2 \quad RSS^w = \sum_{j=1}^{n} w_{ij}(y_i - \hat{y}_j)^2$$

Both approaches are contemplated by myGeoffice© (check Appendix 3 for the GWR code).

A curious study developed by Lu, Charlton, Harris & Fotheringham (2014) was considering distances a non-straight measure, instead. By using a London house price dataset coupled with several hedonic independent variables, these authors indicate that GWR calibrated with a non-Euclidean metric can not only improve model fit but also provide additional and useful insights into the nature of varying relationships within this house price dataset.

The next series of illustrations represents the output of the GWR option based on the 1990 Tokyo, Japan, mortality subset (the complete dataset can be downloaded directly from GeoDa® Center of Arizona State University). The dependent variable is the number of observed deaths (ObsDeaths) while the independent variables considered for this spatial model are as follows: ExpDeaths (expected number of working age deaths, i.e., between 25 and 64 years old – BETA 1), %Workers (proportion of professional workers – BETA 2), %OwnedHouse (proportion of owned houses – BETA 3) and %Elderly (proportion of elderly people, i.e., equal and over 65 years old – BETA 4). The first three fields, Sample ID, XX Coordinate and YY Coordinate, represent the sequential areal ID and the (x,y) centroid coordinates of each area (municipality), respectively (see Figures 5, 6 and 7).

The following maps display the intercept (BETA0) and the first independent variable (BETA 1) spatial distribution according to three exclusive classes. The second (BETA 2), third (BETA 3) and fourth (BETA 4) independent

Figure 5.

Geographically Weighted Regression (Part II)

Sample ID	XX Coordinate	YY Coordinate	Dependent Variable
1	378906,83	17310,41	189
2	385191,03	13064,39	48

(A)Spatial Regression

Figure 6. Besides the basic descriptive statistics for ObsDeaths variable, three basic nearest neighborhood parameters (regarding distances between samples) are shown with the purpose to give clues for the Kernel bandwidth constraint and to answer the Figure 1 issue (30000 units was the chosen one, in this case)

199	378200,19	70677,05	70
200	282095,21	20283,2	95

Dependent Variable Basic Statistics

Conventional Average:	171,04	Maximum:	1215
Minimum:	7	Range:	1208
Variance:	53411,5963819095	Standard Deviation:	231,109490029963
Coeficient of Variation:	1,35120141504889	Skewness:	2,34543760153761

Nearest Neighborhood Basic Statistics

Sampling Mean Distance: 60808,9295804087
Sampling Minimum Distance: 2421,32810426427
Sampling Maximum Distance: 150697,077576471

Kernel Bandwidth of the Gaussian weight function: 30000

Reset Submit

Figure 7. As an interpretation exercise, the Japanese Municipality whose ID equals 200 holds 95 real deaths in 1990 although GWR only estimated 94.57 = 43.28 + 0.93 × ExpDeaths - 281.82 × %Workers - 54.11 × %OwnedHouses + 238.88 × %Elderly. Notice that the total residual and the corrected AIC are useful for the comparison of different Kernel bandwidths (the lower, the better)

Geographically Weighted Regression (Part III)

Check if the executable program was accomplished with success. Afterwards, analyze the output results and press the bottom button.

c:\inetpub\wwwroot\executable\gwr.exe d:\data\657014042_gwr.txt
Dependent variable: ObsDeaths ; First independent variable: ExpDeaths ; Second independent variable: %Workers ; Third independent variable: %OwnedHouses ; Fourth independent variable: %Eldery
The Kernel Bandwidth (KB) of the GWR Gaussian function, W(x)=EXP(-0.5(Distance/KB)^2, equals 30000.000000

GEOGRAPHICALLY WEIGHTED REGRESSION [Sample 1 at (378906.830000,17310.410000) location with a 189.000000 value]
ExpDeaths: 194.572000 ; %Workers: 0.126000 ; %OwnedHouses: 0.606000 ; %Eldery: 0.104000
BETA0: 136.111524 BETA1: 0.935902 BETA2: -549.194631 BETA3: -134.397277 BETA4: 214.752112
GWR estimation: 189.902740 ; Residual (real-estimation): -0.902740 ; Local weighted RSS: 52.157288

GEOGRAPHICALLY WEIGHTED REGRESSION [Sample 2 at (385191.030000,13064.390000) location with a 48.000000 value]
ExpDeaths: 52.392000 ; %Workers: 0.075000 ; %OwnedHouses: 0.767000 ; %Eldery: 0.140000
BETA0: 130.857215 BETA1: 0.937509 BETA2: -517.447422 BETA3: -130.122696 BETA4: 205.472685
GWR estimation: 70.128686 ; Residual (real-estimation): -22.128686 ; Local weighted RSS: 28992.747801

GEOGRAPHICALLY WEIGHTED REGRESSION [Sample 199 at (378200.190000,70677.050000) location with a 70.000000 value]
ExpDeaths: 83.235000 ; %Workers: 0.106000 ; %OwnedHouses: 0.733000 ; %Eldery: 0.110000
BETA0: 186.765687 BETA1: 0.966090 BETA2: -942.240895 BETA3: -145.868114 BETA4: 147.290377
GWR estimation: 76.581275 ; Residual (real-estimation): -6.581275 ; Local weighted RSS: 1986.447495

GEOGRAPHICALLY WEIGHTED REGRESSION [Sample 200 at (282095.210000,20283.200000) location with a 95.000000 value]
ExpDeaths: 97.526000 ; %Workers: 0.107000 ; %OwnedHouses: 0.671000 ; %Eldery: 0.111000
BETA0: 43.283076 BETA1: 0.935559 BETA2: -281.822093 BETA3: -54.113070 BETA4: 238.884657
GWR estimation: 94.575805 ; Residual (real-estimation): 0.424195 ; Local weighted RSS: 8.400455

Total Residual (Weighted) Sum of Squares (RSS): 7922831.555900
Corrected Akaike Information Criteria, AICc=N*LOG(RSS)+2*K+(2*K*(K+1))/(N-K-1), where N and K equals 200 and 5 respectively: 3187.051844

Press here for the GWR mapping...

Weighted RSS (Residual Sum of Squares) = $\sum_{j=1}^{n} w_{ij}(y_j - \hat{y}_j)^2$

Figure 8. The top-left map displays the GWR estimation of the dependent variable at each location while the top-right one stresses the positive and negative residuals (real-estimation) throughout the study area. Certainly, the spatial residuals should be laid out in a random pattern for a good GWR model with a global spatial autocorrelation Moran I close to zero. The below illustration visualizes GWR final estimation mapping of the dependent variable, according to three different categories.

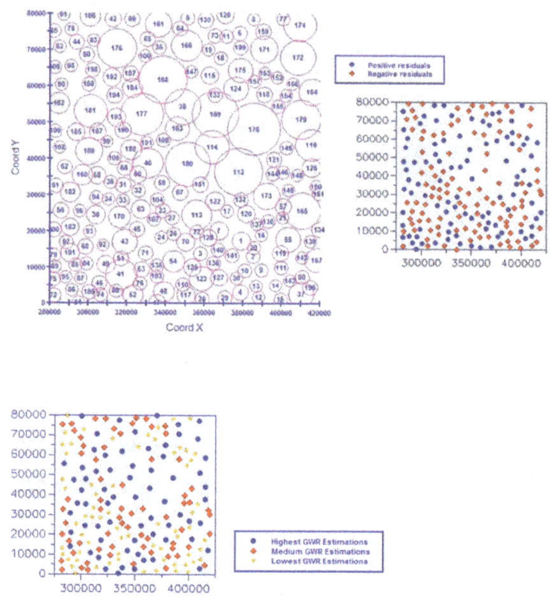

variable mapping are presented afterwards, as well. By means of these maps, it is possible to interpret the parameter's estimates. For example, the linear model Y=0.5-0.7×X tells that when X becomes zero, a Y of 0.5 units can be expected and, for each unit increase in X, Y will decrease correspondingly by 0.7.

The situation is analogous within GWR, except that we have a surface of parameter estimates (be aware that myGeoffice© does not estimate everywhere on the surface, only at the regression points). Depicted in Figure 9, the spatial variation of the independent elderly factor (BETA 4) shows a high impact on the observed deaths (dependent variable) in the bottom-left corner of the map (quite probably, more elderly people live in this particular vicinity). At the same time, the same neighborhood is influenced by a high level of %Workers (BETA 2) and a high %OwnedHouses (BETA 3), which contribute negatively to the real number of deaths.

Under the Sherlock Holmes view, this sub-region might be characterized by a pleasant urban zone environment with a high concentration of office jobs (high BETA 2) and where the available local housing is occupied by their titleholders (high BETA 3) with old ages (high BETA 4). The lowest values for BETA 1 also confirm this interpretation in the sense that the mortality expectancy for people between 25 and 65 years old is quite low. On the other hand, the top-right corner seems to be characterized by a rural area, for instance, due to the low influence of BETA 2 and BETA 3 factors. Still, the strong values of BETA 1 (ExpDeaths) may suggest, for instance, the existence of coal mines, where job accidents and health issues surround this atypical environment. The spatial variation of depth towards the intercept BETA 0 presumably uncovers other attributes that might be added to the model in such a way that would reduce the spatial variation of this factor. Then again, it is up to local expertise to analyze and explain the possible reasons for the intensity of each variable throughout space (see Figure 9).

Spatial Regression: Residuals

Residuals are perhaps the only available data values that can be considered as a reasonable reality test. The ability to verify existing patterns among residuals provides a key information source on possible assumption violations, variable transformations, outliers, surface trends and inappropriate formulations in the raw data. In theory, error assumptions are as follows: (1) The sum of the residuals equals zero with a constant variance; (2) To hold a Gaussian distribution (the most desirable model for every statistics researcher); (3) Errors are independent of each other.

The complete absence of autocorrelation among residuals would result in a value of 0 while a close value to +1.0 or –1.0 suggest the presence of a directly or inversely close relationship between successive residuals. If so, the inference based on the usual T and F statistics would be misleading and the R^2 goodness-of-fit would be wrong (Ebdon, 1998). With regard to predicted values against the associated error (Figure 10) in a spatial linear framework (check that the global mean is basically the same), situation (a) is the most desirable one with homogeneous variance. The prediction intervals associated within interpolation are correct. However, (b) involves a WLS (Weighted Least Squares) approach to model the continuous spatial drift. Situation (c) needs a linear structure. Case (d) requires a quadratic model ($Y=b_0+b_1X+b_2X^2$) or a data transformation of the independent variable prior to regression ($Y=Log(b_0)+b_1Log(X)$).

Figure 9. Confirmed by Fotheringham, Brunsdon and Charlton (2002), the spatial magnitude changes of the estimates' parameters across the surface reveal the local changing influence (highest, medium, lowest) of the independent variables on the dependent one. As expected, its influence is much stronger in some areas than in others.

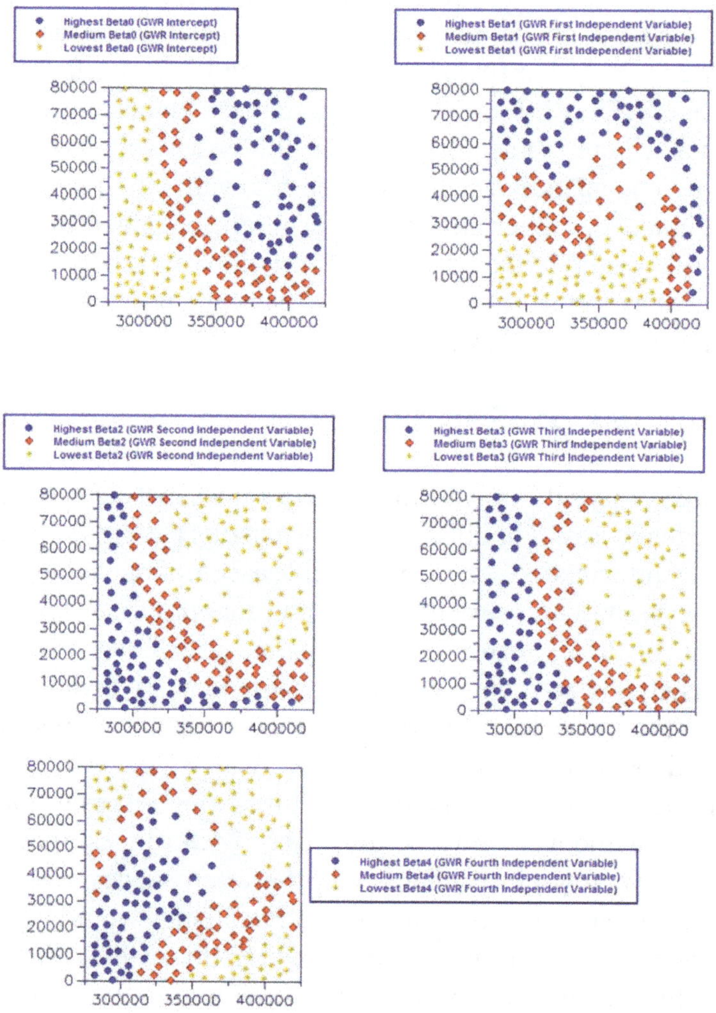

Within multiple regression models, according to Ebdon (1998), error assessment equals the equation shown in Figure 11.

Residuals can also be tested by the binary presence-absence of the join-count statistic. These dual state variables are based on the contiguity test of negative (N) and positive (P) residual patterns, a joint measurement between

Figure 10. Possible shapes of residual plots. Spatial autocorrelation among residuals can thus be viewed as a diagnostic process tool and a post-correction factor

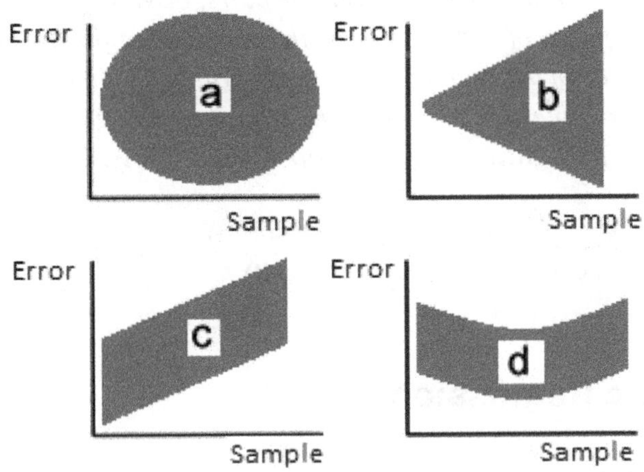

Figure 11. $S_a = \sqrt{\frac{\sum a^2}{n} - \bar{a}^2}$ *and* $S_b = \sqrt{\frac{\sum b^2}{n} - \bar{b}^2}$ *, in which a equals the first n-1 residuals, b is the last n-1 residuals and ab represents the product between two adjacent errors.*

$$\frac{\frac{\sum ab}{n} - \bar{a}\bar{b}}{S_a S_b}$$

neighboring points with non-zero boundaries. A further downgrading possibility for this binary form is the cutoff level. Thus, unit joins can be PN, NN or PP (if similar locations are clustered then there will be few NP joins; if locations were dispersed, there would be relatively many). Figure 12 gives the three standard indices that represent the number of links between negative, positive and negative-positive units. If any of these normal indices are greater or less than 1.96, for a 95% confidence level, then the respective null hypothesis is rejected, residual contiguity emerges and further investigation for missing independent variables is probably required.

Figure 12. J(NN) is the number of links between negative neighbor pairs, J(PP) stands for the number of links between positive neighbor pairs, J(PN) equals the number of links between positive and negative neighbor pairs, σ(NN) denotes the variance of negative neighbor pairs, σ(PP) represents the variance of positive neighbor pairs, σ(PN) is the variance of between positive and negative neighbor pairs while μ(PP), μ(NN) and μ(PN) denotes the expected link average among positive, negative and between positive and negative neighbor pairs, respectively.

$$Z(NN) = \frac{J(NN)-\mu(NN)}{\sigma(NN)} \quad Z(PP) = \frac{J(PP)-\mu(PP)}{\sigma(PP)} \quad Z(PN) = \frac{J(PN)-\mu(PN)}{\sigma(PN)}$$

OLS Bivariate Regression

Ebdon (1998) presents three reasons for classical statistics and OLS use. First, in some of the statistical analysis executed by geographers, the spatial component is of little or no significance. This would mean that spatial data is actually a-spatial. Second, summary information can always be compared to different locations and different times since the interest is not the study variability over space. Third, score variability includes a random error component; therefore, statistical models can be applied. Additionally, the potential gains of simple computing and estimation efficiency are attractive, mostly as a diagnostic tool at an exploratory stage and for the analysis of residuals. After all, non-spatial operations can often lead to significant spatial discoveries.

The conventional first order linear regression is used with two aims: (1) To construct a simple formula that will predict values for a variable given the value of another variable; (2) To test whether and how a given variable is related to another one. In the OLS bivariate regression, the linear model is $Y_i=\alpha+\beta X_i+\varepsilon_i$ with Y_i and X_i being the response and explanatory variables, respectively, α for the intercept, β for the slope and ε_i for the residual or error term. Mathematically, the β slope equals $\frac{\Sigma\left(X_i-\bar{x}\right)\left(Y_i-\bar{Y}\right)}{\Sigma(X_i-\bar{x})^2}$ while $\alpha=\bar{Y}-b\bar{X}$. The total sum of squared deviations ($TSS=\Sigma\left(Y_i-\bar{Y}\right)^2$) is the explained regression of the sum squared deviations ($RSS=\Sigma\left(\hat{Y}_i-\bar{Y}\right)^2$) plus the unexplained error of the sum squared deviations ($ESS=\Sigma\left(Y_i-\hat{Y}_i\right)^2$). Meanwhile, the R^2 coefficient equals RSS/TSS. It is also crucial to check the individual significance of the Student's T statistical test (quite often, at 0.05 level) of the slope. Notice that if the independent variable presents a highly significant T-test with a low R^2, the regression model under study has no statistical value because the independent variable does not explain the dependent variable's behavior.

The global (a-spatial) linear bivariate regression of Figure 13 equals Y=a+bX, in which a=26351.2 and b=0.13 while the Pearson product moment correlation coefficient becomes 0.76 (variance percentage of the dependent variable captured by this regression). Therefore, the R-square equals 0.57. Regarding the significance of the b slope parameter, its Student´s T-test (df=9) equals 3.34. As expected, if this statistic (do not consider the sign if negative) is larger than the critical right tail threshold for a particular level of significance (commonly 2.5%), the computed regression slope is thus significantly different from zero, which means H0 rejection (which is the case in this example). Concerning the spatial autocorrelation possibility among the regression errors, the Durbin-Watson test output equals 1.67 (No autocorrelation -> close values to 2; Positively correlated -> [0...1]; Negatively correlated -> [3...4]). However, it is essential to underline that these results adopt the following assumptions (although spatial data, quite common, does not respect them): (1) Samples are selected randomly from a Gaussian population of interest; (2) The dependent variable is continuous; (3) The error term follows a Gaussian distribution with a zero mean and a constant variance.

Based on the Geostatistical Association of South Africa dataset, Clark and Harper (2000) investigated the possibility of a linear relationship between

Figure 13. Output based on an apartment's financial value (dependent value) against a household income (independent variable) dataset supplied by myGeoffice©

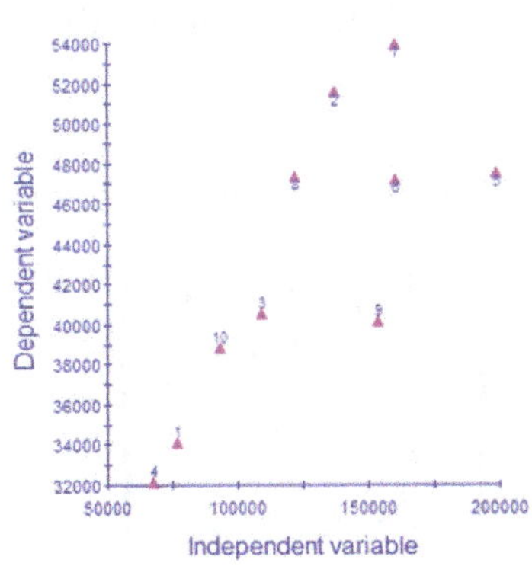

OLS Bivariate Regression Model (Part II)

ID Variable	Dependent Variable	Independent Variable
1	34135	76700
2	51571	137400
3	40558	109400
4	32182	67400
5	47493	198900
6	47203	160100
7	53935	160600
8	47381	122000
9	40127	153500
10	38819	93200

the gold grade in grams per ton of rock (parts of millions) and the thickness of the reef intersection (cm) in the borehole. After they normalized the initial data (27 samples) by applying a lognormal Napierian transformation, ln(width) and ln(grade+0.23), the estimated correlation coefficient became 0.0004. Centered on the null hypothesis of no correlation (H0: $\rho=0$) with a 90% level of confidence (± 0.3365), both authors concluded that it was not possible to predict the value of one variable from the other.

OLS MULTIPLE REGRESSION

Frequently, the relationship of interest involves more than two variables. For example, in an agricultural context, crop yield may depend on the amount of fertilizer applied, average daily sunshine, pH of soil and so on (Clark & Harper, 2000). It is also well-known that grasshoppers are a common agricultural pest, which can be found in a variety of crops and ecosystems. Researchers Negreiros, Damásio, Painho, Oliveira, Aguilar and Aguilar (2008) question if these infestations be predicted with spatial statistics methods.

Although multi-variable least squares regression cannot represent the involved variables spatially, the produced coefficients may allow us to understand the relationship (if any) between the dependent variable and the independent ones. Statistically, ordinary least squares (OLS) regression is a set of techniques that allows to relate k-1 independent variables ($X_1 \ldots X_{k-1}$) to a dependent one (Y) in the following form: $Y=X_0+X_1\times\beta_1+X_2\times\beta_2+\ldots+X_{k-1}\times\beta_{k-1}+\varepsilon$, where Y represents the (n×1) vector, X symbolizes the (n×k) matrix, β embodies the (k×1) regression coefficients vector of each independent variable and ε denotes the (n×1) residuals vector (Figure 14).

In this matrix approach, the least squared estimator for the multivariate regression linear model becomes $\beta=(X\,X^T)^{-1}\,X^T\,Y$, where X^T represents the transpose of X matrix. As an extension of the previous bivariate linear

Figure 14. n equals the number of sub-regions considered in the spatial problem

$$\begin{vmatrix} Y1 \\ Y2 \\ \ldots \\ Yn \end{vmatrix} = \begin{vmatrix} 1 & X11 \ldots X1k\text{-}1 \\ 1 & X21 \ldots X2k\text{-}1 \\ \ldots & \ldots \\ 1 & Xn1 \ldots Xnk\text{-}1 \end{vmatrix} \times \begin{vmatrix} \beta_0 \\ \beta_1 \\ \ldots \\ \beta_{k-1} \end{vmatrix} + \begin{vmatrix} \varepsilon_1 \\ \varepsilon_2 \\ \ldots \\ \varepsilon_n \end{vmatrix}$$

model, this approach allows the researcher to determine how much of the dependent variable variance would be explained by the variables of the model as a group (how much is explained by the model as a whole) and as separate variables (how much of the dependent variable variance is explained by each independent variable). It should be observed that the value of the b_1 coefficient in a multivariate regression ($Y=a+b_1X_1+b_2X_2+\varepsilon$) will not be the same as the b_1 from the bivariate regression ($Y=a+b_1X_1$). This happens because bivariate regression carries out the analysis of one variable in isolation whereas multivariate analysis controls the explanatory effect in a global context.

Equivalent to OLS bivariate regression, the Student T-test for each variable (by default, the researcher should not interpret non-significant variables given by the individual T-test) and the global R^2 (note that adding independent variables will always inflate the R^2, which means the investigator should always look for the Adjusted R^2 instead because it adjusts the results downward by factoring the number of variables in the model) are key. Moreover, the global F-test allows the researcher to see whether at least some of the variables in the model are statistically significant, i.e., the F-test is used to test the probability that the correlation coefficients for all included model variables are zero (acceptance of H0). Empirically, if the p-value significance of the F-test is below 0.05, for example, the researcher may infer that at least one of the independent variables explain the dependent variable's behavior (check Appendix 1 for main statistical distributions).

For reference, the Durbin–Watson statistic test is used to detect the presence of first-order autocorrelation in the residuals of a regression analysis whose index always ranges between 0 and 4. Although the researcher should look upon the Durbin–Watson table (the d_U and d_L factors depend on the sample size and on the number of independent variables considered), as a rule of thumb, if this statistic is substantially less than 1 then there is evidence of positive serial correlation (GLS should be applied, instead). If it is greater than 3, successive error terms are much more different in value from each other (negative correlation can be found). Otherwise, if no autocorrelation is found among the residuals, the regression model can be considered correct.

Padeiro (2015) examined the local factors of influencing urban expansion in the Lisbon Metropolitan Area (LMA), Portugal. Based on 2011 Census data, an OLS multiple linear regression was set up to identify the determining factors for the increase in the number of buildings created between 1991 and 2011. His results showed that 50% of the recently constructed buildings are concentrated in five municipalities (the majority were located in the LMA's second ring) while densely populated areas and sub-regions served by public transportation did not attract new construction projects.

In contrast, Bilginol, Denli and Seker (2015) presented another encouraging study of OLS usage regarding the distribution channels in the finance sector. Since most customers consider ATM accessibility as a key criterion for customers' bank preferences, banks are making investments to increase the number of ATMs to serve more customers and improve ease-of-use access. Hence, it becomes crucial to install new ATMs at the most appropriate locations. In their study, canonical correlation analysis (for ATM site criterion selection) and OLS regression analysis were utilized to find the optimum ATM locations and to predict the efficiency of those ATMs.

Figure 15. In this example, the F-test results showed that the overall model of acceptance is null (acceptance of H0), i.e., the cottage age and the cottage area do not explain the final price of the cottage itself

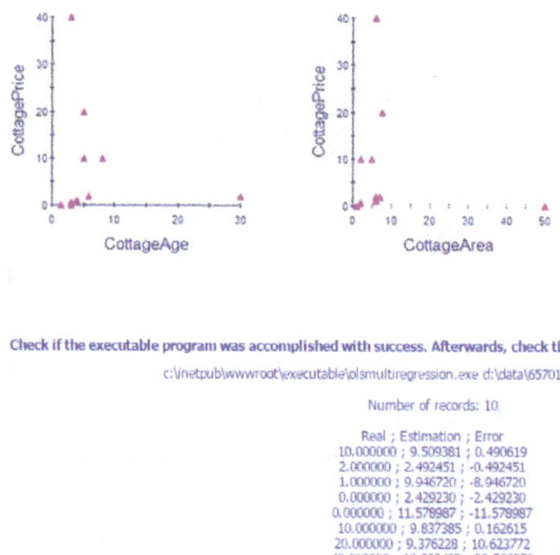

Check if the executable program was accomplished with success. Afterwards, check the output results and press the bottom button.

c:\inetpub\wwwroot\executable\olsmultiregression.exe d:\data\657014042_multilinearOLS.txt

Number of records: 10

Real ; Estimation ; Error
10.000000 ; 9.509381 ; 0.490619
2.000000 ; 2.492451 ; -0.492451
1.000000 ; 9.946720 ; -8.946720
0.000000 ; 2.429230 ; -2.429230
0.000000 ; 11.578987 ; -11.578987
10.000000 ; 9.837385 ; 0.162615
20.000000 ; 9.376228 ; 10.623772
40.000000 ; 10.233422 ; 29.766579
2.000000 ; 9.253287 ; -7.253287
0.600000 ; 10.942894 ; -10.342894

Dependent variable (Y): CottagePrice
Parameters estimation of the multiple linear regression model:
Beta0 = 12.157739
Beta1 = -0.286703
Beta2 = -0.177368

Total sum of squares (TSS): 1476.624023
Explained sum of squares (ESS): 97.597214
Residual sum of squares (RSS): 1379.026978

Mean absolute errors: 8.208715
Estimated mean (computed values of Y): 8.559999
Observed mean (real values of Y): 8.559999

Coefficient of multiple determination (R2): 0.066095

F-test for the overall model significance: 0.247704
Degrees of freedom: 2 (df1), 7 (df2). If F-test is greater than the critical value (see below table) then there is evidence that at least one independente variable affects the CottagePrice (rejection of H0)

Durbin-Watson : 1.589817
This statistic is a summary measure of the amount of autocorrelation in the error terms of the present regression (it is assumed that the present data is collected in a sequence time): No autocorrelation -> Close values to 2; Positively correlated -> [0...1]; Negatively correlated -> [3...4]

Nonlinear Models

Occasionally, there are relationships that do not hold a straight line such as the association between drug dosage for an illness and its effect on patients. As dosage rises, severity of illness goes down. However, at some point, the patient begins to experience negative side effects associated with high dosages and the severity of illness begins to increase.

One way to deal with nonlinear relationships between one Y dependent variable and one Z independent (see Figure 16) one rely on the use of polynomial regressions (functions of higher powers of Z such as Z^2 and Z^3). By definition, a polynomial regression model equals $Y = ß_0 + ß_1 \times Z + ß_2 \times Z^2 + \ldots + ß_m \times Z^m + \varepsilon$, where m is the degree of the polynomial.

Another way to deal with nonlinear relationships is to convert them to a linear form by using a variable transformation of the dependent and/or independent variable. The general multiplicative model between two variables, Y and Z, equals $Y = ß_0 \times Z^{ß1} \times \varepsilon$, where ßs are the unknown conventional parameters to be estimated while ε represents the multiplicative random error. This equation can be transformed from a logarithmic one by using the close mathematical

Figure 16. Graphical plot of linear and nonlinear relationships between two variables

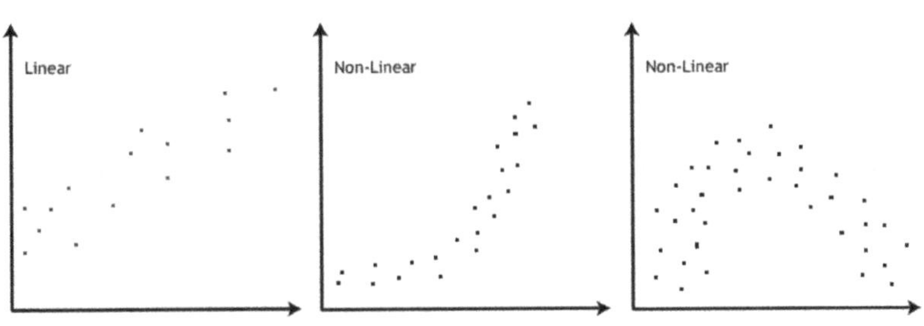

Figure 17. Graphical view of a linear, second and third polynomial models

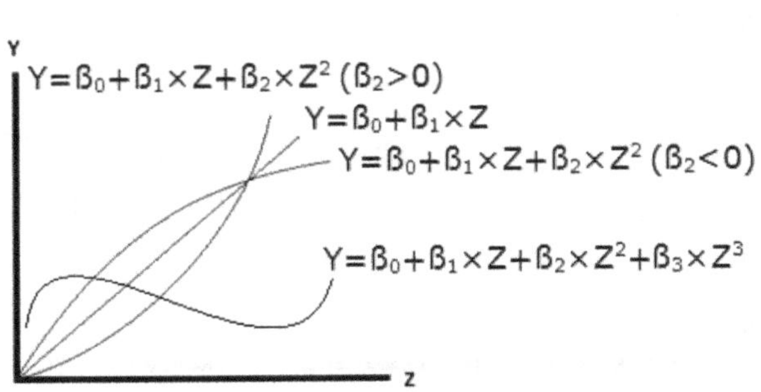

association between exponentials and logarithms operations: log Y=log $ß_0$+$ß_1$×log Z+log ε. Do not forget that the linearized log ε errors should be considered normally distributed with a mean of zero and with an equal constant variance. This also means that the user must take logs of the Y and Z original variables first and run an OLS regression test on the transformed logarithm variables. As expected, the statistical software regarding predictions and confidence intervals should calculate the back transform of the results for visualization purposes by taking antilogs (Sounderpandian, 2008).

Do we have to take logs of both X and Z or can we take the log of one of the variables only? A possible way to answer this pertinent question is to plot the different combinations of Y in terms of Z and check the linearity between them. Based on the layout in Figure 17 and 18, one can do the fitness regression test with some freedom of choice. For instance, an alternative of the previous transformed equation could be Y=log$ß_0$+$ß_1$×logZ+ε. Positively, the coefficient of determination (R^2), the significance of each individual T-test for each ß parameter and the global F ratio becomes mandatory in order to make comparisons between the several models produced by the researcher.

Using the sales data of Table 1, several nonlinear regression models were assessed with MS-Excel® (under the data/data analysis/regression option) whose results are displayed in Table 2. Again, it is essential to look carefully at the regression errors plot and their (supposedly) random distribution. Can

Figure 18. Graphical view of the power curves family $Y=ß_0 \times Z^{ß1} \times \varepsilon$

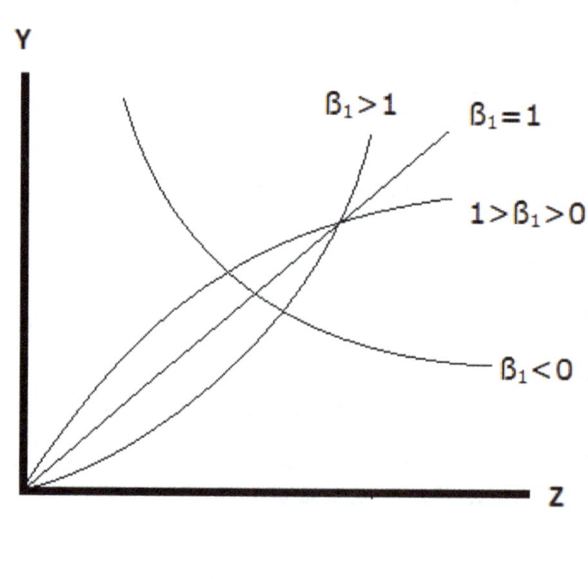

Table 1. Sales response to advertisement data for exemplification purposes only

Quarter	Sales	LogSales	Advert	SquareAdvert	LogAdvert
1	5	1.61	1	1	0.00
2	6	1.79	1.8	3.24	0.59
3	6.5	1.87	1.6	2.56	0.47
4	7	1.95	1.7	2.89	0.53
5	7.5	2.01	2	4	0.69
6	8	2.08	2	4	0.69
7	10	2.30	2.3	5.29	0.83
8	10.8	2.38	2.8	7.84	1.03
9	12	2.48	3.5	12.25	1.25
10	13	2.56	3.3	10.89	1.19
11	15.5	2.74	4.8	23.04	1.57
12	15	2.71	5	25	1.61
13	16	2.77	7	49	1.95
14	17	2.83	8.1	65.61	2.09
15	18	2.89	8	64	2.08
16	18	2.89	10	100	2.30
17	18.5	2.92	8	64	2.08
18	21	3.04	12.7	161.29	2.54
19	20	3.00	12	144	2.48
20	22	3.09	15	225	2.71
21	23	3.14	14.4	207.36	2.67

(Sounderpandian, 2008)

the reader justify why the last computed model (Sales=3.67+6.78×LogAdvert) should be considered the best for this set of data?

Regression With Dummy Variables

Dummy variables allow the researcher to include qualitative, binary or dichotomy independent variables (such as gender, where only male and female status is allowed and are mutually exclusive) in the typical OLS model. Let's suppose the following hypothetical regression: AgeFirstMarriage=23+2.9×Gender. If gender is coded as 0 for women and 1 for men, it is possible to affirm that the average age of men marrying for the first time equals 23+2.9×1=25.9. In addition, it can be asserted that women get married on average 2.9 years earlier than men.

Table 2. Observe that when the dependent variable does not suffer any data transformation, the need of back-transform (particularly for prediction and confidence intervals purposes) becomes redundant. Nonlinear models that cannot be transformed into linear ones are difficult to analyze and, hence, are outside of the scope of this book

Sales=f(Advert)	
Multiple R	0.95
R Square	0.89
Adj R Square	0.89
Stand Error	1.91
Observations	21

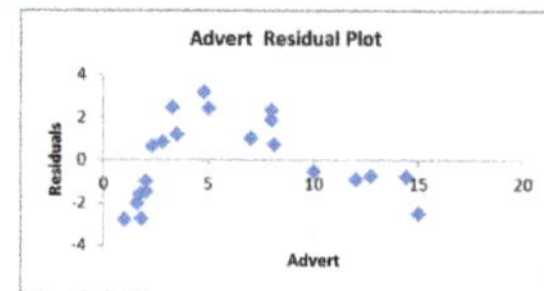

ANOVA

	df	SS	MS	F	Significance F
Regression	1	588.30	588.30	161.76	0.00
Residual	19	69.10	3.64		
Total	20	657.40			

	Coefficients	Standard Error	t Stat	P-value	Lower 95%	Upper 95%
Intercept	6.59	0.70	9.38	0.00	5.12	8.06
Advert	1.19	0.09	12.72	0.00	1.00	1.39

Sales=f(SquareAdvert)	
Multiple R	0.86
R Square	0.74
Adj R Square	0.73
Stand Error	3.00
Observations	21

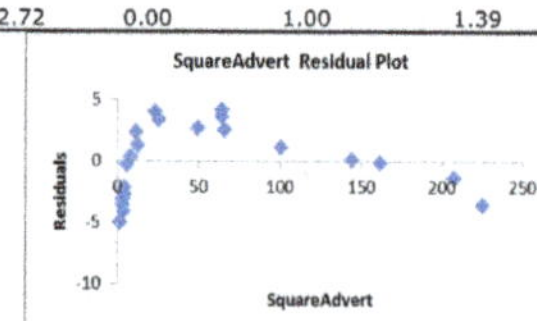

ANOVA

	df	SS	MS	F	Significance F
Regression	1	486.95	486.95	54.28	0.00
Residual	19	170.45	8.97		
Total	20	657.4			

	Coefficients	Standard Error	t Stat	P-value	Lower 95%	Upper 95%
Intercept	9.88	0.84	11.72	0.00	8.11	11.64
SquareAdvert	0.07	0.01	7.37	0.00	0.05	0.09

LogSales=f(LogAdvert)	
Multiple R	0.97
R Square	0.95
Adj R Square	0.94
Standard Error	0.11
Observations	21

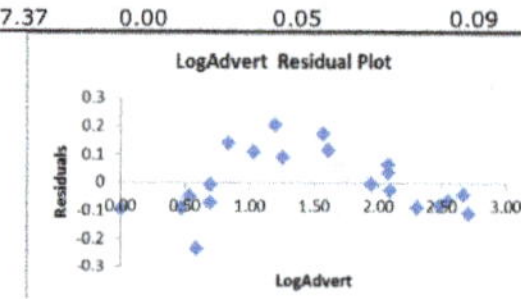

ANOVA

	df	SS	MS	F	Significance F
Regression	1	4.27	4.27	337.56	0.00
Residual	19	0.24	0.01		
Total	20	4.51			

	Coefficients	Standard Error	t Stat	P-value	Lower 95%	Upper 95%
Intercept	1.70	0.05	33.20	0.00	1.59	1.81
LogAdvert	0.55	0.03	18.37	0.00	0.49	0.62

Sales=f(LogAdvert)	
Multiple R	0.99
R Square	0.98
Adj R Square	0.98
Stand Error	0.88
Observations	21

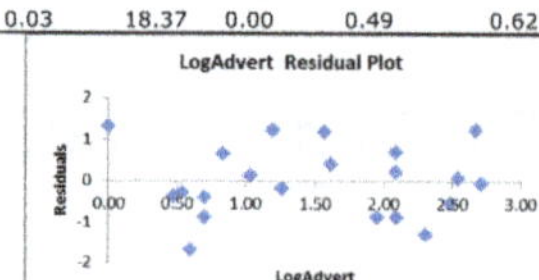

ANOVA

	df	SS	MS	F	Significance F
Regression	1	642.62	642.62	826.24	0.00
Residual	19	14.78	0.78		
Total	20	657.4			

	Coefficients	Standard Error	t Stat	P-value	Lower 95%	Upper 95%
Intercept	3.67	0.40	9.13	0.00	2.83	4.51
LogAdvert	6.78	0.24	28.74	0.00	6.29	7.28

Quite often, certain independent variables hold three or more internal exclusive categories such as race (Europeans, Africans and Asians, for instance). In such a case, the golden rule to follow is as follows: The total number of dummy independent variables required for the OLS regression equals the total number of categories minus one. This means two dummy variables would need to be set up in this case: EuropeansDummy (yes=1, no=0) and AfricansDummy (yes=1, no=0). Furthermore, the omitted status regards the third option, Asians, where EuropeansDummy=0 and AfricansDummy=0. Based on this coding system, consider the next theoretical OLS regression: Income=27.7-6.5×EuropeansDummy +1.3×AfricansDummy. Henceforth, it is possible to make the next three predictions for each ethnic group: (A) Europeans: Income=27.7-6.5×1+1.3×0=21.2; (B) Africans: Income=27.7-6.5×0+1.3×1=29; (C) Asians: Income=27.7-6.5×0+1.3×0 =27.7.

Let's say that the user would like to find out whether the house location (North/South, East/West) of a development area and whether if the house was built before or after the year 2000 affects its sale price. As demonstrated in Tables 3 and 4, the HouseMarketPrice is a numerical dependent variable. The Before2000 dummy variable represents a binary independent one whose content follows the codification rule: 1 if a house was built before the year 2000 or 0 otherwise. Two other dichotomy variables were setup: West (yes=1, no=0) and North (yes=1, no=0). Although four different categories for each house can be setup (NW, NE, SW and SE), it should be stressed that these categories are not mutually exclusive (a house on the North region can belong to both West and East, for instance), a condition not corroborated by the Before2000 binary variable.

The regression on HouseMarketPrice of these three dummy and the amenity (desirable facilities and features of a building such as appliances, services, convenience, accessibility or building construction type) variables yields the following model: HouseMarketPrice=1382695+8.3×Amenity Value+1723500×West-812873×North-71385×Before2000. The intercept constant of 1382695 indicates that houses in this neighborhood start at this sale price irrespectively of location and year built. The coefficient of Before2000 indicates that all things being equal, houses in this neighborhood built before 2000 command a $71385 lower-grade over those built after 2000. Similarly, houses on the West side cost $1723500 more than houses on the East side while houses on the North side cost $812873 less than houses on the South side.

Table 3.

HouseMarketPrice	AmenityValue	West	North	Before2000
$9,321,820	$679,400	1	0	1
$8,374,930	$598,900	1	0	1
$8,172,030	$560,100	1	0	0
$9,939,350	$760,600	1	0	1
$1,973,810	$100,000	0	1	1
$2,141,350	$76,700	0	1	0
$1,801,270	$93,500	0	1	0
$1,088,190	$63,200	0	1	0
$5,115,710	$437,400	1	1	1
$4,987,987	$390,000	1	1	1
$6,000,100	$450,000	1	1	1
$4,906,540	$420,000	1	1	0
$3,105,580	$319,400	0	0	0
$3,700,200	$356,000	0	0	1
$3,650,010	$320,700	0	0	1
$4,004,500	$351,000	0	0	1

Table 4. According to the amenity value ($350000, for example), it is possible to estimate the house market price for a house built before the year 2000 and located on the North/East region from the given OLS regression by substituting Before2000=1, North=1 and West=0, giving 1382695.7+8.3× 350000+1723500.5×0-812873×1-73385×1=$3401437.56.

Regression Statistics
Multiple R 0.98
R Square 0.95
Adjusted R Square 0.94
Standard Error 703280.00
Observations 16
ANOVA
df SS MS F Significance F
Regression 4 1.1088E+14 2.77E+13 56.05 0.00
Residual 11 5.4406E+12 4.94E+11

Total 15 1.1632E+14
Coefficients Standard Error t Stat P-value Lower 95% Upper 95%
Intercept 1382695.7 1475102.5 0.94 0.37 -1863983.0 4629374.5
AmenityValue 8.30 4.7 1.76 0.11 -2.0 18.6
West 1723500.5 1530719.1 1.13 0.28 -1645589.5 5092590.5
North -812873.0 1134944.4 -0.72 0.49 -3310869.0 1685122.8
Before2000 -73385.0 440953.8 -0.17 0.87 -1043917.9 897147.7

A further possibility (but not fully analyzed here) is the interaction examination of two attribute dichotomy variables (gender and marriage status, for instance) which is represented by a third dummy variable as the product of the two individual binary variables ($Y=\beta_0+\beta_1\times Z_1+\beta_2\times Z_2+\beta_3\times Z_3+\varepsilon$, where $Z_3=Z_1\times Z_2$).

Logistic or Logit Regression

Income, hours of TV watched or the grade point average of students are ratio variables. However, there are lots of things that we may want to explain that involve dichotomies. According to Linneman (2011), there are things in life that the outcome process results in a binary value: Do or do not, did or did not, have or have not, believe or believe not, for or against. If a researcher is interested in explaining how many times someone has been married, the outcome could be 0, 1, 2, 3 or 4, for instance. Another plausible possibility is to determine whether someone has been married or not. The same logic pattern follows if you want to explain why people get involved in car accidents. This measured dependent variable (DV) could be zero accidents, 1 accident and so on. Again, it also makes sense to ask if in the past five years, a driver has been involved in any serious car accidents. In this case, the outcome could only be yes (1) or no (0).

Logistic regression is the appropriate maximum likelihood estimation (MLE) regression analysis to conduct when the dependent variable (DV) is dichotomous or binary. Used with predictive purposes, logit regression describes data and tries to explain the relationship between one dependent dualistic variable and one or more metric (interval or ratio scale) independent variables (IV). The output is actually the percentage chance that a particular case gets a yes on the DV. To determine this, the probability computation is $p=e^z / \left(1+e^z\right)$, where $e=2.71828$ while z denotes the MLE logit regression equation. If $z=0$ on this p mathematical expression then the yes probability totals 50%. If $z<0$ then $p<0.5$, otherwise $p>0.5$. Regarding the assumptions of this model, there should be no high intercorrelations (multicollinearity) among the predictors. Meanwhile, Tabachnick and Fidell (2012) suggest that as long correlation coefficients among independent variables are less than 0.90, this assumption is met.

A helpful illustration of this technique is given by Linneman (2011) regarding the factor that determines whether or not American people own a house based on the following three IVs: (A) Living in the city versus living in the country side; (B) Born in the USA as opposed to immigrants; (C) Political party identification (Democrat/Republican). Based on 1,131 samples from a GSS 2006 survey, this author reached the following logistic regression: HomeOwnership=-0.55 +1.20×BornUSA-0.36×LiveCity+0.15×PoliticalParty (the codification of this last IV ranges from 0 to 6 in a continuum scale from being completely a Democrat to a completely Republican). The Nagelkerke R^2 equals 0.10 (still it remains a lot of variation that is unexpected) while the significance T two-tailed test for each IV became 0.000, 0.063 and 0.000, respectively. By comparing a strong Democrat ($Z_{Democrat}$=-0.55+0.15(0)–0.36(0)+1.20(1)=0.65) to a strong Republican ($Z_{Republican}$= -0.55+0.15(6)–0.36(0)+1.20(1)=1.55), the probability of both categories owning a house equals $P_{Democrat} = e^{0.65} / \left(1 + e^{0.65}\right)$ =66% and $P_{Republican} = e^{1.55} / \left(1 + e^{1.55}\right)$ =82%, i.e., strong Republicans are more likely to own their homes than strong Democrats.

To check the effect of city dwelling does not make much sense since it is not statistical significant (p-value>0.05). Yet, the effect of being an immigrant is as follows: $Z_{BornUSA}$=-0.55+0.15(3)–0.36(0)+1.20(1)=1.06, $Z_{Immigrant}$=-0.55+0.15(3)–0.36(0)+1.20(0)=0.10, $P_{BornUSA} = e^{1.06} / \left(1 + e^{1.06}\right)$ =75% and $P_{Immigrant} = e^{0.10} / \left(1 + e^{0.10}\right)$ =48% (immigrants are quite a bit less likely to own their homes). The same appraisal logic can be applied between strong Democrats (city-dweller immigrants) and strong Republicans (non city-dweller Americans citizens):

- Z_1=-0.55+ 0.15(0)–0.36(1)+1.20(0)=-0.91 -> $P_1 = e^{-0.91} / \left(1 + e^{-0.91}\right)$ =29%
- Z_2=-0.55+ 0.15(6)–0.36(0)+1.20(1)=1.55 -> $P_2 = e^{1.55} / \left(1 + e^{1.55}\right)$ =82%

Another fascinating example supplied by Moro et al. (2104) is closely related to direct marketing campaigns (phone calls) of a Portuguese financial institution (see Figure 19). The goal was to predict if a client would subscribe (or not) to any term deposit plan. The input independent variables of this 673 cases are as follows: Age (numeric), job type (admin, blue-collar, entrepreneur, housemaid, management, retired, self-employed, services, student, technician, unemployed, unknown), marital status (divorced, married, single, unknown), education (basic.4y, basic.6y, basic.9y, high.school, illiterate, professional.

course, university.degree, unknown), credit (yes, no, unknown), housing loan (yes, no, unknown), personal loan (yes, no, unknown), contact communication type (cellular, telephone), last contact month of year (Jan ... Dec), last contact day of the week (Mon … Fri), last contact duration (numeric), number of contacts performed during this campaign (numeric), number of days that passed by after the client was last contacted from a previous campaign (numeric, where 999 means client was not previously contacted), number of contacts performed before this campaign and for this client (numeric), outcome of the previous marketing campaign (failure, non-existent, success), employment variation rate (numeric), consumer price index (numeric), consumer confidence index (numeric), Euribor three month rate (numeric) and number of employees (numeric). Again, the binary dependent output variable concerns the subscription (or not) of any term financial deposit of the bank to measure the success (or not) of direct marketing campaigns.

Path Analysis

Regression can be used to establish the possibility of cause and effect relationships among a set of variables (Sprinthall, 2011) in a one-way cause flow technique. Path analysis and structural equation modeling (SEM) are the two main practices that fall under this category. Unfortunately, there is no statistical test that will determine the misspecification (or not) of the proposed model of the researcher.

Path analysis allows the testing of a global model including both direct and indirect effects on some outcomes by exploring the intercorrelation among variables. It begins with the investigator developing a path diagram with arrows connecting variables and depicting the cause flow by using the traditional OLS regression standardized parameters. After all, these beta weights reflect the direct explanation effect of an independent variable (IV) on a dependent one (DV). However, if the path coefficient is not statistically significant, one should consider dropping it from the model unless there is a strong theoretical support for its inclusion (Tate, 1992). Under this semantic, the DVs that are being explained by the IVs are named endogenous variables (have incoming arrows) while the external causes are referred to as exogenous (no arrows going to them).

Bryman and Cramer (1990) provide a clear example using four variables from a job survey: Age, income, autonomy and job satisfaction (see Figures 20 and 21). They propose that age has a direct effect on job satisfaction.

Figure 19. Based on the available dataset (archive.ics.uci.edu/ml/datasets/Bank+Marketing) and on the output of PSPP® free software (www.gnu.org/software/pspp/), this singular form of regression hold a special kind of R2, the Nagelkerke one. Grounded on the B and Sig. columns, the logit regression is then Buy=-17.91+1.65×Is_Female+1.13×Minors+1.86×White+1.56×Prev_Child_Mag

LOGISTIC REGRESSION

```
LOGISTIC REGRESSION Buy WITH Income Is_Female Is_Married Has_College Is_Professional Is_Retired Unemployed
Residence_Length Dual_Income Minors Own House White English Prev_Child_Mag Prev_Parent_Mag
   /CRITERIA = CUT(0.5) ITERATE(20)
   /PRINT = CI(95)
   /NOORIGIN.
```

Dependent Variable Encoding

Original Value	Internal Value
0	0
1	1

Case Processing Summary

Unweighted Cases	N	Percent
Included in Analysis	673	100.00
Missing Cases	0	.00
Total	673	100.00

note: Estimation terminated at iteration number 9 because parameter estimates changed by less than 0.001

Model Summary

Step 1	-2 Log likelihood	Cox & Snell R Square	Nagelkerke R Square
	182.33	.50	.81

Classification Table

	Observed		Predicted Buy 0	Predicted Buy 1	Percentage Correct
Step 1	Buy	0	526	22	95.99
		1	21	104	83.20
	Overall Percentage				93.61

Variables in the Equation

		B	S.E.	Wald	df	Sig.	Exp(B)	95% CI for Exp(B) Lower	95% CI for Exp(B) Upper
Step 1	Income	.00	.00	73.02	1	.000	1.00	1.00	1.00
	Is_Female	1.65	.47	12.53	1	.000	5.19	2.08	12.90
	Is_Married	.57	.59	.93	1	.334	1.76	.56	5.56
	Has_College	-.28	.44	.40	1	.529	.76	.32	1.80
	Is_Professional	.23	.46	.23	1	.628	1.25	.50	3.12
	Is_Retired	-1.16	.93	1.54	1	.214	.31	.05	1.95
	Unemployed	.99	4.69	.04	1	.833	2.69	.00	26373.71
	Residence_Length	.02	.01	3.20	1	.074	1.02	1.00	1.05
	Dual_Income	.45	.52	.75	1	.386	1.57	.57	4.37
	Minors	1.13	.46	5.97	1	.015	3.10	1.25	7.70
	Own	1.06	.56	3.57	1	.059	2.88	.96	8.61
	House	-.93	.62	2.22	1	.136	.40	.12	1.34
	White	1.86	.55	11.68	1	.001	6.45	2.21	18.78
	English	1.53	.84	3.31	1	.069	4.62	.89	24.00
	Prev_Child_Mag	1.56	.71	4.79	1	.029	4.75	1.18	19.15
	Prev_Parent_Mag	.48	.62	.59	1	.444	1.61	.47	5.48
	Constant	-17.91	2.22	64.93	1	.000	.00		

Figure 20. In this model, age is classified as an exogenous variable while the remaining ones are considered endogenous
(source: people.exeter.ac.uk/SEGLea/multvar2/pathanal.html)

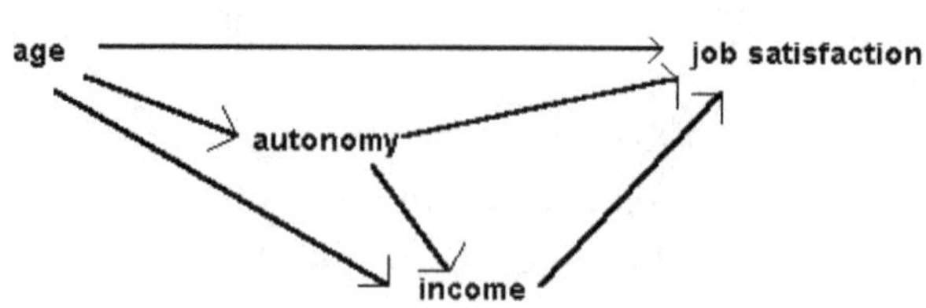

However, indirect effects of age on job satisfaction are also suggested: Age affects income, which in turn affects satisfaction; Age affects autonomy, which in turn affects satisfaction; Age affects autonomy, which affects income, which affects satisfaction. Autonomy and income have direct effects on satisfaction.

As expected, the standardized beta regression coefficients must be assessed through a set of structural equations: (A) satisfaction=$ß_{11}\times$age+$ß_{12}\times$autono my+ $ß_{13}\times$income+ε_1; (B) income=$ß_{21}\times$age+$ß_{22}\times$autonomy+ε_2; (C) autonomy=$ß_{31}\times$age +ε_3. Notice that the intercept constant of each equality is not used in this methodology. The disturbance terms (reflecting unexplained unmeasured variables variance plus measurement error) of each equation are $\varepsilon_1+\varepsilon_2+\varepsilon_3$, respectively. To obtain each path coefficient, three regression analyses need to be run with job satisfaction, income and autonomy being the DVs. In the meantime, the standardized beta weights are obtained by multiplying the original slope with the ratio between the standard deviation of the independent variable and the standard deviation of the dependent one ($\text{StandardizeSlope} = \text{OriginalSlope} \times \text{SD}_{\text{IV}} \;/\; \text{SD}_{\text{DV}}$).

The indirect effects must be calculated in order to compute the total effect of each variable on the system. Hence, the total indirect effect on job satisfaction equals 0.45, which corresponds to the sum of the following three parcels (path multiplication rule):

- age -> income -> satisfaction: 0.57x0.47=0.26
- age -> autonomy -> satisfaction: 0.28x0.58=0.16
- age -> autonomy -> income -> satisfaction: 0.28x0.22x0.47=0.03

Figure 21. Notice that path analysis cannot be used in situations where feedback loops are included in our hypothesis (source: people.exeter.ac.uk/SEGLea/multvar2/pathanal.html). In addition, each of IV must be on an interval scale

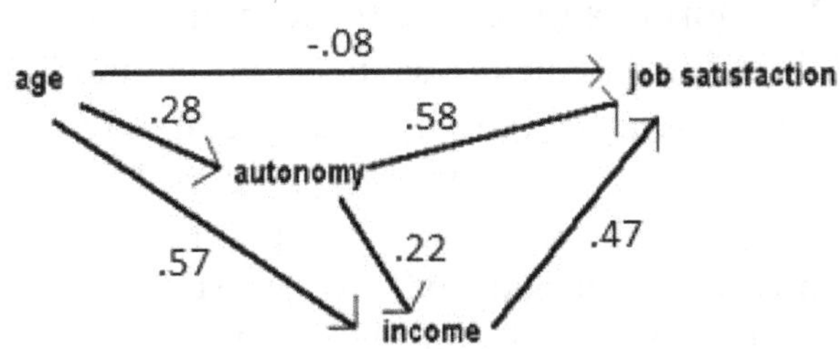

On the other hand, its total effect is -0.08+0.45=0.37, where the first item matches the direct effect of age on job satisfaction. Curiously, 0.45/ (0.45+0.08)= 84.9% of the effect of age on job satisfaction is an indirect one.

Because path analysis is an extension of multiple regression, three major assumptions should be stressed: (1) The relationship between the IVs and the DV is linear; (2) the DV errors are independent among themselves whose total mean tends to be zero; (3) the adequate sample size, according to Kline (1998), should be 10 times as many cases as parameters.

As a reference, SEM is an alternative to OLS, where all coefficients are calculated simultaneously (AMOS module of SPSS® enables this feature), not individually like OLS (check seoul-humanlab.com/web/bbs/board.php?botable=83&wr_id=24&page=2 for a magnificent example of SEM path modelling regarding graduate schools). Confirmed by Nazim and Ahmad (2013), SEM is a much more accurate and efficient way to make estimations for multiple items variables (latent construct). Finally, keep in mind that this technique does not confirm causation in the model. It merely illuminates which of two or more competing models is most consistent with correlation patterns found in the data (Garson, 2008).

SPEARMAN RANK CORRELATION

A well-known measure of correlation is the Pearson coefficient, which can be calculated if the input data type is interval or ratio. The Spearman Rank Correlation Coefficient, Pearson's twin brother, is applied when the original data comes in terms of order (first, second, third, fourth…). This index is a non-parametric measure of correlation, which uses ranks to calculate the relationship between two variables but without making any other assumptions towards the particular nature of their distribution. This technique can thus be used to summarize the strength and direction (negative or positive) of an association between variables (from +1 to -1).

The Spearman index equals -0.73 of Figure 22. In mathematical terms, the Spearman's Rank formula looks like the formula presented in Figure 22, in which d represents the rank difference between sample pairs and n equals the number of measurement sites (in this example, it is 10). As expected, the closer this index is to +1 or -1, the stronger the correlation can be assured.

An example involving the usage of the Spearman rank correlation for geographical purposes is described next. Zoonotic Cutaneous Leishmaniasis (ZCL), an important public health problem in Iran, is sensitive to climate

(A)Spatial Regression

Figure 22. The only requirement for this test is data should be ranked or, at least, converted into ranks

ID Variable	First Variable (X)	Second Variable (Y)	Order Rank of X	Order Rank of Y	Ranks Difference	Square Ranks Difference
1	50	1,8	1	9	-8	64
2	175	1,2	2	7,5	-5,5	30,25
3	270	2	3	10	-7	49
4	375	1	4	5	-1	1
5	425	1	5	5	0	0
6	580	1,2	6	7,5	-1,5	2,25
7	710	0,8	7	2	5	25
8	790	0,6	8	1	7	49
9	890	1	9	5	4	16
10	980	0,85	10	3	7	49

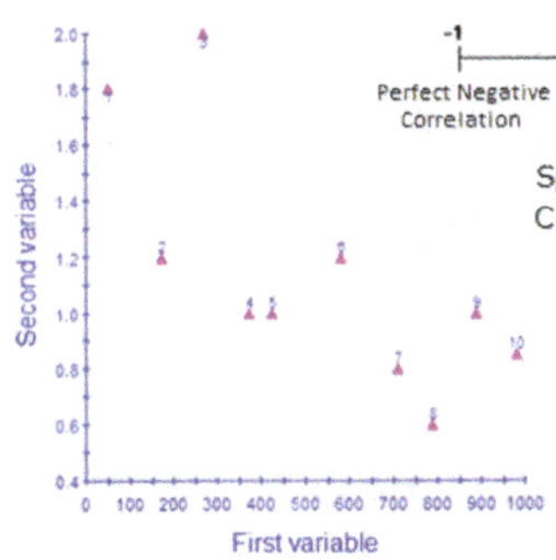

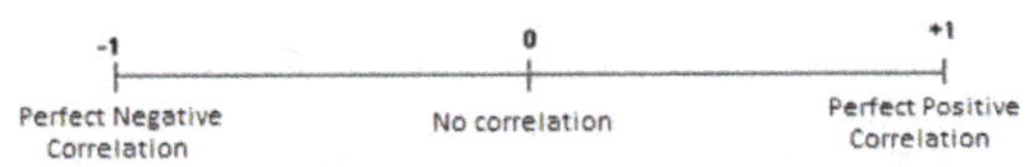

$$\text{Spearman's Rank Correlation Coefficient} = 1 - \frac{6\sum d^2}{n^3 - n}$$

conditions. Shirzadi, Mollalo and Mohammad (2015) examined the dynamic relations between weather and incidence of ZCL in Golestan Province between 2010 and 2012. Data of monthly climatic factors (including temperature, relative humidity, evaporation, total rainfall and number of freezing/rainy days) together with monthly ZCL incidence were tested. Spearman rank correlation was carried out to explain associations between monthly ZCL incidence rate and climate factors at 0, 1, 2, 3 and 4 months lagged intervals. Pearson's correlation analysis was conducted to examine the type and strength of relationships between the spatial averaged climate factors and the ZCL incidence rate at district level. According to Shirzadi et al. (2015), results of both Spearman rank and Pearson correlation analyses showed that ZCL incidences in Golestan Province tend to be more prevalent in areas with higher temperatures, lower relative humidity, lower total rainfall, higher evaporation and lower number of rainy days. It seems that weather factors are major determinants of ZCL in this endemic Iran area.

REFERENCES

Bilginol, K., Denli, H., & Seker, D. (2015). Ordinary least squares regression method approach for site selection of automated teller machines (ATMs). *Procedia Environmental Sciences, 26*, 66-69. doi:10.1016/j.proenv.2015.05.026

Bryman, A., & Cramer, D. (1990). *Quantitative data analysis for social scientists*. Retrieved from people.exeter.ac.uk/SEGLea/multvar2/pathanal.html

Clark, I., & Harper, W. (2000). *Practical geostatistics*. Ecosse North America.

Ebdon, D. (1998). *Statistics in geography*. Blackwell.

Fotheringham, A., Brunsdon, C., & Charlton, M. (2002). *Quantitative geography*. London: SAGE Publications.

Garson, D. (2008). *Path analysis*. Lecture notes, North Carolina State University. Retrieved from hbanaszak.mjr.uw.edu.pl/Wnioskowanie%20Statystyczne/09_ZastosowaniaPrzyklady/Garson_28_PathAnalysis.pdf

Linneman, T. (2011). *Social statistics, the basics and beyond*. Routledge, Taylor & Francis.

Lloyd, C. (2007). *Local models for spatial analysis*. CRC Press.

Lu, B., Charlton, M., Harris, P., & Fotheringham, A. S. (2014). Geographically weighted regression with a non-Euclidean distance metric: A case study using hedonic house price data. *International Journal of Geographical Information Science, 28*(4), 660–681. doi:10.1080/13658816.2013.865739

Mennis, J. (2006, July). Mapping the results of geographically weighted regression. *The Cartographic Journal, 43*(2), 171–179. doi:10.1179/000870406X114658

Nazim, A., & Ahmad, S. (2013). A comparison between ordinary least square and structural equation modeling methods in estimating the influential factors of 8th grades student's mathematics achievement in Malaysia. *International Journal of Scientific & Engineering Research, 4*(7).

Negreiros, J., Damásio, M., Painho, M., Oliveira, T., Aguilar, F., & Aguilar, M. (2008). Grasshopper outbreaks in Colorado: A spatial decision support with SAKWeb©. *Proceedings of the Third Iberian Conference of IT, 2*, 811-824.

Negreiros, J., Painho, M., Aguilar, M., & Aguilar, F. (2009). *Technical outline of a W3 spatial decision support prototype. In Handbook of Research on Social Dimensions of Semantic Technologies and Web Services*. Hershey, PA: Information Science Reference.

Padeiro, M. (2015). Os factores determinantes da edificação recente na AML (1991-2011). *Finisterra Revista Portuguesa de Geografia, 50*(99), 5–29. doi:10.18055/Finis3072

Shirzadi, M. R., Mollalo, A., & Mohammad, R. Y. (2015). Dynamic relations between incidence of zoonotic cutaneous leishmaniasis and climatic factors in golestan province. *Iranian Journal of Arthropod-Borne Diseases, 9*(2), 148–160. PMID:26623427

Sounderpandian, A. (2008). *Complete business statistics* (7th ed.). McGraw-Hill/Irwin.

Springthall, R. (2011). *Basic statistical analysis* (9th ed.). Pearson.

Tabachnick, B., & Fidell, L. (2012). *Using multivariate statistics* (6th ed.). Pearson.

Yu, D., & Wei, Y. (2011). *Geographically weighted regression.* Retrieved in 2014 from the Department of Geography – UWM: pages.csam.montclair.edu/~yu/GISDay_GWR.ppt

Chapter 5
Event Analysis

ABSTRACT

The goal of event analysis is to understand (and perhaps to predict) spatial occurrences that happened in a particular time-frame. Thus, the following five aspects will be covered in this chapter: Binomial and geometric for two outcome type of events (any phenomenon that can be observed such as rain or not rain), Knox and Mantel indicators, moving averages, point patterns, and Poisson distribution.

BINOMIAL DISTRIBUTION

Binomial distribution (check Box 1) analyses discrete random variables. It is appropriate when an event has only two possible outcomes: Success (p probability of first outcome) or failure (q=1-p, i.e., the probability of second outcome). Recalling that probabilities of all outcomes result in a sum of one, it is mandatory that p+q=1 (p represents the probability of a car being robbed in a particular metropolis, for example, while q equals the complementary probability of it not being stolen). Other appealing features for GIS of this peculiar distribution are: (1) All trials are independent events; (2) All trials are subject to the same probability distribution; (3) The probable sum of all possible combinations (n=1,2,3...N) for a given success probability of p with a size population of N equals one; (4) The N number of trials is considered fixed.

DOI: 10.4018/978-1-5225-3270-5.ch005

Box 1. Statistically, the binomial distribution gives the discrete probability distribution, $P_p(n|N)$, of obtaining exactly n successes out of N Bernoulli trials

$$P_p(n|N) = \frac{N!}{n!\,(N-n)!} p^n (1-p)^{N-n}$$

Returning to our car example, this means that the number of the sampled vehicles for that metropolis must be known. Obviously, when the number of trials changes, the number of all possible outcomes and their distributions across the values of the random variable will change too (if we survey six cars instead of four, we will have $64=2^6$ possible outcomes instead of $16=2^4$). Furthermore, the considered n number of successes (stolen cars) must be a positive integer below or equal to N (see the equation in Box 1).

Figure 1. What is the probability of having two or more days of rain in Montevideo, Uruguay, for one single week if the chance of rain is 5% on each day?

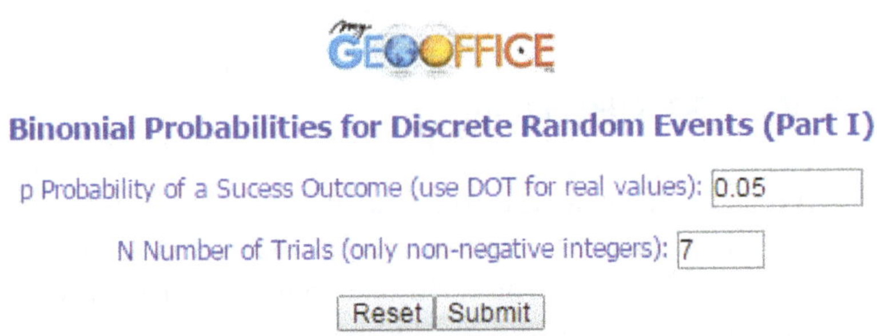

Figure 2. Using this example of myGeoffice© output, the answer becomes P(x=2)+P(x=3)+P(x=4) +P(x=5)+P(x=6)+P(x=7)=4.4%. It is also possible to say that the probability of it not raining during the whole week is 69.83%

Binomial Probabilities Distribution for Discrete Random Events (Part II)

P(X=0) = 0,69833729609375
P(X=1) = 0,25728216171875
P(X=2) = 0,04062349921875
P(X=3) = 0,00356346484375
P(X=4) = 0,00018755078125
P(X=5) = 0,00000592265625
P(X=6) = 0,00000010390625
P(X=7) = 0,0000000078125

Brites-Neto and Roncato Duarte (2015) tried to develop maps to identify environmental suitability for the two core species of scorpions of medical importance in São Paulo, Brazil, and to develop spatial configuration parameters for epidemiological surveillance of these venomous animals. By using 54 georeferenced points for Serrulatus, 86 points for Bahiensis and 8 environmental indicators, both authors generated a species distribution model (ESRI® ArcGIS 10.2.2 for Desktop) using omission error and binomial probability for evaluation purposes. Under the emergency warning actions for prevention and surveillance from scorpion stings, the resulting map showed a high environmental suitability of the north, central and southeastern parts of this Brazilian state, confirming the increasing spread of these species. Moreover, the environmental variables that most greatly affected the scorpion distribution model were rain precipitation (28.9%) and tree cover (28.2%) for the Serrulatus; and temperature (45.8%) and thermal amplitude (12.6%) for the Bahiensis.

As a reference (not included in myGeoffice©), the negative binomial distribution concerns to the probability of k successes in a sequence of g independent trials (if we define 1 as tails and 0 as heads, for instance, and we flip a coin repeatedly until the sixth time 1 appears then the probability distribution of the number of 0s that had appeared will be a negative binomial).

GEOMETRIC DISTRIBUTION

Consider a sequence of trials where each trial has only two possible outcomes. The probability of success is assumed to be the same for each trial. In such a sequence of trials, the geometric distribution is useful to model the probability of the number of trials needed before the first success appears. Basically, the researcher is searching for the number of Bernoulli trials, each having p probability of success, required to achieve just one success. For instance, a newly-wed couple plans to have children and will continue this plan until the first girl arrives (Wikipedia - Geometric Distribution, 2016). What is the probability that the first child is a girl, the second child is their first girl, the third child is their first girl and so on? As expected, for the first success to occur on the xth trial, the first x-1 trials must be failures. Regarding assumptions, the following three conditions must be met: (A) The phenomenon being modelled is a sequence of independent trials; (B) There are only two possible outcomes for each trial, often designated success or failure; (C) The probability of success, p, is the same for every trial.

Statistically speaking, the probability function for the X variable that follows this specific distribution is as follows: $f(x)=P(X=x)=(1-p)^{(x-1)}\times p$, for $x=1,2,3,\ldots$ The mean or expected value regarding the first success equals $1/p$ while its variance is $(1-p)/p^2$. The plot of a geometric probability distribution always presents a right skewness shape, where the highest probability is at the minimum value ($x=1$). Be aware that there is no maximum value for the x discrete variable although myGeoffice© only computes the first one hundred values (see Figure 3).

INDEX OF KNOX AND MANTEL

By definition, space-time analysis tries to assess the interaction between space and time (the two main fields of study of Albert Einstein). Frequently, the spatial pattern of incidents, for instance, does not occur uniformly throughout the year but instead are often clustered together during short periods. At certain times, a suddenly spike in incidents will occur in certain neighborhoods, i.e., there is clustering both in time as well as in space. Certainly, epidemiology (branch of medicine that deals with the incidence, distribution and possible control of diseases and other related health factors) and crime analysis are

Figure 3. The cumulative values of the above output example assesses the probability sum of that particular event happening on or before the xth event. When x=5, for instance, the probability that a monsoon will happen on the first five days of the year is 0.001249, i.e., 0.1249%=P(x=1)+P(x=2)+ P(x=3)+P(x=4)+P(x=5).

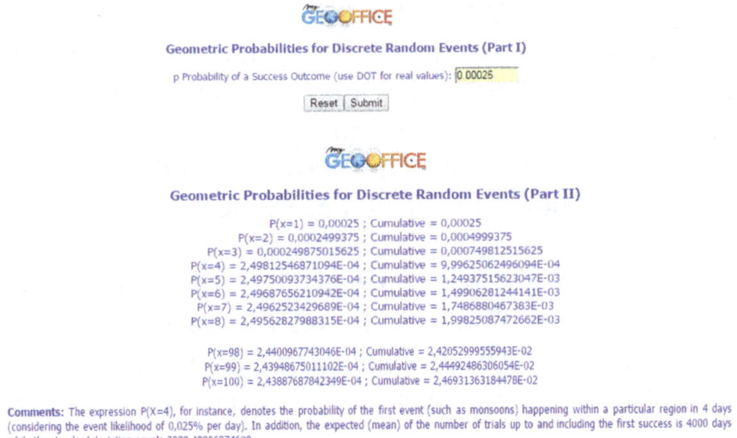

two of the major fields of research under this space-time analysis (Pina, Correia & Negreiros, 2011).

Knox is a comparison index for the relationship between incidents with regard to space and time, where each individual pair is compared in terms of the Euclidean distance and in terms of the time interval. Since each pair of points is being compared, it is possible to compute Nx(N-1)/2 pairs, where N equals the total number of events (six in the particular example case presented in Figure 4). The distance between points is divided into two groups: Close in space and not close in space. Similarly, the time interval axis is also divided into two clusters: Close in time and not close in time.

Regarding myGeoffice©, the definition of 'close' and 'not close' is based on the mean distance/time interval of all paired events since the mean value (the center of the distribution) always denotes a balanced point. Nevertheless, other choices can be made and, as a result, a variety of results can be achieved. Since the Knox Index is a one-tail test that follows a Chi-square distribution (a high value is indicative of spatial interaction), if the result of the Chi-square test (df=1) is larger than the 95% threshold, the null hypothesis of a random distribution between space and time can be rejected (check Appendix 1). It is key to clearly state that observations are considered independent under this framework (sometimes a hard assumption to be fulfilled). Due to this limitation, the Monte Carlo Simulation for the critical Chi-square value should be computed instead (a parameter not covered by myGeoffice©).

Alternatively, the Mantel Index (an alternative assessment that shows the relationship between closeness in time and closeness in distance) is the Pearson product-moment correlation between distance and time interval for event-pairs. It is based on a simple cross-product of two interval variables (distance and time) as illustrated in Box 2.

As noted by Michigan (2009), this correlation coefficient might be not appropriate since both measures are not independent. Furthermore, extreme values could distort the relationship while the sample size needs to be fairly large to produce a stable estimate. Finally, the correlations found quite commonly tend to be very small. This makes this index less intuitive as a measure than the traditional correlation coefficient (which varies between -1 and +1).

Aldstadt et al. (2012) produced an engaging space-time study of dengue fever in Kamphaeng Phet, Thailand. Between January 1, 2009, and May 6, 2011, two hundred and sixty two people was hospitalized with this infection. Spatial coordinates of each patient's home was captured using GPS (Global

Figure 4. Six ambulance crashes in Maputo, Mozambique, were recorded between 2003 and 2005. According to the present analysis, there was no significant clustering between space and time. Notice that the event's time (yyyymmddhh) should be laid out in an ascending order.

Knox and Mantel Indices for Space-Time Event Analysis (Part II)

Event ID	Coord X	Coord Y	Event Time (Ascending Order)	Format Date
1	23	45	2003030200	2003/03/02 (00H)
2	89	125	2003090909	2003/09/09 (09H)
3	100	250	2004121223	2004/12/12 (23H)
4	2	300	2005031222	2005/03/12 (22H)
5	7	77	2005052820	2005/05/28 (20H)
6	200	8	2005110515	2005/11/05 (15H)

Knox Close Mean Time: 1146487,13333333 Knox Close Mean Space: 181,43114788452 Knox Pairs Comparison: 15 Knox Chi-Square Statistic: 1,60714285714286 (The Chi-square critical value with one degree of freedom for a 0.05 alpha value is 3.841)

Knox Index	Close in Space	Not Close in Space	Horizontal Total
Close in Time	3	6	9
Not Close in Time	4	2	6
Vertical Total	7	8	15

Expected	Close in Time	Close in Space	Horizontal Total
Close in Distance	4,2	4,8	9
Close in Time	2,8	3,2	6
Vertical Total	7	8	15

Figure 5. Mantel index output of the same previous example

Standardize Mantel Index: -0,337035136252096 Standard Deviation (Space): 79,6649585359927 Standard Deviation (Time): 816759,004310405

Box 2. X_{ij} is an index of distance similarity between two observations (i and j) for one variable, Y_{ij} stands for the time similarity index between two observations (i and j) for the second variable, N equals the number of events while SD_x and SD_y represent the standard deviation of each variable.

$$\text{Mantel}=\frac{1}{N\times\frac{N-1}{2}-1}\sum_{i=1}^{N}\sum_{j=i+1}^{N}(X_{ij}-\text{meanX})/SD_x(Y_{ij}-\text{meanY})/SD_y$$

Positioning System). A novel method based on the Knox test was approved to determine the temporal intervals between cases in which spatial clustering occurred. For these authors, these intervals are indicative of the length of time between successive illnesses in the chain of dengue virus transmission. The results suggested that the strongest spatial clustering occurred within a

15-17 days interval. There was also substantial spatial clustering over short intervals (2-5 days). For space, the highest contraction risk was observed to be within 200m of a previous hospitalized case and a significantly elevated risk persisted within this distance for 32-34 days. By relying only on detected hospitalized cases with household locations, according to Aldstadt et al. (2012), it is possible to provide a useful space-time tool for the understanding of dengue virus transmission dynamics and outbreaks.

MOVING AVERAGE

This statistic determines the mean moving center of M-observations across time, where M represents a sub-set of the total sample events, N. This moving concept implies that all N observations are sequenced in order of occurrence (see Figure 6). Hence, it is implicit that there is a time dimension associated with the sequence itself (Michigan, 2009). The M sub-events are called the span (in myGeoffice©, this parameter equals five) and it is centered in each observation so that there are an equal number on both sides. Because there are no data points prior to the first event and after the last event, the first few mean centers will have fewer observations than the rest of the remaining sequence (the first and last mean center will have only three observations, the second and the before last will have four observations while the remaining ones will hold five for its mean assessment). According to Michigan (2009), this measure can be quite useful for detecting changes in behavior by serial offenders, suggesting whether the offender´s behavior is stable or not.

POINT PATTERNS

Point pattern analysis (geographical phenomena that may be reasonably modeled as point data such as clustering vs dispersion) is particularly popular in the fields of biology. Nearest neighborhood and K function density are other possibilities of point pattern analysis (see Exploratory Analysis menu of myGeoffice©). This option tries to quantify the strength of a certain event (oil spills, for instance) within two study areas over a time period (six months, for example). However, the exact spatial location of each event is not important. It only focuses on the number of events in each region, according to each occurrence in time.

Figure 6. Imagine the Spatial-Temporal Moving Average of a lawbreaker who committed eleven offenses such as thefts or residential burglaries before being arrested. The actual incidents are shown in pink triangles with the sequence number displayed. The moving average is represented by green squares, including its moving path. As can be seen, there is a definite shift in the field of operation by this hypothetical offender.

Spatial-Temporal Moving Average (Part II)

Event ID	Coord X	Coord Y	Coord X of Moving Average	Coord Y of Moving Average
1	123	34	145	119
2	78	89	195	175,5
3	234	234	225	170,4
4	345	345	260,4	173,6
5	345	150	284,8	167,8
6	300	50	258	123
7	200	60	189,2	54,2
8	100	10	122,2	60,2
9	1	1	74,2	66,2
10	10	180	42,75	67,75
11	60	80	23,6666666666667	87

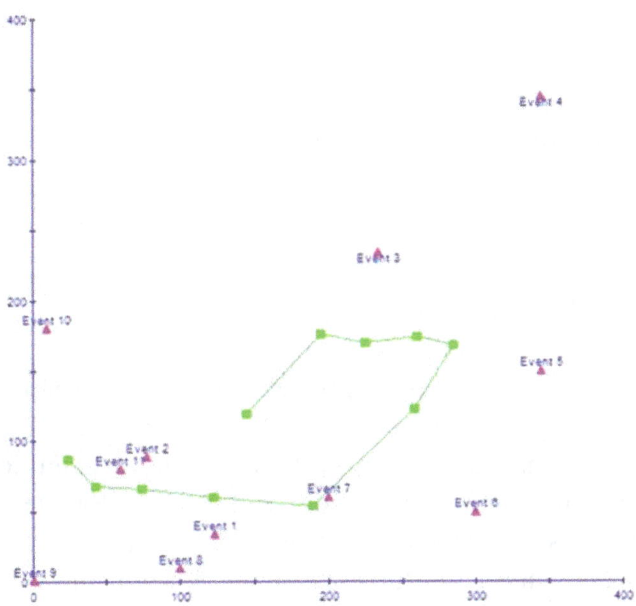

Using an example of Fotheringham, Brunsdon and Charlton (2002), imagine you want to check the correlation between burglaries in two areas over a period of twelve months. When no second-order interaction occurs, if M1 index (the product of the first-order global mean intensities for both regions) is less than M2 index (the ratio between the product average of the events number of region 1 and 2 by the product between both area regions) then this procedure suggests that the burglary counts for those two regions are positively correlated. Furthermore, if those regions are spatially adjacent, this indicates that there is, to some degree, a spatial clustering of the household burglaries. Finally, when M1 and M2 values are approximately equal, no correlation is insinuated. Otherwise, negative correlation is implied (M1 > M2). Mathematically, M1 and M2 equal the following expressions (see Figure 7 as an example implemented by myGeoffice):

$$\text{M1}\left(\text{region 1}\right) = \frac{\text{AVG}(\text{count\_region1})}{\text{Area\_1}}$$

$$\text{M1}\left(\text{region 2}\right) = \frac{\text{AVG}(\text{count\_region2})}{\text{Area\_2}}$$

$$\text{M1}\left(\text{Product}\right) = \text{M1}\left(\text{region1}\right) \times \text{M1}\left(\text{region2}\right)$$

$$\text{M2} = \frac{\text{AVG}(\text{count\_region1} \times \text{count\_region2})}{(\text{Area\_1} \times \text{Area\_2})}$$

POISSON DISTRIBUTION

The Poisson discrete distribution is applicable for count variables. It expresses the probability of a given number of events occurring in a fixed interval of time and/or space. It is different from the Binomial distribution as it relies on an average estimate, which may be derived from the long term occurrence average of an event such as the number of landslides along a mountain slope over a period of 50 years. Thus, the Poisson distribution allows the researcher to determine the probability of having a given number of occurrences. For instance, after studying hurricane statistics over the past 20 years, it is possible to compute an average expected number of hurricanes across Florida, USA, coastline in a given year.

However, some comparisons can be made with Binomial distribution. For instance, events should be considered random and independent in both models.

Figure 7. Based on an example provided by Fotheringham, Brunsdon & Charlton (2002), the above myGeoffice© output indicates that during a specific twelve month period, burglary counts were slightly positively correlated within both regions.

Second-Order Interaction for Events Count (Part I)

Time:	Region 1	-	Region 2
T1:	19	-	40
T2:	17	-	19
T3:	19	-	37
T4:	8	-	22
T5:	21	-	45
T6:	23	-	43
T7:	32	-	55
T8:	36	-	54
T9:	28	-	46
T10:	15	-	24
T11:	18	-	23
T12:	35	-	49

Region 1 Area: 41 Region 2 Area: 93

Reset Submit

Second-Order Interaction for Events Count (Part II)

Region 1	Region 2
19	40
17	19
19	37
8	22
21	45
23	43
32	55
36	54
28	46
15	24
18	23
35	49

M1 Product= 0,225556939709182 M2= 0,248644986449865

The probabilities sum for all events (X=0,1,2,3...) equals one. However, there is a major underlying difference between both distributions: The number of trials, N, is not known with the Poisson distribution.

Based on the following notation (L is the average number of occurrences, E equals the Napierian constant of 2.71828 while X denotes the random variable, representing the number of occurrences of an event over time or across some geographic extent), the probability of observing the event X times under a given condition, can be computed using the Poisson function shown in Box 3

On average, eight tropical storms make landfall in the Philippines each year. In 2013, only six storms affected the country, yet the worst year was 1993 when 19 cyclones pummeled the coastline. Based on this statistical fact, the Poisson option was assessed within myGeoffice© producing the (partial) output of Figure 8.

As previously highlighted, epidemiology is a major field for event analysis. Malaria, for instance, remains a public health concern despite the significant decrease in malaria incidence over the past decades. In addition, history reveals that malaria transmission is unstable and prone to local outbreaks. Thus, understanding spatial, temporal and spatiotemporal distribution of this disease is needed for its effective control and elimination.

Xia et al. (2015) carried out an investigation of annual malaria incidence from 2004 to 2011 in Hubei Province, China. GIS and spatial statistics were utilized to identify spatial clusters of malaria cases at the county level. Pure retrospective temporal analysis scanning was performed to detect temporal clusters of malaria cases using the discrete Poisson model. The study showed that the overall malaria incidence decreased to a lower level from 2004 to 2011. Meanwhile, the spatial clustering of malaria cases indicated that the disease was not randomly distributed in the study area. A total of eleven

Box 3. As stated previously, the Binomial approach is used to determine the probability of success (or failure) of a given trial. Alternatively, the Poisson distribution is used to determine the probability of having a specific number of occurrences of an event.

$$P(X) = \frac{E^{-L} \times L^{X}}{X!}$$

Figure 8. As expected, the probability of tropical storms of being less than 4 or more than 13 in a single year is quite small. On the other hand, the typhoon probability of being between 5 and 12 is over 83%.

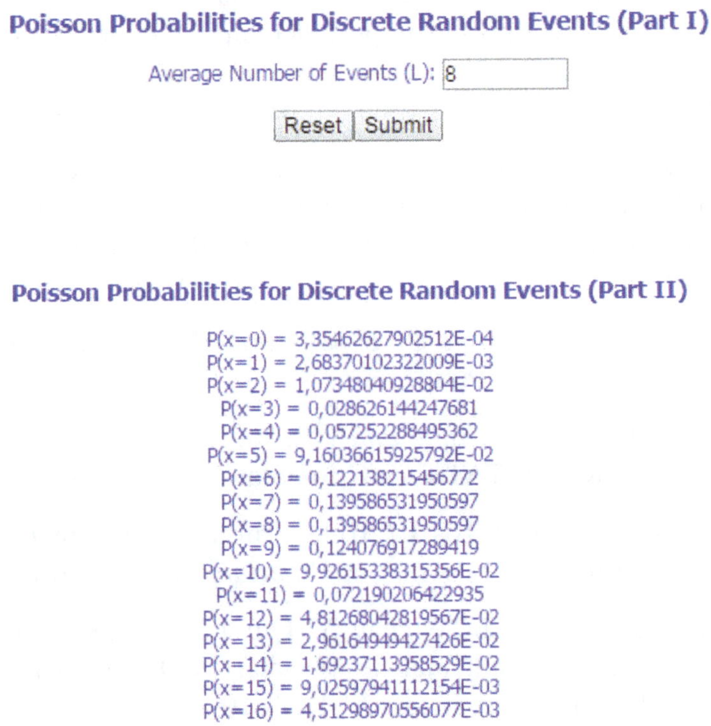

high-risk counties were identified through the use of the Local Moran's I for this time period. Furthermore, the space-time clustering analysis determined that the most likely cluster occurred from April 2004 to November 2007. Again, the detection of spatial, temporal and spatiotemporal distribution of malaria as well as the identification of high-risk areas for malaria contraction will help in prioritizing resource assignment for future malaria control and eradication efforts.

REFERENCES

Aldstadt, J., Yoon, I., Tannitisupawong, D., Jarman, R. G., Thomas, S. J., Gibbons, R. V., & Endy, T. (2012). Space-time analysis of hospitalised dengue patients in rural Thailand reveals important temporal intervals in the pattern of dengue virus transmission. *Tropical Medicine & International Health, 17*(9), 1076–1085. doi:10.1111/j.1365-3156.2012.03040.x PMID:22808917

Brites-Neto, J., & Roncato Duarte, K. M. (2015). Modeling of spatial distribution for scorpions of medical importance in the São Paulo state, Brazil. *Veterinary World, 8*(7), 823–830. doi:10.14202/vetworld.2015.823-830 PMID:27047160

Fotheringham, A., Brunsdon, C. & Charlton, M. (2002). *Quantitative geography.* London: SAGE Publications.

Michigan, U. (2009). *Space-time analysis.* Retrieved in 2014 from www.icpsr.umich.edu /CRIMESTAT/files/CrimeStatChapter.9.pdf

Pina, J., Correia, A., & Negreiros, J. (2011). Criminality management of Praia city, Cape Verde. *Proceedings of the Sixth Iberian Conference of IT, 2*, 29-32.

Xia, J., Cai, S., Zhang, H., Lin, W., Fan, Y., Qiu, J., & Nie, S. (2015). Spatial, temporal and spatio-temporal analysis of malaria in Hubei province, China from 2004-2011. *Malaria Journal, 14*(1), 1–10. doi:10.1186/s12936-015-0650-2 PMID:25557741

Section 2
GIS Process

Chapter 6
Exploratory Data Analysis

ABSTRACT

Exploratory data analysis (EDA) tries to summarize datasets main characteristics such as nearest neighborhood indexes, standard deviation, scatterplots or quadrat analysis. This EDA chapter is divided into several sections to cover myGeoffice© options not forgetting the graphical mode when facing outputs: file data input (after all, any analysis demands data); Descriptive study of the variable (mean, kurtosis, distribution plot, etc.); 2D-3D data posting (spatial location of the data samples); Cutoff layout map (a spatial colorful plot according to the data samples values that are higher and lower against any particular threshold); G and Kipley's K Index (to disclose clustered, uniform and random space sampling); Kernel Gaussian density (a non-parametric way to estimate the probability space density function of a variable); T-Student and F-tests (a parametric approach to check statistical differences between two sub-regions), including a brief section regarding the two-way ANOVA technique; Quadrat analysis (comparison of the statistically expected and actual counts of objects within spatial sampling areas to test randomness and clustering); XX profile scatterplot (silhouette view of the data along XX axis); and YY profile scatterplot (silhouette view of the data along YY axis).

DOI: 10.4018/978-1-5225-3270-5.ch006

INTRODUCTION

Exploratory data analysis (EDA) tries to summarize datasets main characteristics such as nearest neighborhood indexes, standard deviation, scatterplots or quadrat analysis. This EDA covers myGeoffice© ten options, not forgetting the graphical mode when facing outputs:

- File data input (after all, any analysis demands data)
- Descriptive study of the variable (mean, kurtosis, distribution plot, etc.)
- 2D-3D data posting (spatial location of the data samples)
- Cutoff layout map (a spatial colorful plot according to the data samples values that are higher and lower against any particular threshold)
- G & Kipley's K Index (to disclose clustered, uniform and random space sampling)
- Kernel Gaussian density (a non-parametric way to estimate the probability space density function of a variable)
- T-Student and F-tests (a parametric approach to check statistical differences between two sub-regions), including a brief sub-section regarding the two-way ANOVA technique.
- Quadrat analysis (comparison of the statistically expected and actual counts of objects within spatial sampling areas to test randomness and clustering)
- XX profile scatterplot (silhouette view of the data along XX axis)
- YY profile scatterplot (silhouette view of the data along YY axis)

Be reminded that this second menu of myGeoffice© deals with variables that are continuous in nature such as groundwater or air pollution levels. For discrete distributions (the number of diamonds in different sub-regions, for example) should be investigated and evaluated with discrete statistics such as the compound Poisson (check Clark & Harper (2000) bibliography for further information), Poisson, Bernoulli and negative binomial distributions.

PRIMARY FILE INPUT

Every analysis requires data and myGeoffice© is not an exception. The first step of this second menu is the capability to load an ASCII text file that follows the next pattern and separated by one TAB only: Sample_id, Coord_x, Coord_y, Sample_value (use DOT or COMMA for double real values but not for the first ID field). In addition, the first and fourth column cannot be negative. Moreover, the input datasets may vary between 50 and 200 observations (see Figure 1). For practical purposes towards the second part of this writing, the present dataset used for analysis comes from GSLIB® book of Clayton Deutsch and André Journel (1998), which can be downloaded directly from myGeoffice© or gslib.com.

DESCRIPTIVE STUDY

The first step with autocorrelation, Kriging or geostatistical simulation (the statistical approach) concerns the assessment of univariate and bivariate descriptive indexes such as central tendency measures (mean, mode, median), distribution graphs (QQ and PP plot, cumulative frequency table and histogram), spread measures (variance, standard deviation, range and inter-quartile range), skewness, coefficient of variation (useful for lognormal distributions identification), kurtosis and scattergram. For user's reference, the range of an ideal Gaussian dataset should vary between 4 and 6 standard

Figure 1. It is desirable that these input observations sites are posted in a square layout since maps are drawn in a perfect four-sided layout

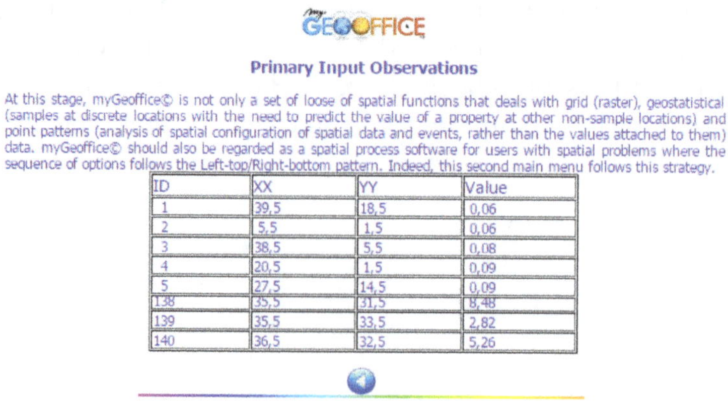
Primary Input Observations

At this stage, myGeoffice© is not only a set of loose of spatial functions that deals with grid (raster), geostatistical (samples at discrete locations with the need to predict the value of a property at other non-sample locations) and point patterns (analysis of spatial configuration of spatial data and events, rather than the values attached to them) data. myGeoffice© should also be regarded as a spatial process software for users with spatial problems where the sequence of options follows the Left-top/Right-bottom pattern. Indeed, this second main menu follows this strategy.

ID	XX	YY	Value
1	39,5	18,5	0,06
2	5,5	1,5	0,06
3	38,5	5,5	0,08
4	20,5	1,5	0,09
5	27,5	14,5	0,09
138	35,5	31,5	8,48
139	35,5	33,5	2,82
140	36,5	32,5	5,26

deviations from the mean with a skewness of zero and a standard kurtosis of three.

Univariate analysis explores each variable in a dataset, separately. Briefly, it describes each variable on its own (for instance, the skew index refers to the shape of the values distribution of any variable). If the values of the observations are distributed symmetrically around the mean, it is called a Normal distribution. In this case, the mean, median and mode will all coincide (see Figure 2). Yet, if the mode is higher than the mean, the distribution is called negatively skewed or left skewed distribution. For Clark and Harper (2000), this last sort of behavior is commonly found in geological applications of limestones and sedimentary iron ores.

Bivariate statistics tries to highlight the possible correlation between two variables but treating data as a set of numbers and ignoring its spatial nature. Hence, spatial description (mathematical techniques that include spatial location and topology of variables) such as nearest neighborhood analysis attempt to fill this gap by exploring the spatial component such as data posting, flux movement maps, cartograms, choropleth mapping and nearest neighborhood analysis. Quoting Isaaks and Srivastava (1989), familiarity with the dataset is an asset in any statistical study.

A major concern regarding random spatial processes is to confirm the Gaussian spatial distribution assumption. Normality of input data is important to obtain interpolation maps with a strong sense of security. If sample values come from a lognormal distribution, for example, the variance assessment from raw untransformed data is synonymous with outright stupidity (Clark & Harper, 2000). The lognormal distribution variance is a skewness measure, not variability, and if the goal is to measure changeability, some transformation

Figure 2. Graphical relationship between the skew type and the mean/mode/median (Sethi, 2015)

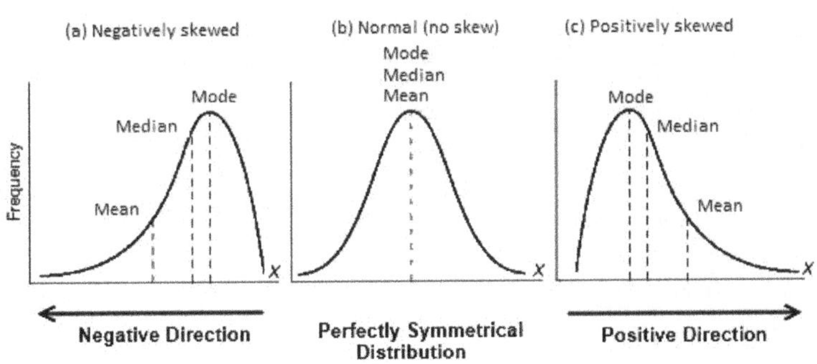

should be produced for a better behaved distribution, particularly for a very small sample size.

The easiest way to check if samples respect the Gaussian curve is given by the cumulative probability plot or fractal diagram: The closer the data values are to the 45° line, the better. An additional method is to verify the overall similarity shape between the Gaussian distribution and the sample histogram. Another classical goodness-of-fit-test is the χ^2 statistic based on

Figure 3. If this resulting statistic is less than χ2[p=95%, df=nc-3] critical value, the acceptance of the null hypothesis becomes true and the observed distribution does not differ from the Normal distribution. The Anderson-Darling, Ryan-Joiner, Shapiro-Wilk or Kolmogorov-Smirnov statistics are other possibilities.

$$\chi^2 = \sum_{j=1}^{nc} \frac{(O_j - E_j)^2}{E_j}$$

Figure 4. Typical descriptive measures supplied by myGeoffice© produced by the present dataset (a highly positive skewed distribution can be found)

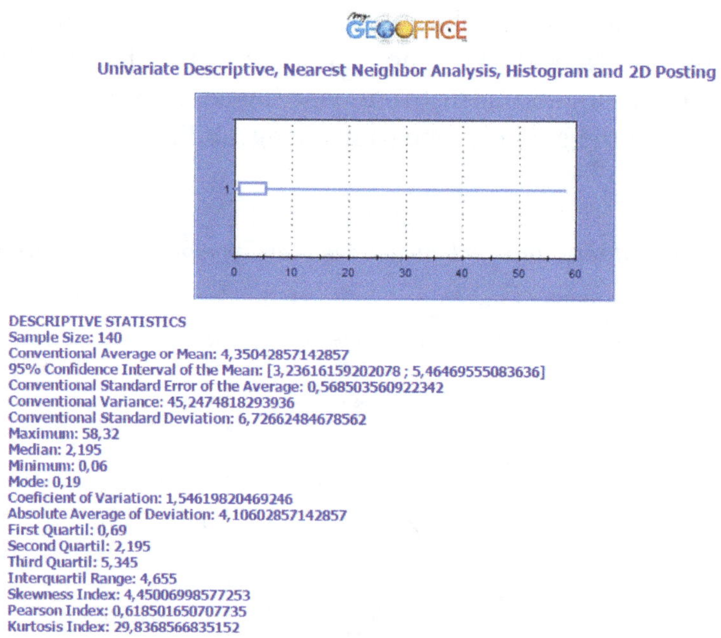

the frequency distribution of classes, as shown in Figures 3 and 4, being Oj the observed value of class j and E_j the expected one while nc equals the number of classes considered (check Appendix 5 for the basic classical statistical formulas).

If the data is skewed, the probability logarithm transformation plot will be curved, kinked or just weird revealing the presence of two or more populations. Therefore, the straight line of the cumulative probability plot on a logarithmic scale requires a logarithmic transformation adjusted by an additive constant, if necessary (EcoSSE® software is a robust option to handle lognormal distributions). Yet, the issue is that GIS users do not really care what the logarithm mean is. Researchers are only concerned about knowing what the average reef thickness is (Clark & Harper, 2000). However, the lognormal distribution estimation can be a complex procedure and based on the Sichel and Finney tables (it includes a bias correction factor). It is imperative to stress that the direct exponential back-transform is biased since the logarithm mean is the logarithm of the median and not the mean. When anti-logged, large values will become exponentially larger while small values will become exponentially smaller and cannot go below zero (Clark & Harper, 2000). At present, myGeoffice© requires the external transformation of the initial skewed dataset into logarithms in order to get a Gaussian distribution. ESRI® GIS software presents the Normal Score Transformation (NST) alternative whose goal is to make all random errors for the whole population normally distributed.

A key issue on spatial sampling regards cluster samples in one single area versus lonely samples in other areas, a representativeness issue (certainly, isolate samples incorporate more information than a sample that lies on a region surrounded by close neighbors). In order to assign dissimilar weights of importance, different techniques can be used, particularly for the mean and variance computation of the input dataset. Conventional cell declustering assigns samples weights, according to the cell size, but since each sample receives an inversely proportional weight to the number of points that fall within the same cell then several outcomes may emerge. With Geostatistical Analyst®, the cell size is determined from the maximum value of Morisita´s index. Furtado and Negreiros (2010) at Santiago, Cape Verde, island used this default cell size on their temperature spatial modulation study, for instance. As confirmed by Goovaerts (1997), when sampling size does not suggest a natural cell dimension, several cell sizes and origins must be tried.

For the sampling global mean and variance estimation, myGeoffice© uses nearest neighborhood analysis, where each sample weight relies on the nearest neighbor distance among all samples and the estimated one (nearest samples receive higher representativeness weights). This happens because political boundaries or delimitation areas of coastlines do not affect this methodology (it only depends on the samples coordinates). However, it is imperative to stress that none of both approaches is considered the best in terms of final results. Still, it can be guaranteed that any of these weighting declustering methods will always lead to a tremendous improvement in local and global distribution estimates when compared with the traditional statistical approach, where all observations have the same weight (Isaaks & Srivastava, 1989).

This sampling representativeness issue will be reflected on the estimated global mean (EGM), and on the estimated global variance (EGV). As expected, the sum of the weights attached to each sample should add up to one. Moreover, the closer the EGM and EGV to the traditional mean and variance becomes, respectively, the more the spatial evenly distribution are.

Sampling spatial direction preference can be uncovered by the standard deviational ellipse, a centrographic measure. Positioned on the mean center (with its long axis representing the maximum dispersion path while the short one highlights the minimal direction), the standard deviational ellipse reveals directional process bias of sampling.

Figure 5. According to the present results, it looks that these sampling locations hold a directional distribution close to 52 degrees (for reference, zero degrees means North)

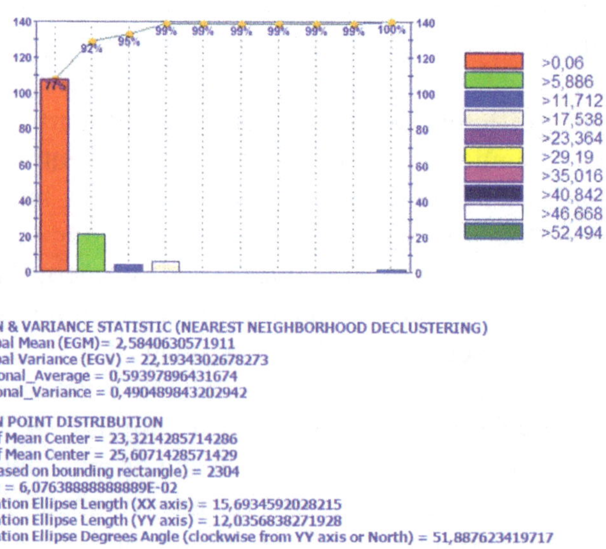

The mean center is a point constructed from the x and y coordinates average of all input samples. This arithmetic mean center (AMC) identifies the point with the minimum squared distance deviation and, if weighted by the magnitude of each variable, it represents the population centroid. On

Figure 6. myGeoffice© output: Extension of Figure 4 and Figure 5

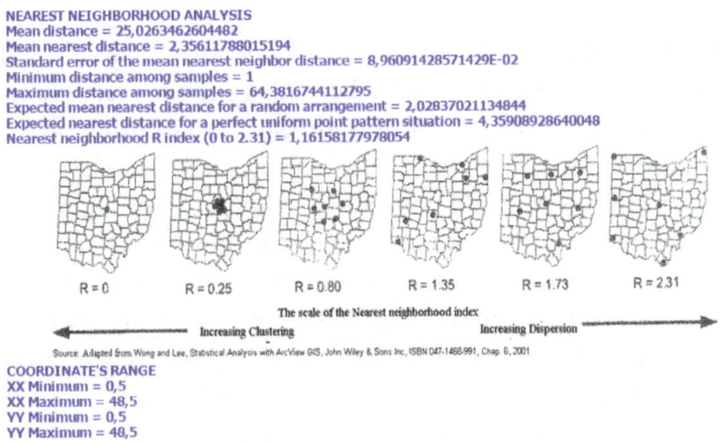

Figure 7. Data posting plotted with its corresponding ID sample

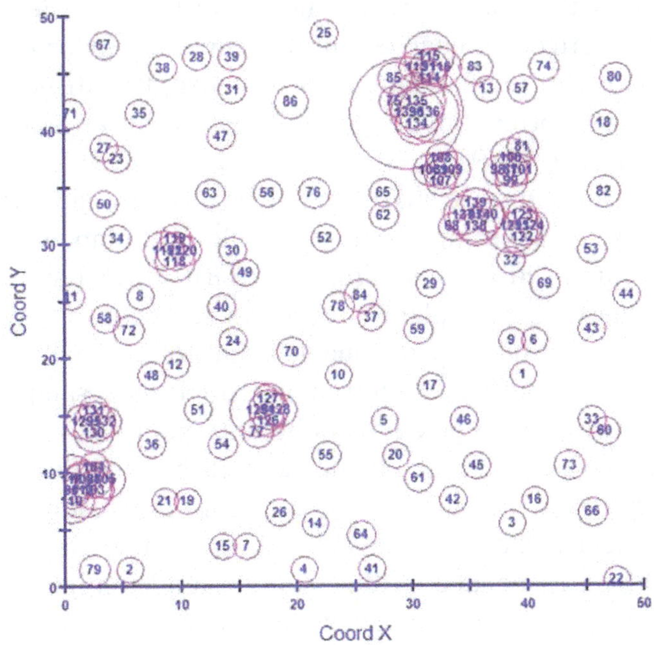

the other hand, the $R_{nearest}$=observed/expected ratio reveals a cluster zone, if a close zero value is found. In the presence of a perfectly dispersed region, this index equals 2.149. In regard to the present output, the dataset falls into a slightly ordered pattern (see Figures 6 and 7).

2D-3D DATA POSTING

Data sampling can be a difficult matter and some considerations may be raised. Statistically, samples should be independent, representative of a very large population and truthful. Regarding probabilities properties with two independent events, $P(e_1 \cap e_2)=P(e_1)\times P(e_2)$ and $P(e_1 \cup e_2)=P(e_1)+P(e_2)$, geographical sampling is rarely concerned with events that are not mutually exclusive. If sampling with replacement constitutes a random sample, it is a common practice not to evaluate the same element twice for time and financial reasons and to evaluate as many elements as possible. In addition, samples like people or animals are not stationary in space but, when surveyed, they are considered to be for a particular instance of time.

If spatial variability and process scales do not match, it is complicated to combine various datasets for spatial analysis because sampling interval measurement should capture the same distribution nature. Quite common, spatial variation varies inversely with sampling support size: Samples hold greater variances than blocks while their distribution is less symmetric, making support effect another conditional bias for incorrect thresholds. Regarding variogram regularization, for instance, the affine correction or indirect lognormal corrections are possibilities for this up-scaling operation (Isaaks & Srivastava, 1989).

Concerning sampling size, no rule could be provided although, regularly, time and money dictate sample size. As a rule of thumb, a sample size of at least 30 is employed in most situations (Griffith & Layne, 1999). But in a spatial context, the problem is even worse because spatial sampling distribution can be critical for the analysis that follows. Clustered samples will not improve reliability as dispersed ones do. Hence, sample density can turn out to be only a rough guide to quality interpolation.

Isaaks and Srivastava (1989) demonstrated that additional sampling is most necessary in anomalous areas since it improves estimates where the proportional effect makes them less accurate. Even so, variograms with a proportional effect associated with clustered samples might not work well either due to the apparent hole-effect, a dip variance at distances greater than

Figure 8. Under this option, myGeoffice© plots the 2D and 3D view of the samples by dividing the whole dataset into three exclusive subsets. Its aim is to give a first glance of the location samples according to their magnitude

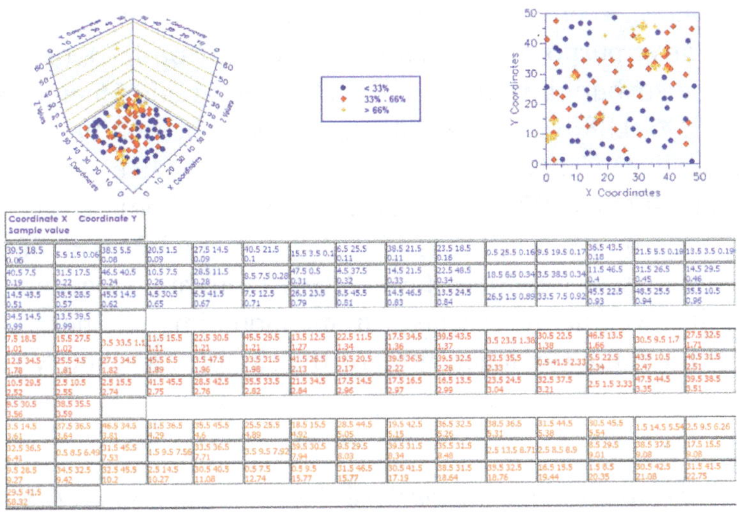

the range. Certainly, a major variogram hard assumption is that only one single dominant correlation scale is assumed. In addition, heterogeneity on a larger scale is not evident while heterogeneity on smaller scale is reflected into the nugget-effect. In this sense, the variogram is a short- and farsighted middle-aged man (with difficulty seeing objects both in close proximity and further apart).

CUTOFF LAYOUT MAP

The estimation of a complete cumulative distribution can be accomplished either by estimating the proportion of values that fall within particular classes or by estimating the proportion of values that fall above or below certain thresholds (Isaaks & Srivastava, 1989). This implies that it is possible to estimate the global proportion below or above any particular cutoff value.

The naive approach means to forget samples location and draw the cumulative distribution function in a non-spatial context. Yet, clustered and isolated samples may misrepresent this ascending distribution curve because samples should hold a corresponding weight since information attached to isolated samples are greater than to clustered ones. Only in a perfect grid layout, all samples hold the same geographical information, i.e., the same weight. As long as the sum of all samples weights are one, polygonal or cell declustering, for instance, can be used to estimate the cumulative proportion global distribution to produce a better representation of the unknown reality.

For illustrative purposes and based in Figure 9 (top-left), the proportion of the global distribution values that falls between 5 and 15 units is close to F(15)-F(5)=0.96-0.88=0.08=8%, if the nearest neighborhood weight approach is used (pink line). Under the conventional perspective, this probability becomes F(15)-F(5)=0.92-0.7=0.22=22% (orange line). Using the 2500 samples of the true dataset, F(15)=0.973=97.3% and F(5)=0.866=86.6%, which confirms Isaaks & Srivastava (1989) thoughts: In general, the use of declustering weights represents a tremendous improvement over naive weighting. In a similar manner, 90% of the samples are below seven (the

Figure 9. For illustration purposes, the three chosen thresholds were 0.5 (top-right), 2 (bottom-left) and 10 (bottom-right), where green square dots represents samples above the chosen cutoff value (the label attached to each triangle denotes their sample ID)

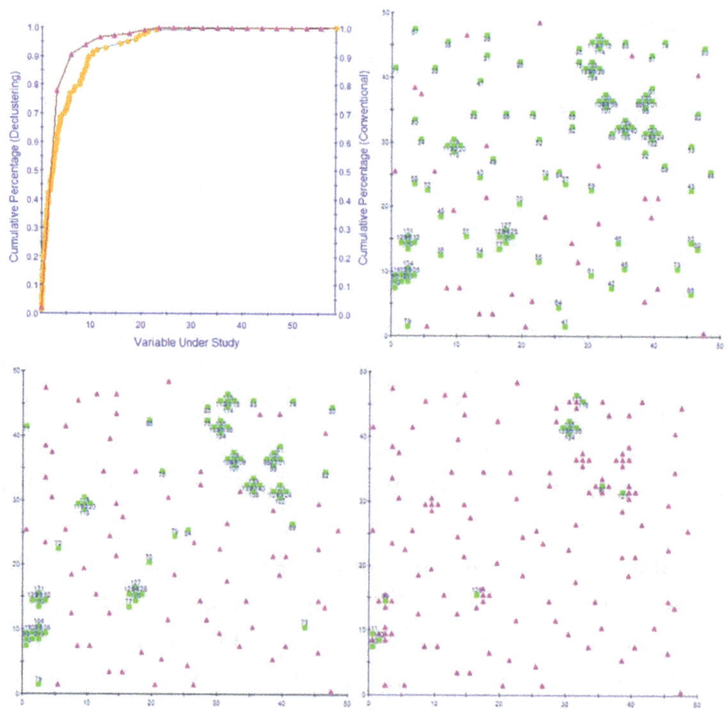

true percentage of samples below seven is 91.6%) but this percentage drops to 78%, if the traditional method is applied (orange line).

G AND KIPLEY'S K INDEX

The main goal of point pattern analysis is to determine if there is a tendency of points that may reveal a clustered (concentration of points) and uniform (every point is as far from all of its neighbors as possible) pattern over an area, as opposed to being randomly (the location of any point is not affected by the position of any other point) distributed.

The G(h) cumulative distribution, based on nearest neighborhood analysis, can be used as an exploratory method to check if there is any evidence of events (samples) interaction. Mathematically, the G(h) function equals the ratio between the number of samples for a particular h distance and the total sample numbers. As expected, there is a close relationship between the considered h distance (for myGeoffice©, the one hundred successive h increment equals 10% of the mean nearest distance) and the ratio numerator. For instance, G(h) curves with faster growth for closer distances suggest interaction between samples (clustered scale presence). On the other hand, smaller values for closer distances and faster growth for greater ones imply a more regular or even spatial distribution.

The major limitation of nearest neighbor distance is that it only uses the nearest distance. On the other hand, Ripley's K-function uses more points by comparing spatial patterns that are produced by a homogeneous Poisson point process. Yet, it assumes isotropy over the region. Essentially, K(h) counts the number of samples that lie in an area within h distance of a randomly chosen sampling. The estimation of the K function could, hence, be summarized as follows: (1) An observation is visited; (2) Under the observation vicinity, a set of concentric circles is constructed; (3) The cumulative number of observations in each distance is counted; (4) The remaining observations are visited in the same way; (5) The cumulative number of samples in different distance bands around all observations are counted and scaled by the study_area/n^2 ratio, where n equals the total number of observations. This is the estimate of K(h) and, as expected, it can be plotted against different ranges of h. Typically, this statistic is very sensitive to the size of the study area since identical site arrangements can exhibit different clustering or dispersion depending on the size of the study area.

Figure 10. Both G(h) and L(h) confirm a dispersed pattern, according to G(h), with some clustered sampling spots, according to L(h), for the given dataset

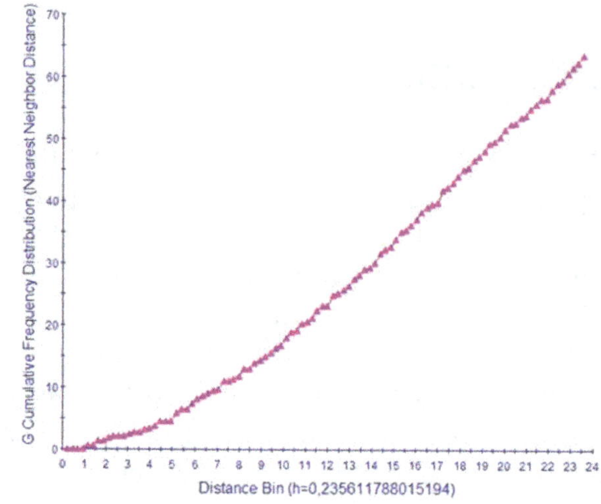

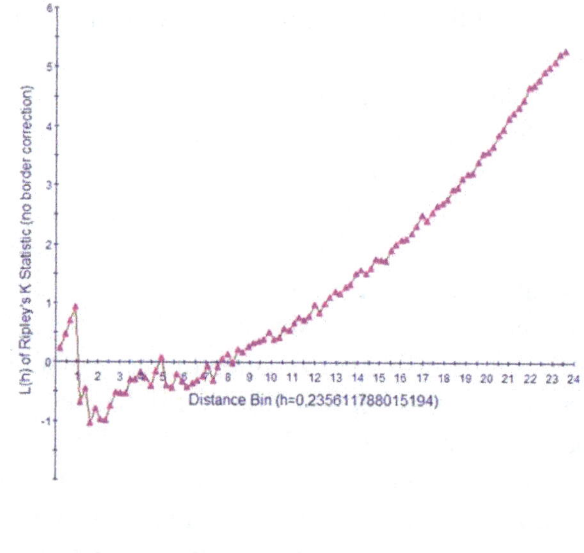

The second graphic of Figure 10 is based on the transformed values of K(h), the L(h) standardized version of K(h) and computed as sqrt(K(h)/3.14159)-h. A 45 degree line indicates a random distribution while a clustered one is identified by a sudden jump of L(h) for distances that are close to zero. Inversely, a slow L(h) increase suggests an uniform pattern.

KERNEL GAUSSIAN DENSITY

This selection creates a raster density map (houses or population, for instance) from the available point data using a moving kernel function (under myGeoffice©, the Gaussian model was used) to produce a continuous tapered surface and to identify hot spots. Conceptually, the produced surface value becomes higher at the point location and contracts with increasing distance from the point. In fact, larger values of search radius yields a smoother and more generalized density raster map while small radius generates a detailer raster output. In other words, a small d bandwidth implies a small influence region while, as d increases, a larger and smooth influence region will be produced.

By definition, the Kernel Gaussian function for any d distance becomes the average number of events found in a circle of d radius around an event divided by the mean intensity of the process (the number of events divided by the study area). Under myGeoffice©, this d radius search parameter must be setup by the user based on the average distance between samples (the locally-adaptive window approach). It should be highlighted if the neighborhood number of samples becomes zero for a particular estimation site, the Kernel function will compute zero for that particular cell. Notice that this intensity estimator only takes into account the distribution of the samples in space but forgetting their values.

T-STUDENT AND F-TESTS

The parametric T-test gives an indication of the difference of behavior or characteristics of two sets of samples whose result should be examined on a T-table (df=number_samples-1) to determine the existence (or not) of a significant difference between two sets of measures (check Appendix 1). If the T-test outcome is greater than the T-table critical value for a particular level of confidence (in general, statisticians use a 95% level of confidence, i.e., 0.025=2.5% for each tail), both subsets of data hold a difference to be claimed. Otherwise, no dissimilarities of behavior can be stressed.

It is significant to highlight that the T-Student (for mean comparison) and the F-test (for variance comparison) are based on the premise of a complete lack of spatial autocorrelation between samples. According to Fotheringham, Brunsdon, and Charlton (2002), this holds some key consequences for any

Figure 11. myGeoffice© output generated by four different searches for the next radius distance: 0.25 and 1 (top row), 1.75 and 3 (bottom row) times of the nearest average distance among samples (four major hot spots of samples density can be confirmed as the radius is increased)

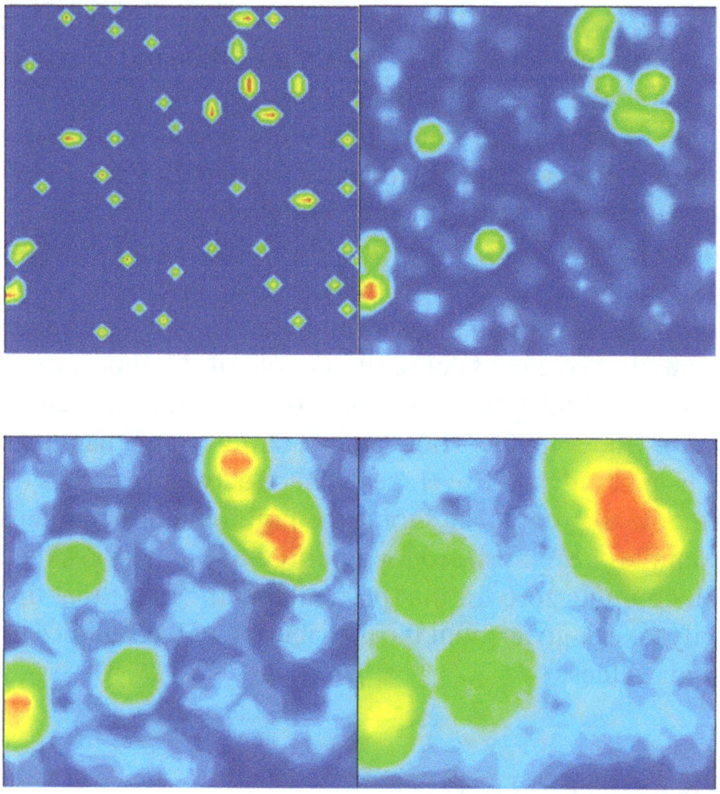

interpretation you may decide afterwards, particularly if the researcher wants to compare means for two spatial groups. If both groups are positively spatially auto-correlated, the variance of the mean becomes wider (if contrasted when spatial autocorrelation is null). In that case, the researcher may decide that the two sample averages are not sufficiently different. However, if the user ignores this autocorrelation factor then s/he runs an increasing risk of making the wrong decision - concluding that the population means are different whereas in fact there is insufficient evidence to support that decision. If data exhibit negative spatial autocorrelation (the calculated confidence interval becomes narrow) and by ignoring it, the researcher may also conclude incorrectly that there is no difference between the two population averages (Negreiros, Costa, Painho & Lopes, 2008).

Developed by Ronald Fisher, the one-way analysis of variance (ANOVA) is a technique used to compare means of two or more datasets by using the F non-symmetrical distribution, where the null hypothesis clearly states that samples of two or more groups are drawn from populations with the same mean. This one-way F-test equals the variance ratio calculated among the means between the different regions (explained variation) over the variance within the samples of each region (not explained variation). If both sub-regions means are drawn from populations with the same mean, the variance among the region means should be less than the variance of the samples (Wikipedia-ANOVA, 2015). A higher ratio, therefore, implies that samples were drawn from populations with different averages.

As illustrated in Figure 12, a pre-defined sub-region was setup so that it could be compared with the leftover area. In this case, (20,50) and (50,25) were the top-left and bottom-right rectangle coordinates of the chosen sub-region, respectively.

In geographical analysis, it is often assumed that the levels of variability across the sub-regions are relatively uniform or stationary, i.e., there is only one spatial process across the entire region. According to Wong and Lee (2001), the differences in variance among the sub-regions are due to sampling errors (H0) or non-stationary processes over space (H1). Formally, under the ANOVA

Figure 12. myGeoffice© outlook to set up the sub-region boundaries from the remaining area. Notice that each region must have, at least, five samples

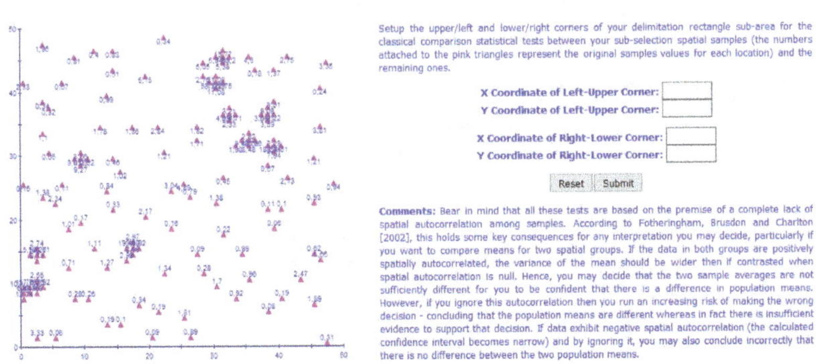

Figure 13. Partial myGeoffice© results (the T-Student's table for different degrees of freedom and levels of confidence were omitted in this writing, for instance) for the comparison of two subsets. By considering the common 95% level of confidence, the 2.91 T–test is greater than the critical value of 2 (df=63), which means both areas hold different mean populations.

T-Student & F Tests

Sample ID (Group 1)	Coordinate X	Coordinate Y	Sample Value	Rank Order
13	36,5	43,5	0,18	13
18	46,5	40,5	0,24	18
25	22,5	48,5	0,34	25
29	31,5	26,5	0,45	29
32	38,5	28,5	0,57	32
44	48,5	25,5	0,94	44
52	22,5	30,5	1,21	52
53	45,5	29,5	1,21	53
57	39,5	43,5	1,37	57
62	27,5	32,5	1,71	62
65	27,5	34,5	1,82	65

Sample ID (Group 2)	Coordinate X	Coordinate Y	Sample Value	Rank Order
1	39,5	18,5	0,06	1
2	5,5	1,5	0,06	2
3	38,5	5,5	0,08	3
4	20,5	1,5	0,09	4
5	27,5	14,5	0,09	5
6	40,5	21,5	0,1	6
7	15,5	3,5	0,1	7
8	6,5	25,5	0,11	8
9	38,5	21,5	0,11	9
10	23,5	18,5	0,16	10
11	0,5	25,5	0,16	11

The parametric Student´s T Test (according to David Ebdon, Statistics in Geography, 1998, p. 62, see formulae below) equals 2,91972714858969 while its degrees of freedom is 63. Notice that the first group holds 52 samples while the second one has 88 samples.

$$t = \frac{|\bar{x} - \bar{y}|}{\sqrt{\frac{(\Sigma x^2/n_x) - \bar{x}^2}{n_x - 1} + \frac{(\Sigma y^2/n_y) - \bar{y}^2}{n_y - 1}}}$$

n_x = *Number of samples of group 1*

n_y = *Number of samples of group 2*

$$df = \frac{(SX + SY)^2}{SX^2/(n_x - 1) + SY^2/(n_y - 1)}$$

x = *Samples values of group 1*

y = *Samples values of group 2*

$$SX = \left[\left(\sum x^2/n_x\right) - \bar{x}^2\right]/(n_x - 1)$$

$$SY = \left[\left(\sum y^2/n_y\right) - \bar{y}^2\right]/(n_y - 1)$$

$\bar{x}$ = *Average of group 1*

$\bar{y}$ = *Average of group 2*

T-Student & F Tests

Sample ID (Group 1)	Coordinate X	Coordinate Y	Sample Value	Rank Order
13	36,5	43,5	0,18	13
18	46,5	40,5	0,24	18
25	22,5	48,5	0,34	25
29	31,5	26,5	0,45	29
32	38,5	28,5	0,57	32
44	48,5	25,5	0,94	44
52	22,5	30,5	1,21	52
53	45,5	29,5	1,21	53
57	39,5	43,5	1,37	57
62	27,5	32,5	1,71	62
65	27,5	34,5	1,82	65

Sample ID (Group 2)	Coordinate X	Coordinate Y	Sample Value	Rank Order
1	39,5	18,5	0,06	1
2	5,5	1,5	0,06	2
3	38,5	5,5	0,08	3
4	20,5	1,5	0,09	4
5	27,5	14,5	0,09	5
6	40,5	21,5	0,1	6
7	15,5	3,5	0,1	7
8	6,5	25,5	0,11	8
9	38,5	21,5	0,11	9
10	23,5	18,5	0,16	10
11	0,5	25,5	0,16	11

The parametric Student´s T Test (according to David Ebdon, Statistics in Geography, 1998, p. 62, see formulae below) equals 2,91972714858969 while its degrees of freedom is 63. Notice that the first group holds 52 samples while the second one has 88 samples.

$$t = \frac{|\bar{x} - \bar{y}|}{\sqrt{\frac{(\Sigma x^2/n_x) - \bar{x}^2}{n_x - 1} + \frac{(\Sigma y^2/n_y) - \bar{y}^2}{n_y - 1}}}$$

n_x = *Number of samples of group 1*

n_y = *Number of samples of group 2*

$$df = \frac{(SX + SY)^2}{SX^2/(n_x - 1) + SY^2/(n_y - 1)}$$

x = *Samples values of group 1*

y = *Samples values of group 2*

$$SX = \left[\left(\sum x^2/n_x\right) - \bar{x}^2\right]/(n_x - 1)$$

$$SY = \left[\left(\sum y^2/n_y\right) - \bar{y}^2\right]/(n_y - 1)$$

$\bar{x}$ = *Average of group 1*

$\bar{y}$ = *Average of group 2*

Figure 14. Both F and T-test confirm that both sub-regions present a statistical difference regarding their local mean. Local variography (see section "Variogram Setup" and "Variogram Fitness") can be a solution for this local non-stationarity situation.

The F index of ANOVA (Analysis of Variance) equals 12,066472434218 (check the below formulae). The degrees of freedom between both groups are 1 while the degrees of freedom within samples is 138.

$$\sigma^2_{WITHIN} = \frac{\sum_k \sum_n (x-\bar{x})^2}{N-k}$$

$$F = \frac{\sigma^2_{BETWEEN}}{\sigma^2_{WITHIN}}$$

$$\sigma^2_{BETWEEN} = \frac{\sum_k n(x-\bar{x}_{Global})^2}{N-k}$$

K – Number of groups

N – Total number of individuals

$\bar{x}_{Global}$ - Overall mean of all samples

n – Number of samples of each group

x – Sample value

$\bar{x}$ - Mean of each group

Below, examine the 95% Confidence level table for the F distribution with 1 degree of freedom (Between) to confirm the acceptance of H0 (null hypothesis) or H1 (alternative hypothesis). Notice that if the F test is smaller than the chosen critical value then the null hypothesis (both set of groups are from two identical normally distribution populations with the same mean) should be accepted. Otherwise, rejection of H0 (acceptance of H1) should take place.

DF (Within)	1	2	3	4	5	6	7	8	9	10
	161.0	18.5	10.1	7.71	6.61	5.99	5.59	5.32	5.12	4.96
DF (Within)	11	12	13	14	15	16	17	18	19	20
	4.84	4.75	4.67	4.60	4.54	4.49	4.45	4.41	4.38	4.35
DF (Within)	21	22	23	24	25	26	27	28	29	30
	4.32	4.30	4.28	4.26	4.24	4.23	4.21	4.20	4.18	4.17
DF (Within)	35	40	50	60	70	80	100	150	300	1000
	4.08	4.03	4.00	3.98	3.96	3.94	3.90	3.87	3.85	4.12

In geographical analysis, it is often assumed that the levels of variability across the subregions are relatively uniform or stationary, that is, there is only one spatial process across the entire region. According to Wong and Lee [2001], the differences in variance among the subregions are due to sampling errors (H0) or nonstationary processes over space (H1). Formally, the null hypothesis is formulated as (Variance_region_1/Variance_region_2) ratio that follows an F distribution (df=(NumSamplesGroupOne-1,NumSamplesGroupTwo-1)). According to the present data, the variances of the left and right subregions are 84,2924797511312 and 17,0663042711599, respectively, which signifies a variance ratio of 4,93911736318776. The degrees of freedom (df) for this one tail test is (52,88). Using the below 95% F-table (click here for other confidence levels), H0 should be rejected if the calculated variance ratio is greater than the chosen F critical value (the variance differences between both groups does not exist by chance, that is, there is a variance difference about their means between both areas). Otherwise, the initial null hypothesis (H0) should be accepted (the minor variance difference between both groups is due to sampling error).

perspective, the null hypothesis is formulated as a ratio (Variance_region_1/ Variance_region_2) that follows an F distribution and whose degrees of freedom equal (NumSamplesGroupOne-1, NumSamplesGroupTwo-1). Using the typical 95% F-table, H0 should be rejected if the calculated variance ratio is greater than the chosen F critical value (the variance differences between both groups do not exist by chance, i.e., there is a variance difference between both areas). Otherwise, the initial null hypothesis (H0) should be accepted (the minor variance difference between both groups is due to sampling error).

Not assessed by myGeoffice©, an alternative possibility to test two independent dataset means (H0:$\mu_1 = \mu_2$) is based on the pooled variance ($s_p = \sqrt{\frac{\left(n_1 - 1\right)s_1^2 + \left(n_2 - 1\right)s_2^2}{n_1 + n_2 - 2}}$). If both variances are considered identical, it is possible to standardize the T-test ($\frac{\mu_1 - \mu_2}{s_p\sqrt{\frac{1}{n_1} + \frac{1}{n_2}}}$) for rejection or acceptance of the initial hypothesis (df=n_1+n_2-2). If both variances are believed unequal (unpooled scenario), the T-test equals $\frac{\mu_1 - \mu_2}{\sqrt{\frac{s_1^2}{n_1} + \frac{s_2^2}{n_2}}}$ whose df=

$$\frac{\left(\frac{s_1^2}{n_1} + \frac{s_2^2}{n_2}\right)^2}{\frac{1}{n_1 - 1}\left(\frac{s_1^2}{n_1}\right)^2 + \frac{1}{n_2 - 1}\left(\frac{s_2^2}{n_2}\right)^2}.$$

Two-Way ANOVA

Although myGeoffice© does not cover this technique, a brief explanation of two-way ANOVA is presented here. Under this procedure, the effects of two factors or treatments, A and B, can be investigated simultaneously, allowing the investigation of the effects of either factor alone (main effect) and of the two factors together. An interaction effect between two treatments occurs if the total effect at some pair of levels of the two treatments differs significantly from the simple addition of the two main effects (Sounderpandian, 2008). In theory, factors that do not interact are called additive. Basically, three questions can be answer by the two-way ANOVA: Are there any factor A main effects? Are there any factor B main effects? Are there any interaction effects between factors A and B?

The following example concerns the effective lifetime of batteries against two factors: The material type (1, 2 and 3) and the operating temperature (low, medium and high). The research question becomes, indeed, quite simple: Is there any difference in the batteries mean lifetime for differing temperatures level and material type? Under the two-way ANOVA, it is then possible to test the following two statements:

- **$H0_a$:** There are no differences in the average battery lifetime for the three different types of batteries under study.
- **$H1_a$:** There are differences in the average battery lifetime among the three different types of batteries.

  In addition, it is possible to test the next null and alternative hypothesis:
- **$H0_b$:** There are no differences in the mean battery lifetime for the three different temperatures levels studied.
- **$H1_b$:** There are differences in the mean battery lifetime by the three different temperatures levels.

Finally, it is also feasible to check the interaction effect between battery types and temperatures levels:

- **$H0_c$:** There are no interactions of the battery types and the temperatures levels under study.
- **$H1_c$:** At the very least, there is one interaction of battery types and temperatures levels.

According to the results of Tables 1 and 2, the three null hypothesis can be rejected (p-value or Sig. < 0.05). So both material and temperature explain the behavior of battery lifetime. As well, the presence of the interaction between material and temperature means that the way battery lifetime changes for different materials depends on the temperature. Similarly, the way battery lifetime changes for different temperatures depends on material (see Figure 5).

Table 1. Life in hours of batteries by material type and temperature

	Temperature (°C)		
	Low (-10°C)	Medium (-10°C)	High (-10°C)
Type 1	130, 155, 74, 180	34, 40, 80, 75	20, 70, 82, 58
Type 2	150, 188, 159, 126	136, 122, 106, 115	25, 70, 58, 45
Type 3	138, 110, 168,160	174, 120, 150, 139	96, 104, 82, 60

(Montgomery, 2001)

Table 2. SPSS® (Analyse/General Linear Model/Univariate) results for the two-way ANOVA

Source	Type III Sum of Squares	df	Mean Square	F	Sig.
Corrected Model	59416.222[a]	8	7427.028	11.000	.000
Intercept	400900.028	1	400900.028	593.739	.000
Material	10683.722	2	5341.861	7.911	.002
Temp	39118.722	2	19559.361	28.968	.000
Material * Temp	9613.778	4	2403.444	3.560	.019
Error	18230.750	27	675.213		
Total	478547.000	36			
Corrected Total	77646.972	35			

Figure 15. In general, the battery life decreases with high temperatures although type 3 batteries remain high at medium temperature

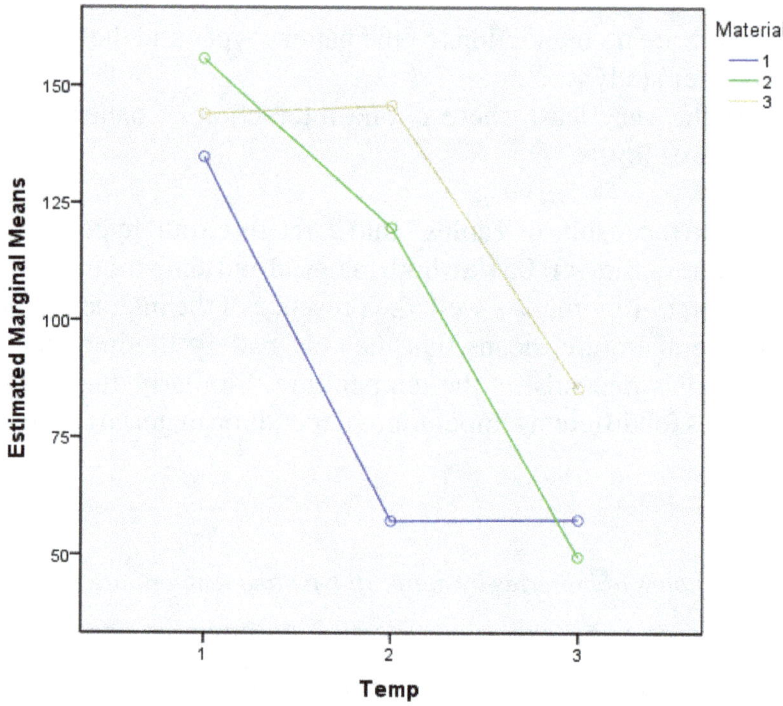

QUADRAT ANALYSIS

Spatial dependence may describe different species of plants at different locations, for instance. To visualize a pattern that is random or aggregated (clustered), quadrat or plot analysis becomes a possibility. Based on the availability of spatial point data and confirmed by Wong and Lee (2001), it allows us to determine if the distribution of a set of points is similar to a random pattern by using a spatial sampling framework, i.e., it examines the density changes over space (expressed as the number of observations per quadrat).

Computationally, this concept relies on the following procedure: Overlay of the study area within a regular square grid (for myGeoffice©, the total number of quadrats may vary from 36 to 196) and the count of the number of points falling in each square. As expected, some squares will contain no points while others will have one, two, three, four or more points. Quadrat analysis

Figure 16. According to the present output, the 140-sample arrangement (10 was considered as the input number of grids) follows a cluster pattern (the D index equals 1.27 while the critical value for a 95% level of confidence is 0.11, a situation already confirmed by the Ripley's K function (section G & Kipley's K Index)

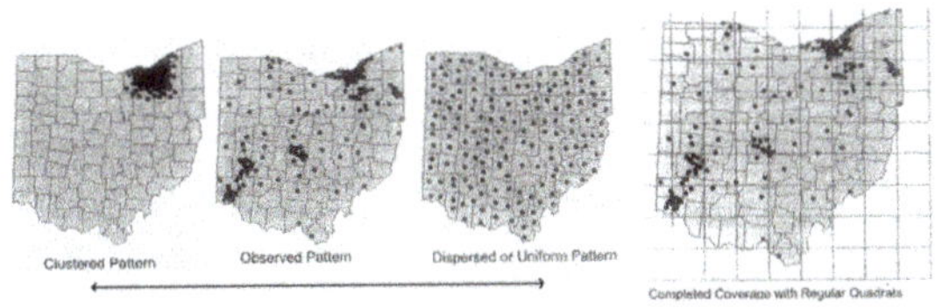

In an extremely clustered point pattern situation, one can expect all samples to fall inside one or a few squares only (check the two above left illustrations). On the other hand, in an extremely dispersed pattern, one would expect all squares to contain a relatively similar number of observations. For myGeoffice© and in this particular case, the quadrat size equals 4,80048 and 4,80048 for the XX and YY axes, respectively, while the number of total points is 140. The number of quadrats becomes 100, a 10 by 10 grid.

Number of points in each quadrat	Observed frequency	Cumulative observed frequency	Observed proportion	Cumulative observed proportion	Poisson proportion	Cumulative Poisson proportion	Poisson frequency	Cumulative Poisson frequency	Absolute difference of cumulative proportions
0	14	0	0,0000	0,0000	0,2466	0,2466	34,5236	34,5236	0,2466
1	61	61	0,4357	0,4357	0,3452	0,5918	48,3330	82,8566	0,1561
2	15	91	0,2143	0,6500	0,2417	0,8335	33,8331	116,6897	0,1835
3	1	94	0,0214	0,6714	0,1128	0,9463	15,7888	132,4785	0,2748

Regarding the above table, the "Observed frequency" column represents the number of quadrats for a particular number of points in each quadrat while the following three columns involve derived further computations (cumulative observed frequency, observed ratio proportion and cumulative one) needed for the Kolmogorov-Smirnov (K-S) statistical test. To achieve this, it is crucial to compare the observed point pattern with a point pattern generated by the random Poisson process. The average number of occurrences of the Poisson distribution is defined as the average number of points in a quadrat. By assuming n quadrats and r points of the entire study area, hence, this estimate mean becomes L=r/n (1,4 in this case). Based on the next notation (L is the average number of samples in each quadrat, E equals the Neperian constant of 2.71828 whereas X denotes the random variable which represents the number of samples in each quadrat), the probability of X observations in any quadrat can be computed using the following Poisson function:

$$P(X) = \frac{E^{-L} \times L^{X}}{X!}$$

The "Poisson proportion" column reflects this calculation. The next three columns denote the analogous previous intermediate assessments. Finally, the farright column ("Absolute difference of cumulative proportions") equals the absolute difference between the two sets of cumulative probabilities, that is, the difference between the fifth ("Cumulative observed proportion") and seventh ("Cumulative Poisson proportion") columns. The largest of these differences is the Kolmogorov-Smirnov D statistic (0,274846178203353). Based on the one-sample test framework with a 0.05 level of significance (in this particular case, this critical value is 0,114940978643078), if the K-S test is greater than the critical value then the observed point pattern rejects the initial null hypothesis (the present sampling layout should be considered a cluster instead of a random pattern). Otherwise, the observation layout is considered not significantly different from a random Poisson distribution (H0 acceptance).

n \ α	0.01	0.05	0.1	0.15	0.2
1	0.995	0.975	0.950	0.925	0.900
2	0.929	0.842	0.776	0.726	0.684
3	0.828	0.708	0.642	0.597	0.565
4	0.733	0.624	0.564	0.525	0.494
5	0.669	0.565	0.510	0.474	0.446
6	0.618	0.521	0.470	0.436	0.410
7	0.577	0.486	0.438	0.405	0.381
8	0.543	0.457	0.411	0.381	0.358
9	0.514	0.432	0.388	0.360	0.339
10	0.490	0.410	0.368	0.342	0.322
11	0.468	0.391	0.352	0.326	0.307
12	0.450	0.375	0.338	0.313	0.295
13	0.433	0.361	0.325	0.302	0.284
14	0.418	0.349	0.314	0.292	0.274
15	0.404	0.338	0.304	0.283	0.266
16	0.392	0.328	0.295	0.274	0.258
17	0.381	0.318	0.286	0.266	0.250
18	0.371	0.309	0.278	0.259	0.244
19	0.363	0.301	0.272	0.252	0.237
20	0.356	0.294	0.264	0.246	0.231
25	0.320	0.270	0.240	0.220	0.210
30	0.290	0.240	0.220	0.200	0.190
35	0.270	0.230	0.210	0.190	0.180
40	0.250	0.210	0.190	0.180	0.170
45	0.240	0.200	0.180	0.170	0.160
50	0.230	0.190	0.170	0.160	0.150
OVER 50	$\frac{1.63}{\sqrt{n}}$	$\frac{1.36}{\sqrt{n}}$	$\frac{1.22}{\sqrt{n}}$	$\frac{1.14}{\sqrt{n}}$	$\frac{1.07}{\sqrt{n}}$

compares this frequency distribution with the theoretically Poisson random distribution. When quadrat data fits a Poisson distribution, it is assumed to be random but when the present data fits a negative binomial distribution, it is said to be nonrandom (aggregated). In an extremely clustered point pattern situation, one can expect all points to fall inside one or a few squares only. On the other hand, in an extremely dispersed pattern, one would expect all squares to contain a relatively similar number of samples.

Figure 17. myGeoffice© samples outline along the X axis

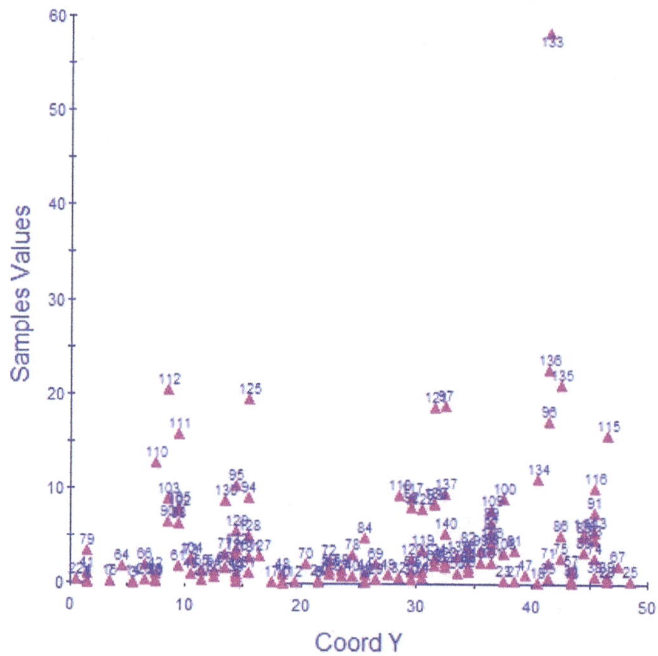

Figure 18. myGeoffice© samples silhouette along the Y axis

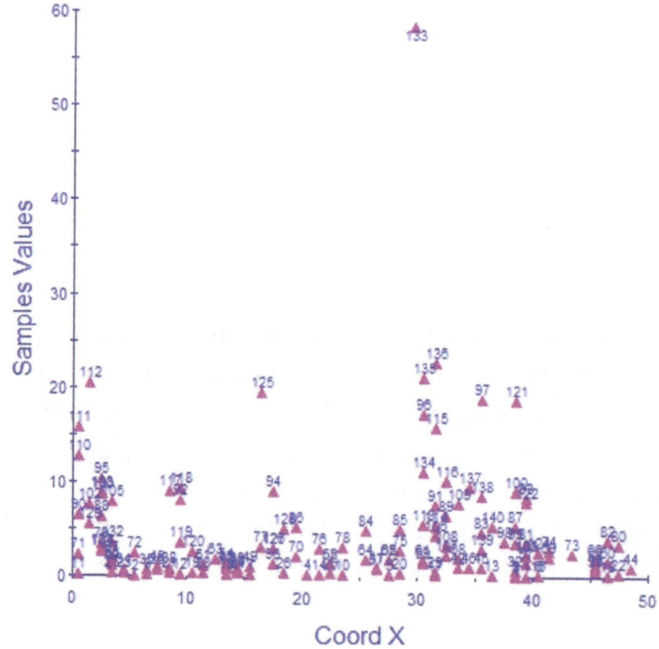

Under myGeoffice©, the Kolmogorov-Smirnov D is assessed to test if the observation layout is not considered to be significantly different from a random Poisson distribution (H0). Based on this one-sample test framework (it gives the probability of X observations in any quadrat), if the Kolmogorov-Smirnov D test is greater than the critical value then the observed point pattern rejects the initial null hypothesis (the present sampling layout should be considered clustered instead of random). Otherwise, H0 should be accepted.

REFERENCES

Clark, I., & Harper, W. (2000). *Practical geostatistics*. Ecosse North America.

Deutsch, C., & Journel, A. (1998). GSLIB. Oxford University Press.

Fotheringham, A., Brunsdon, C., & Charlton, M. (2002). *Quantitative geography*. London: SAGE Publications.

Furtado, A., & Negreiros, J. (2010). Modelação espacial da temperatura na ilha de Santiago, Cabo Verde, com o GeoStatistical Analyst®. *Proceedings of CAPTAR–Environmental Science, 2*(3), 46-54. (in Portuguese)

Goovaerts, P. (1997). *Geostatistics for natural resources evaluation*. Oxford University Press.

Griffith, D. A., & Layne, L. J. (1999). *A casebook for spatial statistical data analysis: a compilation of analyses of different thematic datasets*. Oxford University Press.

Isaaks, E., & Srivastava, M. (1989). *An introduction to applied geostatistics*. New York: Oxford University Press.

Montgomery, D. C. (2001). *Design and analysis of experiments* (5th ed.). New York: Wiley & Sons Inc.

Negreiros, J., Costa, A., Painho, M., & Lopes, I. (2008). *Spatial autocorrelation and association measures*. In G. Putnik & M. Cunha (Eds.), *Encyclopedia of Networked and Virtual Organizations* (Vol. 3, pp. 1502–1508). Hershey, PA: Information Science Reference.

Sethi, S. (2015). *Mode: Merits and demerits*. Retrieved in 2015 from Durofy.com: www.durofy.com/ tutorial-8-mode-merits-demerits/

Sounderpandian, A. (2008). *Complete business statistics* (7th ed.). McGraw-Hill/Irwin.

ANOVA. (2015). In *Wikipedia.* Retrieved from: en.wikipedia.org/wiki/One-way_analysis_of_variance

Wong, D., & Lee, J. (2001). *Statistical analysis with ArcView® GIS*. John Wily & Sons, Inc.

Chapter 7
Spatial Autocorrelation

ABSTRACT

This chapter respects the six procedures of spatial autocorrelation covered by myGeoffice©, including variogram setup and fitness, Moran I correlograms, an innovative version of the conventional Moran scatterplot and the recent Moran variance scatterplot.

INTRODUCTION

If the finding of spatial structures is fundamental for geography then spatial interpolation, spatial autoregressive models and spatial autocorrelation exists as three major techniques for a better understanding of spatial relationships among spatial entities.

Kriging means optimal interpolation of unknown values from observed data at known locations. After the variogram has been defined, this algebraic relationship between values at different distances is used to estimate Kriging weights based on four factors: (A) Closeness to the location being estimated; (B) Redundancy between data values; (C) Anisotropic continuity; (D) Magnitude of continuity. Kriging is BLUP (best linear unbiased predictor), whether or not data is normally distributed. It is linear since estimations are weighted linear combinations of the available data. It is unbiased because the error of the mean tries to be close to zero, i.e., there are no over- or under-estimates. It is best since its goal is to minimize error variance.

DOI: 10.4018/978-1-5225-3270-5.ch007

Spatial autoregressive models expand the standard linear model regression with additional terms that account for local patterns (Banerjee, Carlin & Gelfand, 2015). Essentially, a spatial lag model expresses the notion that the variable value at (x,y) location is related to the values of the same variable measured at nearby locations (the interaction effect and measured by the spatial weights matrix, i.e., W). The basic spatial lagged autoregression equation equals $y_i = \rho \sum_j W_{ij} y_i + \varepsilon_i$, where the spatial autoregression parameter (ρ) should typically be estimated from the available data (Smith, 2015).

Spatial autocorrelation measures the way things are distributed in space (clustering) and it can be seen as a casual process by measuring the degree of influence (correlation) exerted over its neighbors. This means location leads to spatial dependence (correlation or variation that each neighbor holds in relation to a particular point) and spatial heterogeneity (clustering, concentration or proportion of neighborhood average in relation to a specific point) established by Tobler's First Law of Geography. Unfortunately, this situation creates a conflict between the goals of classical statistics and its assumptions (samples are precise, representative, taken randomly from the population and constitute a small portion of a very large one) because classical estimators tend to be inconsistent as the degree of spatial dependence increases (either positive or negative). It appears that Mother Nature does not care about the assumptions of statisticians (Negreiros, Costa, Painho & Santos, 2007).

Supported by Griffith and Layne (1999), spatial autoregression plus spatial autocorrelation provides yet another linkage with Kriging: The spatial interpolation issue. According to both authors, several relationships among these three concepts emerge:

- Spatial autocorrelation is the required condition of Kriging and spatial regression, i.e., spatial autocorrelation is a physical reality and is a necessity for interpolating nearby values.
- Spatial autocorrelation seeks spatial identification while spatial regression and Kriging seek spatial prediction.
- The variance-covariance matrix is included in spatial regression and Kriging.
- Once a variogram is computed, Kriging can be used for spatial interpolation.
- Incorporating spatial autocorrelation within spatial regression results in a 5-10% improvement in the statistical description for a given georeferenced dataset.

- Kriging is primarily applied with continuous regions while spatial autoregression involves aggregations of phenomena into discrete regions such as ward units.
- Spatial autoregressive methods assume that spatial interpolation follows an underlying trend plus random residuals. However, Kriging presumes two other views: (A) If Universal Kriging is assumed then spatial interpolation is interrelated to a background trend; (B) If Ordinary Kriging is chosen, spatial interpolation is consistent with the samples global average plus random residuals.
- Autoregressive residuals can be used for a reasonable reality approximation test to verify existing patterns among residuals and a key information source of possible assumption violations, variable transformations, outliers and trends surface.

As in the data mining process of finding attribute anomalies, spatial autocorrelation measurements look for patterns, outliers, transitions and relationships within vast spatial digital archives. Typically, spatial autocorrelation indices are categorized into two groups:

- Distance view, an approach based on Kriging, where spatial interaction is conceptualized as a continuous distance metric function. It focuses on spatial prediction (variogram).
- Neighboring view is where the interaction is concerned with phenomena aggregations of discrete regions or correlation among adjacent observations. Neighbors, for instance, are defined as spatial units with a common boundary or within a given critical distance. In both cases, no interaction occurs further away. Commonly, the W matrix contents are $W_{ij}=1$ if i and j regions are adjacent to or below the pre-defined cutoff distance and zero otherwise ($W_{ii}=0$). This view enhances statistical description (variogram cloud, Cressie box plot, spatial lag scatterplot, spatial lag pies, Moran I, Geary C, local $G_i(d)$ and global G(d)). If extended with autoregressive models then spatial interpolation becomes real.

The variogram cloud is the square difference between two locations, x_i and x_j, against the distance that separates them in a certain direction. Outliers hold high values in the variogram cloud irrespective of distance and affects computation independently of being a true outlier, skewness datum or wrong observation. Closer distances characterize local outliers. Global ones are considered for larger distances.

The spatial lag scatterplot is distinguished by a graph, where its horizontal axis corresponds to the value at each site while the vertical one shows the corresponding value at locations within a specific d distance apart, representing different contiguity order along different distance bands. Hence, looking for local outliers becomes obvious while a linear relationship along the diagonal indicates a strong association.

Geary C (see Figure 1) is a variant of the Moran I coefficient for positive values only, consisting of the sum within neighbor locations for a specific range divided by the sum of all observations. According to Wikipedia-Geary (2015), Moran's I is a measure of global spatial autocorrelation while Geary's C is more sensitive to local spatial autocorrelation (areal units with more connections are weighted more heavily with Moran I than Geary C). Unfortunately, this ratio is not easy to interpret. Under the random map pattern of a Gaussian distribution, the expected value of Geary C equals one. For positive auto-correlated data, Geary C becomes less than one while a dissimilar cluster tendency reveals a Geary C greater than one.

The spatial lag pies consist of pie charts located on observational coordinates, which compare the magnitude of the sample value relative to its spatial lag. More precisely, if the observation value at i location is x_i then its spatial lag is confirmed by Figure 2. Similar bar heights suggest positive spatial autocorrelation whereas unequal ones tend to indicate a negative spatial autocorrelation or a transition between spatial regimes.

As an alternative for identifying local patterns not revealed by global indexes, a spatial scale issue, Getis and Ord (1992) introduced the local $G_i(d)$ statistic for positive variables. This index measures the degree of clustering that result from the spatial concentration (not correlation) of the sum of

Figure 1. $\bar{x}$ stands for the overall mean, W_{ij} is the weight of W matrix between i^{th} and j^{th}, n represents the total number of observations while x_i and x_j are i^{th} and j^{th} sample value

$$C=\frac{(n-1)\sum\sum w_{ij}(x_i-x_j)^2}{2\sum_i\sum_j w_{ij}\sum_j(x_j-\bar{x})^2}$$

Figure 2. x_j represents i's neighbor and W_{ij} is the weight assigned between site i and neighbor j samples

$$\text{spatial\_lag}_i = \sum_j w_{ij} x_j$$

points within d distance of point i. Therefore, $G_i(d)$ equals the sum of all $x_i x_j$ values within d distance of point i. A large standard negative index implies small values for x_j within i location whereas a large positive one involves larger ones. Accordingly, a spatial pattern has been found. Furthermore, $G_i(d)$ is scale-invariant (but not location-invariant) and is inappropriate with regression residuals (it is centered on positive variables only) and sensitive to transformations such as $x_i = \log(x_i)$.

This chapter respects the six procedures of spatial autocorrelation covered by myGeoffice©, including variogram setup and fitness, Moran I correlograms, an innovative version of the conventional Moran scatterplot and the recent Moran variance scatterplot.

DATA CLASS MAPPING

Classical statistics offer a wide range of inferential methods based on restricted assumptions. The 95% confidence level of the true unknown population mean (tested by the Normal or T-Student distribution) and variance (tested by the Chi-square distribution of Appendix 1) are only valid if the samples are homoscedastic and independent with uncorrelated error. Another

Figure 3. $W_{ij}(d)$ is the weight of W matrix between i^{th} and j^{th} (according to any cutoff d distance), n represents the number of observations while x_j denotes j^{th} neighborhood sample value of ith site

$$G_i(d) = \frac{\sum_{j=1}^{n} w_{ij}(d) x_j}{\sum_{j=1}^{n} x_j} \qquad G(d) = \frac{\sum_i \sum_j w_{ij}(d) x_i x_j}{\sum_i \sum_j x_i x_j}$$

classical possibility is the Chi-square (χ^2) contingency test, which is used to find particular relationships between two categorical variables, where each variable consists of a number of classes. If χ^2 is greater than the expected χ^2 critical value for a certain level of confidence, the initial hypothesis (H0) is rejected and thus a statistical interdependency between regions and groups is determined. To confirm the average difference of two sub-populations is an alternative under the independent random samples and Gaussian distribution assumptions. The linear and multiple regression models, including the global F and individual T-Student, assume linearity, independency of random variables, homoscedasticity (the weak assumptions) and Gaussian distribution (the strong assumption).

However, spatial data does not always follow statistical rules and spatial patterns must be taken into account. As Griffith & Layne (1999) points out, spatial autocorrelation is usually positive with a correlation range between 0.3 and 0.5. Yet, autocorrelation restricts the ability to perform standard statistical hypothesis tests because the estimator standard errors are not minimized and regression coefficients from least squares are unbiased while their variances are underestimated (Ebdon, 1998). This occurs because new observations with a lack of independence do not even allow bring one full degree of freedom since the observer holds some prior knowledge at new locations, reflecting information loss. As spatial autocorrelation approaches 1, the effective degrees of freedom approach 0. As spatial autocorrelation approaches 0, the effective degrees of freedom lean towards the total sampling number. As spatial autocorrelation approaches -1, the effective degrees of freedom increase beyond the number of observations (Griffith & Layne, 1999).

Accordingly, fewer samples are needed as positive spatial autocorrelation increases. If someone records the air pollution levels in downtown Taipei, Taiwan, the monitoring devices will report the same levels, adding almost no new information to the previous samples. Information redundancy is the logical consequence as samples collected for two juxtaposed points in space tend to represent essentially the same information.

With independent data, the two-sided 95% confidence interval for the population mean μ becomes [$-1.96\sqrt{\sigma}$,$+1.96\sqrt{\sigma}$] but with positive n correlated samples, Cressie (1993, 2015) proved that the previous interval equals [$-2.485\sigma/\sqrt{n}$,$+2.485\sigma/\sqrt{n}$], leading to a wider confidence interval. Moreover, the estimated variance of the mean with a population of n samples is not σ^2 / n but σ^2 / n', where $n'=n/[1+2\{\rho/(1-\rho)\}\{1-(1/n)\}-2\{\rho/(1-\rho)\}^2(1-\rho^{n-1})/n]$, which can be interpreted as the equivalent independent observation number. If n=10 and spatial autocorrelation equals 0.26, then n'=6.2, that is, 6 independent observations achieve approximately the same accuracy as 10 correlated samples.

Identifying patterns, transitions and outliers is another aim of most spatial autocorrelation measures. If an outlier is located on the center of the map, for example, only smaller variogram lags will contain it, leading to a variogram decrease for greater distances. However, if the outlier is located in the corner of the field then the variogram will present almost no difference to the one in which the outlier does not exist. Certainly, if heteroscedasticity exists, variability will increase with distance lags, regardless of the variogram choice.

A typical method to check high/medium/low patterns, transition sub-regions and local outliers is to employ visual classification maps based on different thresholds (this is an extension of indicator maps, where only two types of symbols are plotted above or below a specific threshold). To check anisotropy, the property of a spatial process in which spatial autocorrelation changes with both direction and distance between two sites, is also crucial for geographical matters. Thus, the first purpose of this myGeoffice© option is to highlight the main spatial continuity direction (azimuth) of the phenomenon that is under study (a key requirement for the Variogram setup option).

Regarding anisotropy variogram modelling, two perspectives can be stated. Anisotropies modelling should be done by trial and error using a graphic terminal (Soares, 2000). Armstrong (1998) argues that it is not a good idea to apply least squares or other automatic methods to fit variogram models. Firstly, the model must hold a positive definite function to guarantee having one and only one stable solution whilst the Kriging matrix must have an inverse. Polynomials obtained by least square regression will rarely satisfy these conditions. Secondly, least squares assume that sample points are independent observations, which is clearly not true. Thirdly, critical variogram behavior close to the origin is not known and least squares do not take this into account. Outlier adjustments can also be complicated. Further, it does not take into account additional information that may justify the choice of one model instead of another. myGeoffice© follows this philosophy.

For automatic fitting, Goovaerts (1997) uses the minimally weighted sum of the squares of the differences between the experimental variogram and the theoretical one. Clark and Harper (2000) suggest Cressie's modification of the automatic fitting by replacing the scaling division of the total number of pairs. Basically, this formula quantifies the squared difference between

Figure 4. Consistent with this result of myGeoffice©, it seems at first that the present phenomenon embrace a maximum continuity of North-West/South-East or North-East/South-West in space. Contour isoline maps might be another visual possibility to check anisotropy (the state of having different properties or behaviors along different space directions).

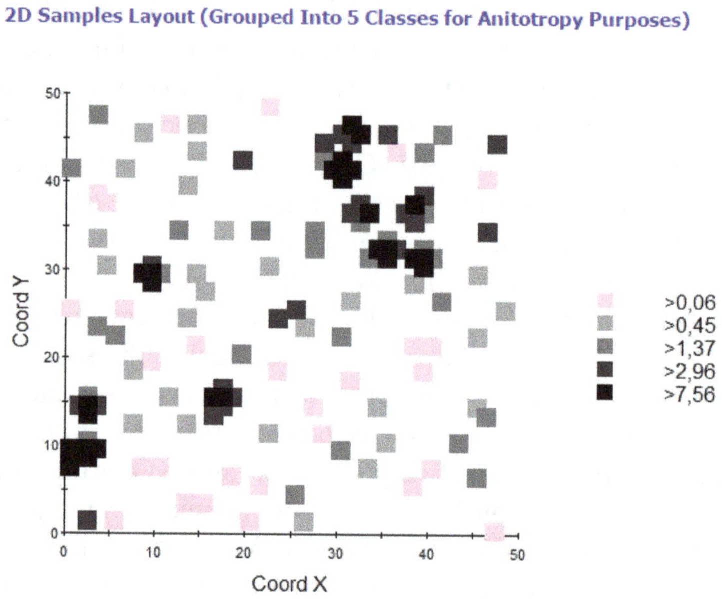

theoretical and real variograms but is scaled by the model itself: The lower this index becomes, the better the model turns out to be. In addition, each point is weighted according to the number of pairs available: If more samples available, errors are minimized.

VARIOGRAM SETUP

The regionalized theory of the French Professor Georges Matheron (1930-2000) states that unknown regions are correlated on the basis of location distance, i.e., there is a variance rate between samples over space in a context of physical continuity. If traditional statistical methods assume that all data is independent, spatial samples taken close to one another may be expected to have similar values to the samples taken farther apart. This phenomenon is usually referred to as spatial dependence, spatial continuity or autocorrelation (Soares, 2000).

According to Matheron (1965), the regionalized theory treats spatial properties as the sum of three components: (1) Deterministic structural component, drift or trend; (2) Stochastic or random term with local variation; (3) Gaussian independent spatial residual. Hence, spatial data equals large-scale variation plus small-scale plus error. Data transformation and removal trend concern the first factor. The variogram, $\gamma(h)$, involves the second issue while error measurement is closely linked to the nugget-effect or nugget-variance, C_0. This section only focuses on the second and the third components.

Under the regionalized assumption, local variation is not generally unstructured but is spatially dependent at some scale because the points within a given distance apart depend on one another in a statistical sense (Armstrong, 1998). For any space lag (h) increment in a particular direction, the distribution of $z(x_1)$, $z(x_2)...z(x_n)$ observations is the same as that of $z(x_1+h)$, $z(x_2+h)...z(x_n+h)$, where z(x) is regarded as the sum of three terms: (1) Mean, drift or trend, (2) random dependent variable and (3) error.

The stationary assumption requires the first two factors (mean and variance) to be constant. By default, spatial homogeneous regions make random functions homogeneous and self-repeating in space. The variance of the random spatial variable equals the covariance at zero lag distance, $\sigma^2=C(0)$, while the covariance between z(x) and z(x+h) becomes $\sigma_{x,x+h}=\sigma^2 f(h)=C(h)$. This means that the stationary covariance only depends on distance and direction between two samples and not on their locations (the second order stationary assumption). Therefore, the expected difference between values at two locations separated by h distance is close to zero.

If $E(\mu(x))=m$ then $z(x)=m+e(x)$ and $E[z(x)-z(x+h)]=0$, the covariance at h lag distance is $E\{[z(x)-m][z(x+h)-m]\}=E[z(x)\times z(x+h)]-m^2=Cov(h)$, z(x) being the value at x site, z(x+h) the value at x site plus h lag distance, m the overall mean and E() the expected value. However, if there was a trend or drift over the area then the expected value would change and the mean difference between any two samples would not necessarily be zero. This is why Matheron developed the intrinsic hypothesis, the stationary assumption of the difference or variance of differences by assuming a weakly stationary situation (stationary of order two): The z(x)-z(x+h) increments exist and they are location independent (Armstrong, 1998). Although the covariance does not exist, the variance of the difference between two locations is assumed to be present and does not depend on the location and $Var[z(x+h)–z(x)]=E[\{z(x+h)-z(x)\}2]=2\gamma(h)$, with $E[z(x)-z(x+h)]=0$. In other words, differences between sites are merely a displacement function between them, the intrinsic hypothesis (Armstrong, 1998).

The $\gamma(h)$ is called the dissimilarity average function (semi-variogram, in a strict sense, or variogram, in a large sense) and is defined as $\gamma(h)=0.5\times Var[z(x+h)-z(x)]$. Since the mean of z(x+h)-z(x) equals zero for stationary and intrinsic variables, the variance is just the mean square difference. Consequently, the variogram is as follows: $\gamma(h)=0.5\times E\{[z(x+h)-z(x)]^2\}$. Mathematically, N(h) being the number of point pairs separated by the h distance lag, the variogram estimator becomes as it is shown in Figure 5 ($z(x_i)$ denotes the sample value at i^{th} location whereas z(x+h) stands for the value at i^{th} site plus h lag). Quoting Clark and Harper (2000), the semi-variogram is a picture of your data spatially and will give you a lot of information about relationships you may not even have thought about.

Two major parameters describe this model: The range or lag distance at which all successive values are independent on each other and the sill, the variogram value corresponding to the range (Negreiros, Painho, Oliveira, Aguilar & Aguilar, 2008). Therefore, the range becomes a length measure of heterogeneity correlation. For Armstrong (1998), there is an equivalence between the variogram, $\gamma(h)$, and the covariance, C(h), but no covariance exists for a non-stationary situation. However, the variogram always exists and this makes it generally more useful. Other useful relationships can be drawn: (1) $C(0)=\sigma^2$, the covariance between the same sample equals the variance; (2) C(h)=C(-h), the covariance between two given samples does not depend on their positions; (3) $|C(h)|<=C(0)$, the module of the covariance is always below or equal to the variance; (4) $C(\infty)=0$, the covariance between samples further apart an infinite distance equals zero; (5) $\gamma(h)=\gamma(-h)$, the variogram computed in both directions is the same; (6) $\gamma(h)=C(0)-C(h)$, the corresponding variogram is obtained by turning the covariance upside down (see Figure 6).

The proof of the relationship between the variogram and the covariance starts out from the variogram definition: $2\gamma(h)=E[\{z(x+h)-z(x)\}^2]=E[\{z(x+h)-m\}^2+(z(x)-m)^2-2\times(z(x+h)-m)\times(z(x)-m)] = E\{Z^2(x+h)\}+E\{Z^2(x)\}-2E\{z(x+h)\times z(x)\} = \sigma^2(z(x+h))+\sigma^2(z(x))-2C(h) =$

Figure 5. Conceptually, the variogram indicates how different the values are as distance increases in a specific direction

$$\gamma(h)=\frac{1}{2N(h)}\sum_{i=1}^{N(h)}[z(x_i)-z(x_i+h)]^2$$

Figure 6. The variogram and the covariogram function relationship with and without the nugget-effect

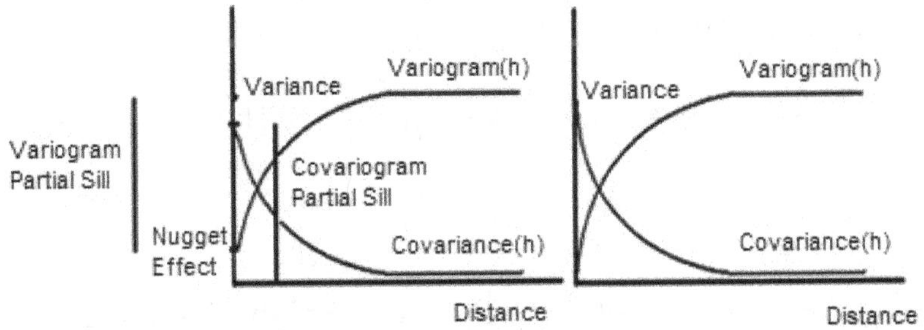

2C(0)–2C(h). Clearly, this relation is only possible with bounded variograms and only stationary regionalized variables have this feature (Armstrong, 1998). Unbounded ones arise from non-stationary situations with infinite dispersions. In this case, the trend has to be defined as a vicinity position function in the field.

A gradual change in C(h) function near the origin indicates strong spatial continuity while a sudden decrease in the variogram points to scale variability. If space is continuous then $E[z(x+h)-z(x)]^2$ should be zero when h=0. The main reason for a nugget-effect greater than zero is a measurement error. Yet, it is conceded that the micro-scale variation also leads to a highly irregular variogram pattern at a very short range. According to Cressie (1993), since nothing can be said about variograms at lag distances smaller than $\min\{\|z(x+h)-z(x)\|\}$, it is not possible to tell whether a micro-scale variation is continuous or not. Consequently, the nugget-effect equals $C_{\text{Micro-scale Variation}}$ plus $C_{\text{MeasurementError}}$. This dramatic short scale variability may have a serious impact on the accuracy of future estimates and myGeoffice© handles it by generating four dissimilar nugget-effect approaches (see Chapter 9). On the other hand, a pure nugget-effect concerns a positive constant value for all lags, which means uncorrelated locations for any h distance. As expected, the residual random noise with zero mean, constant variance and zero covariance also exists within this stochastic approach, as presented in Figure 7.

Fortunately, the variogram is only used up to a certain distance (range). The problem is to decide which series of moving neighborhoods can be found with a constant expected value and enough data to give meaningful estimates. The quasi-stationary assumption thus becomes a homogeneity scale and a sampling density compromise. If there is a spatial trend, the Ordinary Kriging (OK) variogram is no longer the appropriate tool to model spatial patterns.

Figure 7. $\mu(x)$ represents the deterministic function that describes the structural component of Z at x location; $\varepsilon(x)$ denotes the stochastic term whose variance is the nugget-effect; $\gamma 1(h) + ... + \gamma n(h)$ equals the different variogram structures while ξ defines the spatially independent Gaussian noise term.

$$z(x) = \mu(x) + \varepsilon(x) + \xi \text{ or } z(x) = \mu(x) + \gamma_1(h) + ... + \gamma_n(h) + \xi$$

Some alternatives are possible: (1) Universal Kriging, where the trend and the residual local are simultaneously estimated leading, quite often, to the use of a lower order surface. According to Clark & Harper (2000), it gives more realistic confidence intervals for the estimations; (2) Stratified regionalization; (3) Simple Kriging with a varying local mean, where interpolation is based on secondary information and external drift; (4) To model the global trend component using a multi-linear coordinate regression to remove the drift component before OK is carried out on autocorrelated residuals (myGeoffice© only offers the first option for this issue).

The variogram quantifies Tobler's Law at all scales by summarizing the degree of similarity between data values for all possible data pairs as a distance function. Most variograms display a behavior like the one presented in Figure 8. The upper limit is the sill, which implies no spatial dependence between data points because all variances are invariant to the sample separation

Figure 8. Typical behavior of a transitive variogram. At this stage, it is worth noting that a sill rescale by 10 or 100 would not change prediction values. Maps would still look the same although changes would be reflected in prediction variances only: The higher the sill, the higher the prediction Kriging variance.

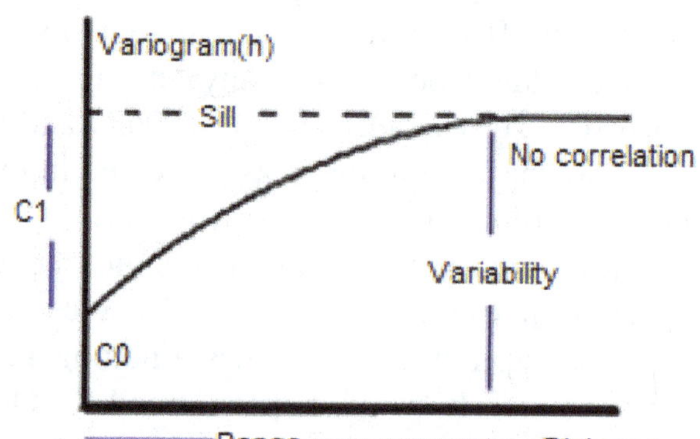

distance. If the variogram has a sill then this behavior tends to support the assumption that samples come from a stationary distribution with a fixed mean and standard deviation (Clark & Harper, 2000). The percentage ratio between the nugget-effect and the total sill is called the relative nugget-effect.

The separation distance at which samples are spatially auto-correlated is the range (how large the search window should be, a local neighborhood stationary issue). If the distance between the estimation and the sample is greater than the range, those samples make no useful contribution to the final interpolation: It is too far and, in theory, its weight becomes zero. As expected, stronger spatial autocorrelation increases the range of the variogram and provides a fairly smooth map since Kriging weights become less powerful with respect to closer samples. Conversely, as the range becomes shorter, more spatial variation is expected.

The nugget-effect represents measurement error variance and spatial variation at distances much shorter than sample spacing. It measures the unreliability of the sample value and the lack of a micro-scale model that the researcher cannot predict because it derives from spatial measurement error (a non-zero nugget-effect indicates repeated measurements at the same point with different outcomes, a precision issue) and/or purely random variation. Theoretically, the variogram should be strictly zero for zero distance. Citing Clark and Harper (2000), if we do not accept zero at zero then we are basically saying that we do not believe in our sample values. Controversially, once a uranium sample has been drawn, it is not there anymore to be replicated (in the statistical sense, replicate sampling is not possible for the majority of spatial studies).

In accordance with the next illustration, the researcher of myGeoffice© must setup the six parameters (lag distance, number of lags, tolerance angle, lag tolerance, bandwidth, azimuth of main direction) required to build a good (perfection does not exist:-) global variogram for future stochastic interpolation and sequential simulation procedures. Analogous to the data class mapping in section 8.1, it is imperative to have a global view of the variability values in all directions. Computationally, the $\gamma(0)=0$ value of the variogram map is plotted below. The directions of anisotropy are usually evident from this map, where dark blue squares stand for low variance, green ones denotes medium variability and yellow and red suggest high dissimilarities.

The number of lags may vary between five and fifteen although a {8,9,10} choice is, in general, a good start. The lag (the distance between successive h-scatterplots) may vary between the nearest mean distance among all observations and 1/3 of the mean distance. A reliable initial value is one

Figure 9. The variogram anisotropy direction must be reflected on the azimuth setting. If the spatial dataset is isotropic, the variogram map will display a perfectly round circle and any azimuth value will do

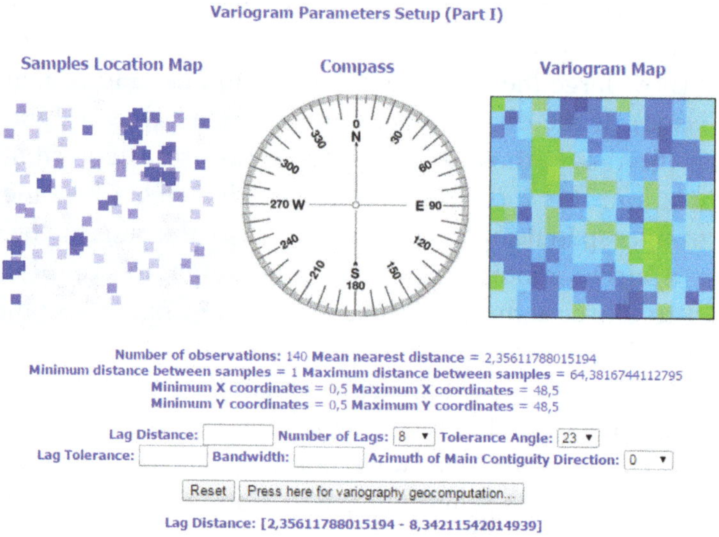

close to the mean nearest distance (the best estimate of the average spacing between neighboring samples). It should be remembered that the number of lags multiplied by the chosen lag distance must not exceed 80% of the maximum distance between all available samples.

Three complementary parameters must be setup (see Figure 10) to create small sub-areas for the computation of the variogram for each lag distance (otherwise, no samples will fall within each particular lag distance):

- An acceptable first default value for the bandwidth is the lag distance itself. Under myGeoffice©, this parameter cannot be negative or greater than the double of the chosen lag distance.
- The lag tolerance of 50% of the chosen lag distance is a decent primary value. Similar to the previous setting, this cannot be negative or greater than the double of the chosen lag distance.
- The tolerance angle may vary between 1 and 90 degrees (omnidirectional) although 23 degrees is the common default starting point to avoid overlapping among the four different variogram directions.

Figure 10. It is the users' responsibility to abide by all the previous rules to get good results (source: spatial-analyst.net/ILWIS/htm/ilwisapp/spatial_correlation_functionality.htm). It should be stressed that the variogram with the greatest range among the four directional variograms is the key factor that decides which direction should be considered as the main azimuth.

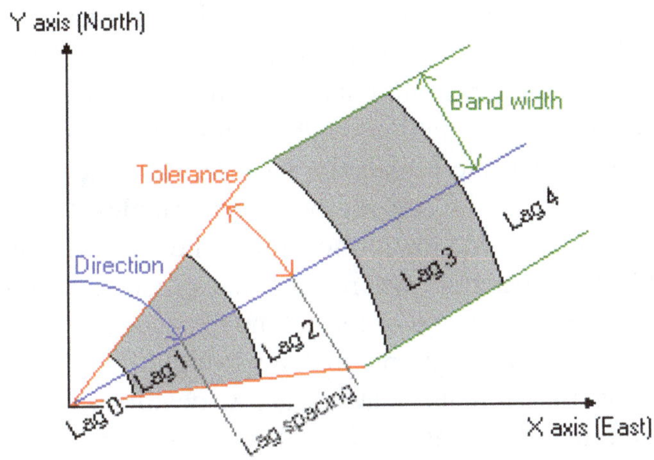

Figure 11. Partial output of the second part of Figure 9 and based on the following parameters: Lag distance=4; Number of lags=8; Tolerance angle=23 degrees; Lag tolerance=2; Bandwidth=4; Azimuth=0 degrees

Check if all executable programs (assessment of traditional, pairwise relative and general variogram, correlogram, madogram, covariance and other types of spatial autocorrelation measures) were accomplished with success

```
c:\inetpub\wwwroot\executable\gamv.exe d:\data\7915763150.par
GAMV Version: 2.905

data file = d:\data\791576315.gamv
columns for X,Y,Z = 1 2 0
number of variables = 1
columns = 3
trimming limits = 0.000000E+00 1.000000E+20
output file = d:\data\7915763150.out
number of lags = 8
lag distance = 4.000000
lag tolerance = 2.000000
number of directions = 1
azm, atol, bandwh = 0.000000E+00 23.000000 4.000000
dip, dtol, bandwd = 0.000000E+00 2.000000 2.000000
flag to standardize sills = 0
number of variograms = 1
tail,head,type = 1 1 1

maxdat = 140
Variable number 1
Number = 140
Average = 4.350429
Variance = 44.924290

Variogram 1 Semivariogram : tail=Main Variabl head=Main Variabl
currently on 14 of 140
currently on 28 of 140
currently on 42 of 140
currently on 56 of 140
currently on 70 of 140
currently on 84 of 140
currently on 98 of 140
currently on 112 of 140
currently on 126 of 140
currently on 140 of 140

GAMV Version: 2.905 Finished

Stop - Program terminated.
```

VARIOGRAM FITNESS

A brief theoretical summary of variography should be addressed here due to its importance on Kriging (chapter 9) and geostatistical simulation (chapter 10). As illustrated in Figure 12, there are four types of behavior regarding the nugget effect (C_0): Quadratic (i) - The regionalized variable is highly continuous and differentiable with the presence of a drift; Linear (ii) - The regionalized variable is continuous but not differentiable and thus less regular; Discontinuous (iii) - The variable is highly irregular at short distances (the higher the nugget-effect, the higher the Kriging variance); Flat (iv) - Pure randomness or white noise, i.e., the regionalized variables z(x) and z(x+h) are uncorrelated for all values no matter how close they are. This is the extreme case of a total lack of structure, where the best estimate for an unknown site would be the population average (if there is no distance relationship between samples, it is not possible to setup a distance weight function).

Although insufficient data may create a deadlock situation for the variogram estimation, the creation of local ones for specific areas according to their local mean (see Figure 13) is a possibility when it is difficult to find a perceptible global variogram. By using local models, Isaaks and Srivastava (1989) demonstrated how stratification reduces interpolation errors based on prior information (qualitative maps, for instance).

Another option presented by both authors is the pairwise relative variogram (Figure 14), where the denominator serves to reduce the influence of very large differences. Notice that if $z(x_i)$ and $z(x_i+h)$ are zero, dropping that calculation for the next pair becomes the natural solution.

Disregarding the variogram option, the removal of the most erratic values can be a plus, even if this might be a questionable solution. In mining excavation, an unusually large value may be a cause for celebration (Goovaerts, 1997). If outlier removal is crucial in generating negative spatial autocorrelation,

Figure 12. The four types of behavior close to the origin
(Armstrong, 1998).

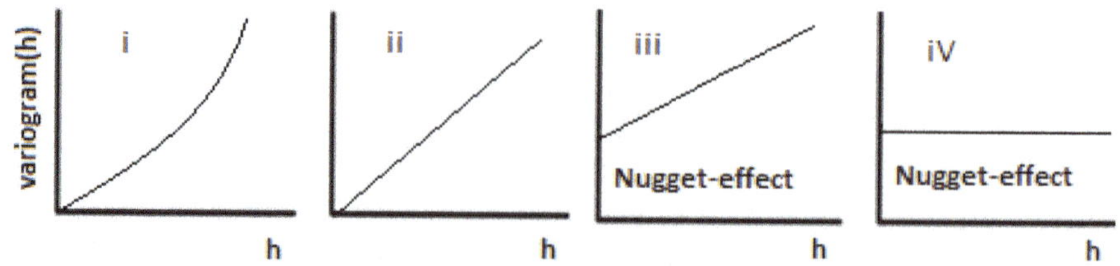

Figure 13. $\gamma 1(h),\ldots,\gamma n(h)$ are the local variograms from n regions, $m_1,\ldots,m_n$ is the local region mean, $N_1(h),\ldots,N_n(h)$ are the number of sample pairs from each region

$$\gamma_W(h)=\frac{\sum_{i=1}^{n} N_i(h)\frac{\gamma_i(h)}{m_i^2}}{\sum_{i=1}^{n} N_i(h)}$$

Figure 14. As an alternative, statisticians may also use the covariance function ($\gamma(h)=C(0)-C(h)=\sigma^2-C(h)$), the madogram or the correlogram model ($\gamma(h)/\sigma^2 = \rho(0)-\rho(h)=1-\rho(h) =1-C(h)/C(0)$)

$$\gamma_{PR}(h)=\frac{1}{2|N(h)|}\sum_{i=1}^{N(h)}\frac{[z(x_i)-z(x_i+h)]^2}{\left[\frac{z(x_i)+z(x_i+h)}{2}\right]^2}$$

small high-grade deposits can equally make the difference between opening a mine and keeping it closed. Moreover, with high skew distributions such as with gold grades, it is not possible to cut them in an arbitrary way.

Data standardization is an alternative although this may create data artifacts. For instance, if a variable holds a zero value and a logarithm transformation is applied, a math problem may then emerge. A plausible solution is to replace 0 by 0.00000001. Even so, depending on the variable range context, either option may or may not change the variogram shape.

If high-grade samples are close together in a specific sub-area, eliminating those samples may not be a good idea either. The model variogram is the spatial averaging result of obscuring regional variations and differences caused by local anomalies (Isaaks & Srivastava, 1989). Therefore, it is essential to examine spatial contiguity and short-range differences before modelling variograms with first-class spatial exploratory software.

The majority of literature recommends computation in four directions to divide the omnidirectional search into four subsets such as 0° (North/South), 45° (North-East/South-West), 90° (East/West) and 135° (North-West/South East). The contiguity ranges of each of these can confirm or refute the assumption of the anisotropy. When plotting a rose diagram, for instance, circular contour lines are indicative of an isotropic surface whereas an elliptic shape points to an anisotropy occurrence. An extension of this procedure is to draw the contour map of the variogram surface. As expected, more weight will be given to samples that lie in the direction of maximum continuity and considerably less, where minimum continuity is involved.

Concerning elliptic or geometric anisotropy, the variogram should hold the same sill for major directions although their ranges might be different (see Figure 15). Given $sill_1=sill_2$, $range_{minor}=a_1=1$, $range_{major}=a_2=a$ and $\gamma_1(h/a)=\gamma_2(h)$, according to Isaaks & Srivastava (1989), the overall 2D variogram becomes

$$\gamma(h)=\gamma(h_x,h_y)=\gamma_1(h_1)=\gamma_1\left(\sqrt{\left(\frac{h_x}{a_x}\right)^2+\left(\frac{h_y}{a_y}\right)^2}\right),$$

where h_x and h_y are the cosine and sine of direction h while a_x and a_y respectively equal the length of major and minor axes of the resulting ellipse.

The last step concerns the creation of a mathematical model that fits the quantitative regionalized variation description so that autocorrelation can be characterized for every distance. The ultimate purpose is to obtain the covariance values to setup the A and B matrices within equation A×W=B (see section "Ordinary Kriging"). Therefore, the variogram formula attached

Figure 15. Occasionally, the variogram exhibits different ranges with the same sill thus creating an overall variogram division of two linear components. myGeoffice© does not cover zonal anisotropy (dissimilar sills for different directions), a spillover effect of the proportional effect (the higher the average of the samples within the block, the higher the variance)

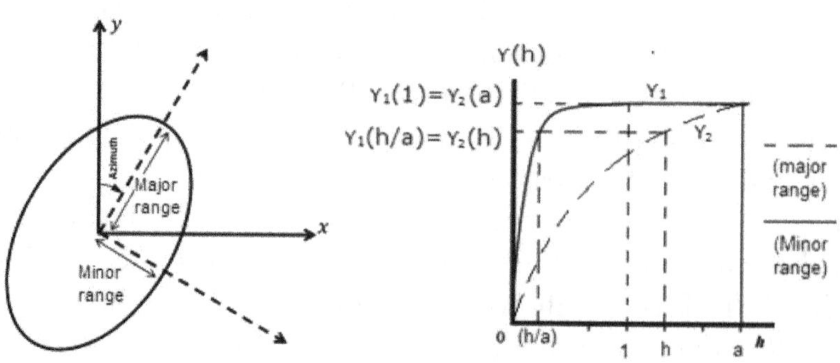

is critical to the interpolation quality. It is imperative that this model fits experimental values well and avoids negative variances, the conditional negative definite property (Armstrong, 1998).

Figures 16, 17 and 18 present the three admissible transitive variograms of myGeoffice©. Not reviewed here, other specialized literature discusses others such as the cardinal sine, wave-hole, intrinsic power function, nugget-effect, prismato-magnetic, Cauchy, Bessel, cubic and prismato-gravimetric model. Non-transition variogram models used to model intrinsic random functions are not enclosed either (Negreiros, Painho, Costa, Cabral & Oliveira, 2008).

The exponential model initially rises more rapidly (see Figure 19), leading to fuzzier images (less structured) than Gaussian because of its linearity near the origin. On the other hand, the Gaussian variogram reaches about 66% of its height at the square-root of its range. Yet, this model does not work well when it has a low nugget-effect with close space data.

Regarding computational issues, any experimental variogram must be aware of observations whose layout follows an irregular network. This is the reason for the associated tolerance angle (see previous section). If the two major directions are orthogonal, the total angular tolerance should be less than 45 degrees to avoid points overlapping (the default 23 degrees for the

Figure 16. The spherical model, where C0 is the nugget-effect while C1+C0 denotes the total sill of the considered variogram, γ(range)

$$\text{Spherical} \quad \gamma(h) = C_0 + \left[C_1\left(\frac{3}{2}\frac{h}{a} - \frac{1}{2}\left(\frac{h}{a}\right)^3\right)\right] \quad \text{if } 0 < h \leq a$$

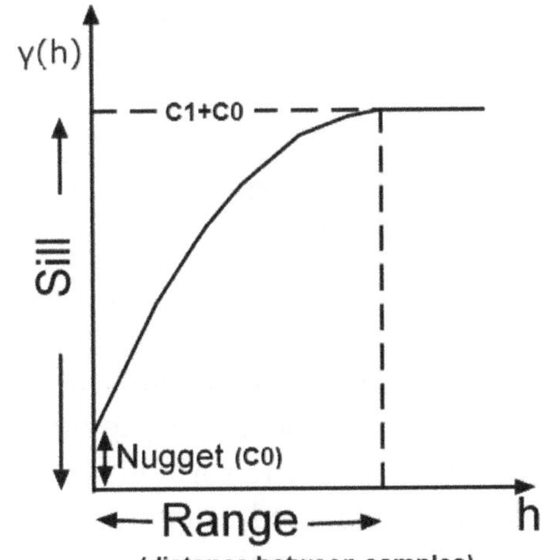

Figure 17. The exponential model

Gaussian $\gamma(h) = C_0 + C_1(1 - e^{-(h/a)^2})$ for $0 < h$ $(a' = a\sqrt{3} \rightarrow \gamma(a') = 0.95 \times (C_0 + C_1)$

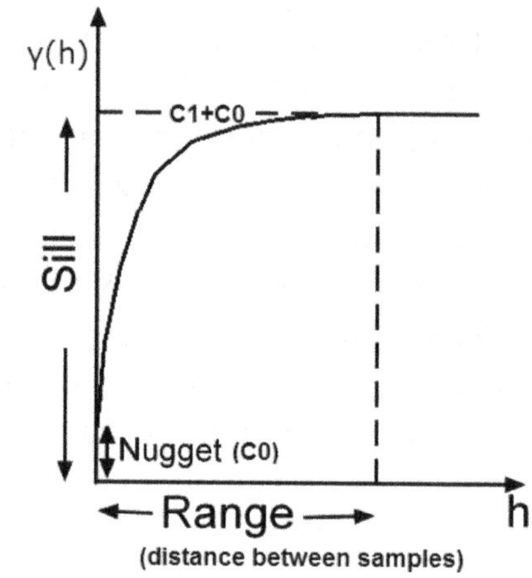

Figure 18. The Gaussian model

Exponentia l $\gamma(h) = C_0 + C_1(1 - e^{-h/a})$ for $0 < h$ $(a' = 3a \rightarrow \gamma(a') = 0.95 \times (C_0 + C_1)$

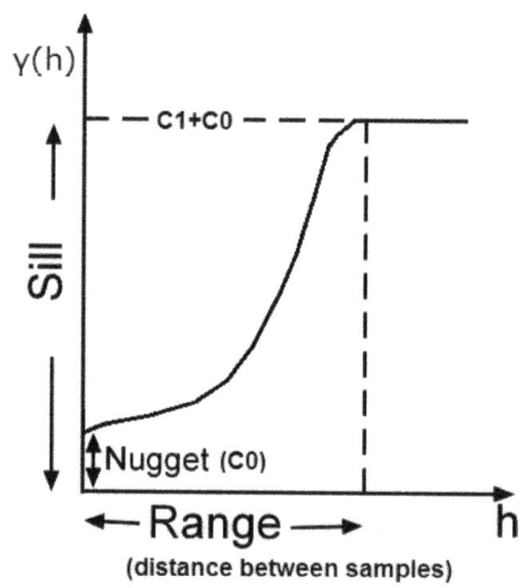

tolerance angle of Figure 9 is the result of this condition). Furthermore, it is crucial that samples with mistaken measurements should not be substituted by zero because they would distort the true variability.

Since variography should be considered an interactive process in finding the best six parameters to fit each spatial reality (see Figure 10), some

Figure 19. Comparison of the three myGeoffice© types of variograms (adapted from ibis.geog.ubc.ca/courses/geog570/talks_2001/kriging.html)

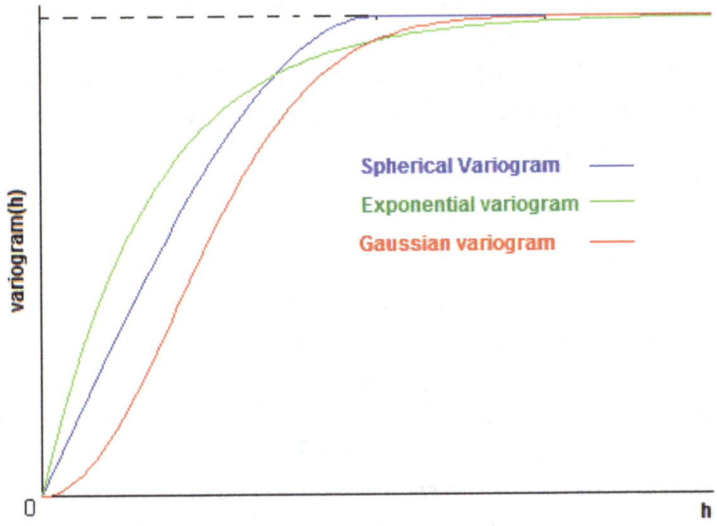

further practical tips to improve a variogram layout are set out next: (A) 30 to 50 pairs at each lag distance; (B) Similar lag space should be applied for different anisotropy axes; (C) Data transformation for skewed distributions; (D) An omnidirectional variogram (angular tolerance of 90° degrees on either side) should be setup before directional ones are calculated. In fact, an omnidirectional variogram contains more sample pairs than any directional one and, therefore, becomes more likely to display a clearly interpretable long-range structure; (E) Try several different lag intervals and bandwidths for plotting the experimental variogram and obtain the maximum detail at small distances without being misled by structural artifacts due to the particular interval used; (F) Large lag tolerance leads to nugget-effect overestimation and masked short-range spatial autocorrelation. In contrast, small lag tolerances may generate many empty bins; (G) If samples are located within a regular grid, the grid spacing is itself the lag.

Figure 20 illustrates the correlogram, madogram, general relative variogram and pairwise relative variogram of myGeoffice© towards the present dataset. The correlogram represents the relationship between the correlation coefficient of an h-scatterplot and h lag while the madogram equals the plot of the mean absolute difference of paired sample measurements as a function of distance and direction (it should not be used with Kriging, according to Isaaks &

Srivastava (1989)). The general relative variogram adjusts the moment of inertia for each h-scatterplot using the mean of the data that appears on that particular h-scatterplot. Finally, the pairwise relative variogram is adjusted by the square mean and done separately for each pair of sample values, using the average of the two values as the local mean. As a general rule, if the traditional variogram appears too noisy, the user may check other measures of spatial continuity presented in Figure 20 and reflected in his/her conclusions in Figure 22.

Figure 20. According to the general/pairwise relative variogram, it seems that the present variable is well behaved in space up to a range of 25 units (when it ceases a growing continuity of behavior). A non-null nugget-effect is highly expected (close to 25% of the total sill). For the researcher, the number next to each triangle represents the number of pairs used to compute each value. In addition, the end-user should perceive that all these spatial variance measures are more resilient to outliers and non-Normal distributions (by presenting a better contiguity behavior) than the variogram itself.

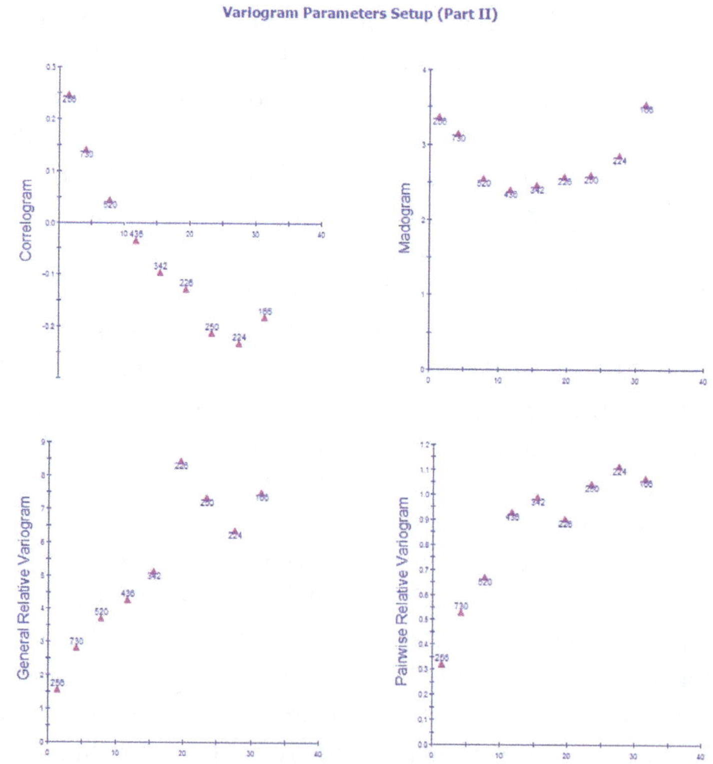

Figure 21.

Check if all executable programs (assessment of traditional, pairwise relative and general variogram, correlogram, madogram, covariance and other types of spatial autocorrelation measures) were accomplished with success

```
c:\inetpub\wwwroot\executable\gamv.exe d:\data\7915763150.par
GAMV Version: 2.905

data file = d:\data\791576315.gamv
columns for X,Y,Z = 1 2 0
number of variables = 1
columns = 3
trimming limits = 0.000000E+00 1.000000E+20
output file = d:\data\7915763150.out
number of lags = 8
lag distance = 4.000000
lag tolerance = 2.000000
number of directions = 1
azm, atol, bandwh = 0.000000E+00 23.000000 4.000000
dip, dtol, bandwd = 0.000000E+00 2.000000 2.000000
flag to standardize sills = 0
number of variograms = 1
tail,head,type = 1 1 1

maxdat = 140
Variable number 1
Number = 140
Average = 4.350429
Variance = 44.924290

Variogram 1 Semivariogram : tail=Main Variabl head=Main Variabl
currently on 14 of 140
currently on 28 of 140
currently on 42 of 140
currently on 56 of 140
currently on 70 of 140
currently on 84 of 140
currently on 98 of 140
currently on 112 of 140
currently on 126 of 140
currently on 140 of 140

GAMV Version: 2.905 Finished

Stop - Program terminated.
```

The four variograms below were assessed with a tolerance defined in Part I. Based in Figure 22, the highest range (25 units) follows a zero degree main direction while the minor contiguity is perpendicular to it (90 degrees). Figure 22 was computed with a tolerance angle of 90 degrees (omnidirectional).

MORAN I CORRELOGRAM

Under the spatial independency of the null hypothesis, the global Moran I is evaluated by measuring the covariance between attributes at each place and near sites to the overall mean (see Figure 25). If both neighboring values are above or below the mean (similarly high-high or low-low values), the multiplication becomes positive, reflecting the presence of a similar spatial autocorrelation. Otherwise, the product of the two mean deviations will be negative (unrelated high-low and low-high values), highlighting a non-positive one.

Anselin (1998) points out the possibility of using the Moran scatterplot to visualize and identify the degree of spatial instability, based on the bivariate

Figure 22. The fitness variography choice can be controversial. Somehow, the variogram fitness is more an art rather than a science, i.e., different researchers may present slightly different solutions for the same spatial case. For instance, to consider the azimuth angle of 135 degrees in Figure 9 could be another plausible solution for this dataset. Certainly, common sense and geographical knowledge about the variable under study can be a plus in this step.

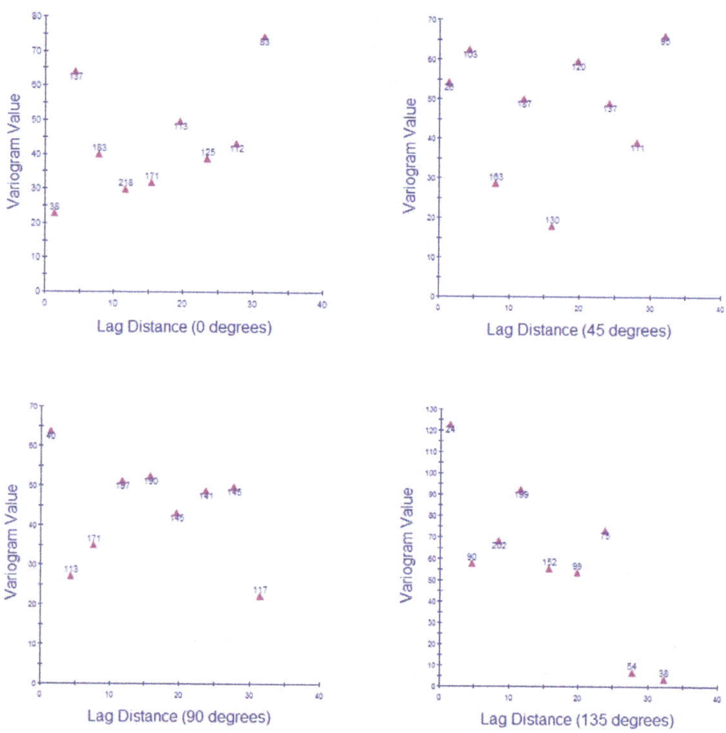

regression coefficient of the spatial lagged variable (Wx) against the original variable (x). The four quadrants centered on the global mean are composed by the x-axis (deviations from the original variable mean) and the y-axis (neighborhood weight average mean deviation). As expected, this scatterplot is divided into four association types: The upper-right quadrant (high values above the mean surrounded by high values), the bottom-left (high values enclosed by low values), the upper-left (low values encircled by high values) and the bottom-right (high values bounded by low ones).

Theoretically, the expected random value of the Moran I is $-1/(n-1)$, where n equals the number of observations. This means that zero cannot be used as a reference point to distinguish positive and negative spatial autocorrelation

Figure 23. After several attempts (by changing the six settings of Figure 9), a Spherical final model was chosen for the present dataset based on the following parameters: (Total) Sill=45, Nugget-effect=10; Major range=25; Minor range=14 (main contiguity azimuth of zero degrees). Interestingly, it is notoriously difficult to find a well-behaved variogram with this highly skewed spatial dataset.

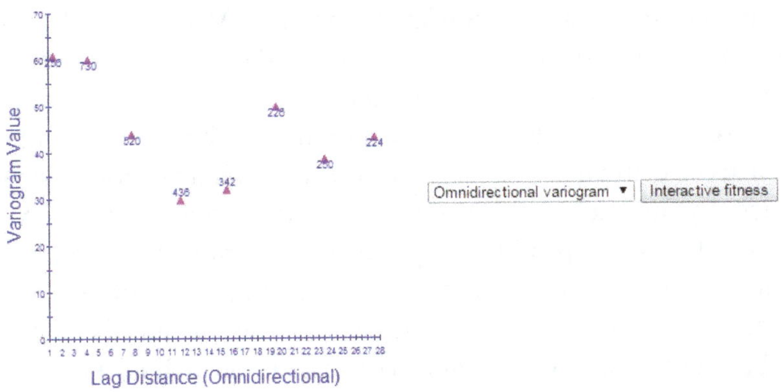

Figure 24. Following the trial and error philosophy, myGeoffice© also allows interactive fitness variography for the omnidirectional and the four chosen directions to help the researcher to set up the estimated parameters of the final variogram.

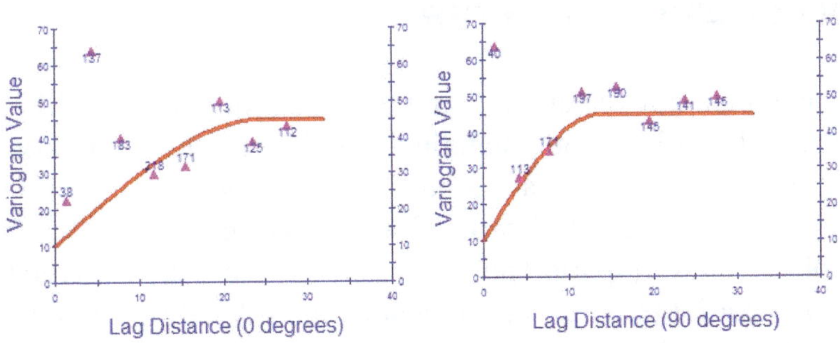

Figure 25. $\bar{x}$ symbolizes the overall sampling mean, W_{ij} is the associate weight between i^{th} and j^{th}, n represents the total number of observations, x_i and x_j are i^{th} and j^{th} sample values and S_0 equals $\Sigma_i\Sigma_j w_{ij}$ (the sum of the spatial weights)

$$I = \frac{n\sum_{i=1}^{n}\sum_{j=1(i<>j)}^{n} w_{ij}(x_i-\bar{x})(x_j-\bar{x})}{S_0\sum_{i=1}^{n}(x_i-\bar{x})^2}$$

(particular in the presence of a small dataset). Yet, since the minimal number of samples for myGeoffice© is fifty, it is possible to consider zero for the expected random value of Moran I. Confirmed by Wong and Lee (2005), the Moran I range varies between -1 and +1 although empirical studies may occasionally obtain a value slightly beyond this range.

In summary, this section confirms that the Moran I falls under the neighborhood view (see the introduction of the present chapter), where interaction involves phenomena aggregations of discrete regions (polygons or lattice). Neighbors are defined as spatial units with a common boundary or within a given critical distance. In both cases, no interaction occurs further away. Since myGeoffice© does not handle lattice data, the vicinity weights of the Moran I index is produced by the covariogram (the inverse function of the variogram as a spatial distance decay function) of the Variogram Fitness option (section "Variogram Fitness"). The Moran I is calculated for five equal incremental range distances (a scale issue) whose largest distance equals the chosen variogram maximum range (see Figure 26). Theoretically, it is expected that the Moran I for this particular last distance should lean towards or be close to zero.

MORAN LOCATION SCATTER

Spatial analysis should be centered on the search for patterns within spatial phenomena and involves understanding, prediction and simulation, where location, time and geometry are significant. Focusing on the concept of spatial autocorrelation and following Getis (2008), given a set S containing n geographical units, spatial autocorrelation refers to the relationship between some variable observed in each of the n localities and a measure of geographical proximity defined for all nx(n-1) pairs chosen from n. So, spatial autocorrelation gives us an idea about the degree to which a set of features tends to be clustered together or evenly dispersed over the Earth's surface. Actually, most real world patterns are somewhere between a random and a positive clustered pattern.

The capability to handle spatial autocorrelation measures is thus the issue of this section whose goal is to propose an enhanced methodology of the Moran conventional scatterplot (called hereinafter the Moran location scatterplot). A second contribution is the introduction of a new exploratory spatial data analysis (ESDA) tool named the Moran variance scatterplot (section "Moran Variance Scatter"), a way to test stationarity in spatial data

Figure 26. Following Tobler's principle, it is predictable that the Moran I will decrease as distance lags increase. According to the present dataset, the distance with the highest Moran I index is 5 units while the 25 units distance holds a Moran I close to zero (the same variogram range distance of Figure 23, a validation issue).

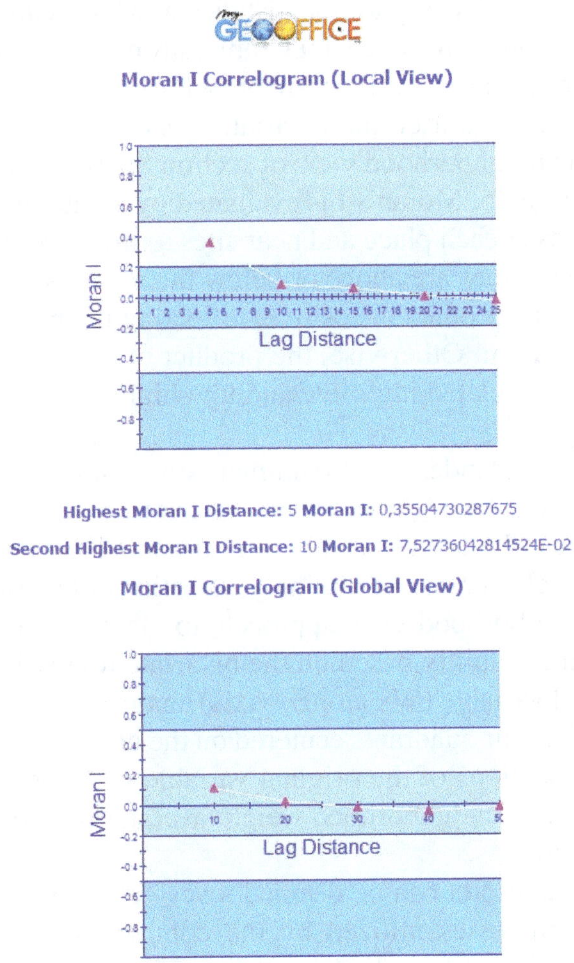

analysis by exploring the neighborhood homogeneity/heterogeneity. This mapping allows the study of spatial patterns, outliers, changeover areas and trends in an ESDA process.

The variogram underlies the keystone and success of Kriging. It quantifies spatial autocorrelation at all scales by summarizing the degree of similarity between data values for all possible observation pairs as a distance function. The upper limit is the sill, which implies no spatial dependence between data points because all variances are invariant with the sample separation distance. The nugget-effect represents error measurement and spatial variation at distances much shorter than sample spacing. For Clark and Harper (2000), this nugget variance of $2C_0$ (not C_0) is seen as a measure of the unreliability of our sample values. The separation distance at which samples are spatially autocorrelated becomes the range. Inevitably, stronger spatial autocorrelation increases the variogram range and provides a fairly smooth map. Conversely, as the range becomes shorter, more variation is expected.

Regarding the neighborhood view of section Spatial Autocorrelation and according to Figure 25, Moran's I is evaluated by measuring the covariance between attributes at each place and near sites to the overall mean. Again, if both neighboring values are above or below the mean (similar high-high or low-low values), the product is positive, reflecting the presence of a similar spatial autocorrelation. Otherwise, the product of the two mean deviations will be negative (unrelated high-low and low-high values), stressing a non-positive situation.

However, Moran's I index tries to summarize a complete spatial distribution in a single number. As a global index, a serious flaw dealing with large spatial datasets can be found: The hiding of spatial heterogeneity or local patterns of spatial clustering. Accordingly, Anselin (1995) points to the Moran scatterplot, a neighborhood view approach, to visualize and to identify the degree of spatial instability based on the bivariate regression coefficient of the spatial lagged variable (Wx along y-axis) against the original variable (x along x-axis). The four quadrants centered on the global mean are composed by the x-axis, deviations of the original variable with respect to the global mean, and the y-axis, neighborhood weight average deviations with respect to the global mean.

The Moran scatterplot can be deemed a very effective visual diagnostic tool for ESDA processes utilized by the common GIS user. But, some shortcomings remain such as the W matrix construction, the fuzzy concept of neighborhood and the samples categorization when posted in the vicinity of both Moran scatterplot axes.

The first element of this enhanced version of the traditional Moran scatterplot addresses the construction of the W matrix. Binary contiguity is well suited for delimited regions. Areal data has, however, two major restrictions: 1) Frequently, the variability between regions relies on political boundaries, i.e., its polygon shape does not depend on the variable under study; 2) The non-variability assumption within the considered region can be considered a hard supposition. In truth, users enjoy lattice data because it is simple and convenient in terms of GIS management. However, much of the traditional sampling used in geostatistics (characterization of the existing sample data and prediction of unsampled locations) does not have defined boundaries. Users are, however, unable to use a lattice spatial autocorrelation measure with the point's data. This also means that it is unfeasible to assess the Moran scatterplot with sample data. To overcome this division barrier with a mixed approach is, therefore, the challenge.

Under this proposal, the W neighborhood matrix is setup by the covariogram contiguity, where each W_{ij} weight is based on the variogram model for that particular dataset (the distance view). To avoid bias, the sum of all neighborhood weights should equal one. At first, the variogram range indicates vicinity boundaries. Thus, if the distance between the location and its neighbor is larger than the variogram range, that particular neighbor is not included in the weighted vicinity matrix.

The neighborhood concept can be considered a fuzzy problem. One must bear in mind that there are a lot of factors that may change the W matrix definition (e.g., travelling time, boundary length, the rush hour factor, migration flows, vehicle speeds). Yet, if spatial distance between samples is the only factor to be taken into account, the variogram can be a tool to remove this constraint and making this methodology suitable for the W matrix definition. Moreover, there are no strict setup rules to compute the W matrix. As stated by Wong & Lee (2005), the choice of geographical weights for spatial statistical modeling is not a clear issue although is largely decided in an ad-hoc manner. In fact, it is determined, primarily, by convenience or convention, if the Rook and Queen strategy is chosen as the W matrix: It is symmetric, its diagonal is zero, the row sum of W is the total number of neighbors for each area unit while its division by two equals the total number of shared joins.

It is clear that these different weights lead to an assorted Moran I and Moran scatterplot results. Fortunately, the W neighborhood matrix is not influenced by the region or sample value. For Negreiros (2004), by choosing different neighborhood limits for the same layout, the Moran scatterplot observations can only exchange vertical positions between quadrants I and IV or II and III whose maximum amplitude is, on average, 15% of the overall mean of the variable (Figure 27). Furthermore, according to Negreiros (2009), the stronger the Moran I becomes, the better the Moran scatterplot highlights the different patterns, outliers and uncertainty/shift regions (the Moran I working as an anti-gravity force against both axes of the Moran scatterplot). Therefore, the neighborhood range for the W matrix construction should be determined in an adjustment mechanism context because spatial autocorrelation is scale-

Figure 27. Overall picture of the Moran scatterplot showing the vertical and horizontal uncertainty sub-regions, high-high, medium-medium and low-low patterns whereas (1), (2), (3), (4), (5) and (6) denotes dissimilar types of outliers.

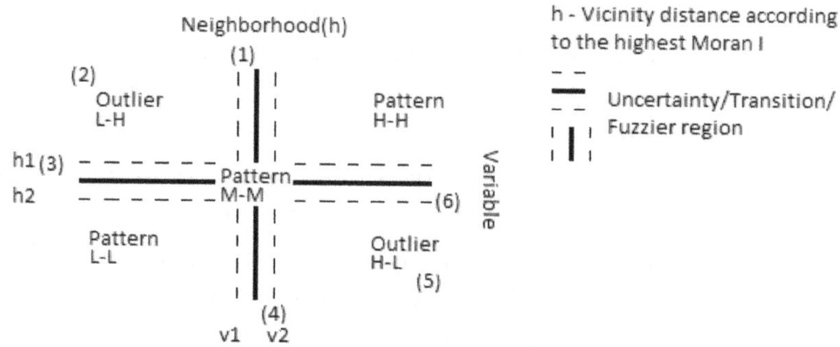

dependent, which is the first cornerstone of this approach: The higher the Moran's I, the more spatial patterns, transitions and outliers are enhanced.

Regarding geocomputation, after the covariogram contiguity (the inverse function of the variogram setup in section "Variogram Fitness") has been defined, myGeoffice© maps the Moran scatterplot. The covariogram range equals the highest Moran's I encountered, according to different lags and where the maximum possible active lag distance equals the initial variogram range. The purpose is to find the interval distance which maximizes Moran's I but does not exceed the variogram range because the distance at which the curve levels off reveals the scale of the spatial pattern.

Since Mother Nature regularly pursues a positive spatial autocorrelation pattern, the prevalence of smaller radii is expected. Nevertheless, the short-term danger of omitting a good percentage of the original samples can be real because of the lack of neighbors for computational assessment. Under myGeoffice©, the initial default value for the first considered covariogram range is one and half times the mean nearest distance. Based on assorted Moran I outcomes for different lags, the range of the covariogram function (to be used on the construction of the W matrix) will be setup according to the highest Moran I distance (Negreiros, Painho, Aguliar, & Aguilar, 2009).

The second enhancement provided by this approach is connected with its layout. The conventional Moran scatterplot plots the number of observations along the four quadrants but it overlooks the true location of each sample. It is frequently difficult to visualize them in space unless an additional location map with brushing and linking capabilities is installed. If each quadrant is plotted on the basis of the coordinates for each sample and highlighted according

Figure 28. The contiguity plot of the covariogram whose range is defined by the highest Moran I distance. Obviously, closer samples hold higher weights than distant ones.

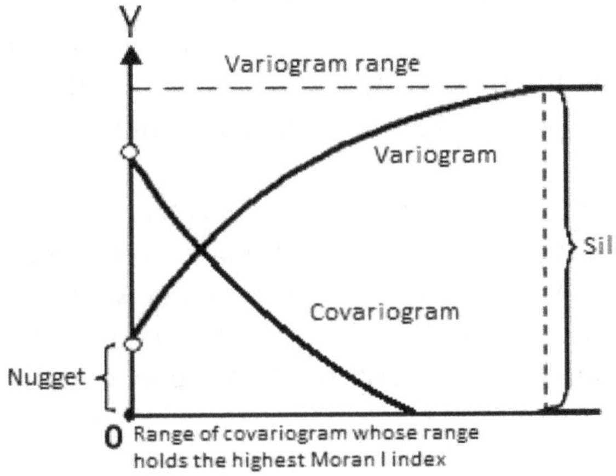

to their categorization, this new design may lead to a more specific ESDA tool than the conventional one. Therefore, the Moran location scatterplot of myGeoffice© plots all sample sites in the same four-quadrant layout while emphasizing each sample according to their classification.

The third improvement to the conventional Moran scatterplot is the introduction of uncertainty regions. This is linked to the cold clear cutoff of both means (vertical and horizontal). For instance, points 1, 2 and 3 in Figure 27 are located in quadrant II and, therefore, the traditional Moran scatterplot identifies them as negative outliers. However, looking carefully at those points, they are characterized by three different situations:

- Case 1 represents an average sample surrounded by high neighborhood values. This means that this sample is a false negative outlier or a shift region between a positive and a negative pattern.
- Case 2 exemplifies a high and true negative outlier.
- Case 3 highlights a very low sample value whose neighborhood is slightly above the overall mean. This indicates a false negative outlier or a shift region between a negative and a positive pattern.

The same can be said of points 4, 5 and 6, located in quadrant IV:

- Case (4) represents an average sample surrounded by very low neighborhood values. This implies that this particular sample is a false positive outlier or a shift region between a negative and a positive pattern.

- Case (5) stresses a high and true positive outlier.
- Case (6) reveals a very high sample value whose neighborhood is slightly below the overall mean, i.e., a false positive outlier or a shift region between a positive and a negative pattern can be found.

This cutoff topic must be converted from a clear cut issue to a fuzzy area for those points close to both axes and reflected within the v1-v2 and h1-h2 limits of Figure 27. After computing these confidence intervals, these fuzzy boundaries (nearby both Moran scatterplot axes) can be setup. Summing up, the myGeoffice© algorithm for constructing the Moran location scatterplot is as follows:

1. A variogram model is setup based on the available data (section "Variogram Fitness").
2. The global Moran's I (section "Moran I Correlogram") is computed for ten lag distances. Their search range is between 3/2 of the average distance between samples (a nearest neighborhood analysis parameter) and the original variogram range. The purpose is to identify the best scale of positive autocorrelation.
3. If the highest Moran's I encountered corresponds to a very small distance that does not process more than 90% of all samples (initial default value) within the Moran scatterplot, the second highest Moran's I is computed (and so on). The corresponding cutoff distance is, thus, defined as the neighborhood boundary of each observation.
4. In accordance with Tobler's Law, the W matrix is based on the best covariogram contiguity of step 1 whose neighborhood range is defined by the distance setup in step 2 and 3.
5. Definition of the four uncertainty limits h1-h2 and v1-v2 (under myGeoffice©, both ranges may vary respectively between 0% and 15% of the variable horizontal mean and neighborhood vertical mean).
6. Optionally, the traditional Moran scatterplot might be plotted.
7. The Moran location scatterplot is drafted on the basis of the Cartesian coordinates of the available data using a pre-defined symbol code:
    a. A red star always symbolizes a medium pattern (the neighborhood and the sample itself are similar to the overall mean).
    b. The top-right hand corner scatterplot corresponds to quadrant I, the dark blue plots represent positive patterns and the green squares and

red diamonds are samples that fall respectively into the horizontal and vertical uncertainty borders.

c. Within quadrant II, the dark blue circles represent negative outliers. Green squares and red diamonds suggest false negative outliers (as in the corresponding previous situation, these points denote shift patterns).
d. In quadrant III, the dark blue circles represent a negative pattern sub-region. Likewise, green squares (h1-h2) and red diamonds (v1-v2) highlight transition local areas.
e. The dark blue symbols of quadrant IV point out true negative outliers whereas green squares and red diamonds suggest false positive outliers.

This iterative algorithm was applied to the present dataset. After setting up a 15% fuzzy boundary (Figure 29), it becomes clear from Figure 31 that two dissimilar patterns emerge from quadrants I and III while quadrants II and IV present a mix between a clustered and a random design. Some of these samples are true outliers in need of inspection while others are false because they belong to changeover regions. Since major interpolation processes are particularly sensitive to true outliers, this is key to locate them as best as possible (removal is a questionable option).

It should be noted that the number of shift observations increases with a lower Moran's I. Hence, the total number of true outlier's decrease as shrinking

Figure 29. A zero percentage signifies no transition region among the four quadrants. As expected, different percentage definitions lead to a slightly diverse sample classification.

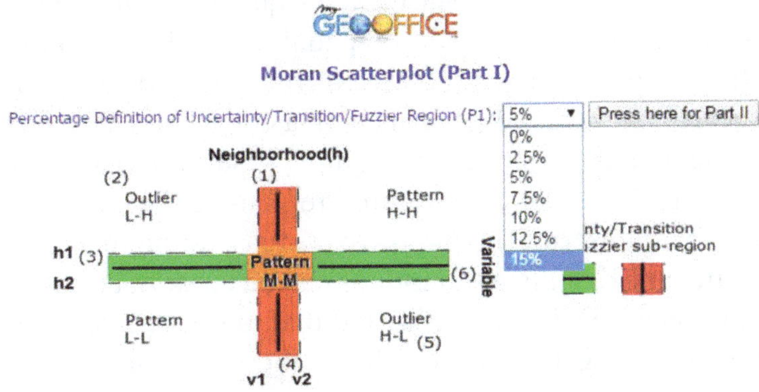

Figure 30. The traditional scatterplot of myGeoffice© for the present dataset

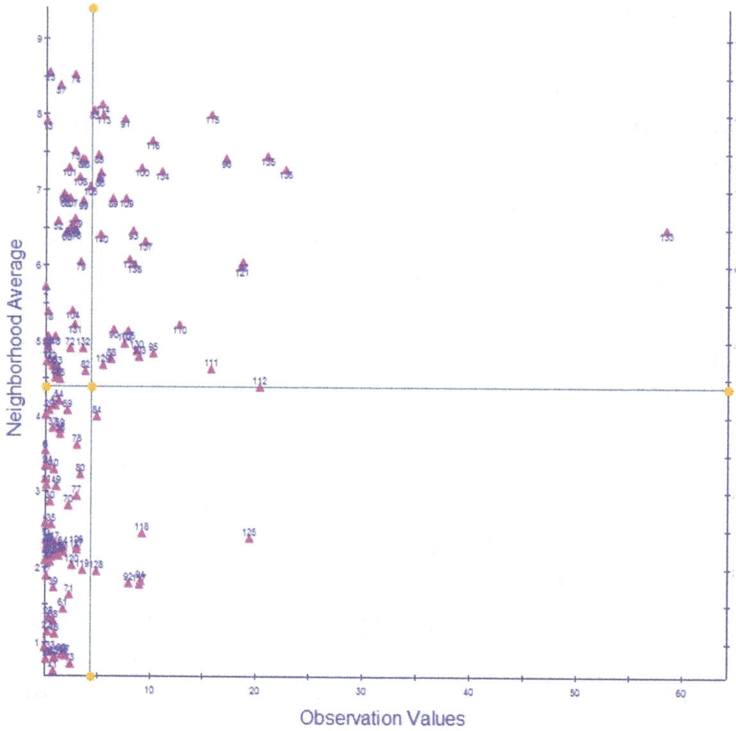

spatial patterns becomes real. Since this algorithm computes the Moran's I with the highest value, this combination of the variogram-Moran's I-Moran scatterplot can reach a deeper level of knowledge concerning the detection of local spatial pockets, trends, true outliers and transition samples (Negreiros, Painho, Aguliar, & Aguilar, 2009).

It would be advisable to highlight that a zero Moran's I produces a random and dispersed outcome throughout the four quadrants, the least favorable circumstance for anyone whose aim is spatial interpolation and simulation. Moreover, the samples that are subtracted from the ones of the fuzzy/shift region will reveal outliers and patterns with a confident level of reliability.

In this sense and as illustrated in Figure 31, several visual findings can be highlighted: 1) Two positive clustered regions; 2) The negative outliers of Quadrat II circumscribe both positive areas, implying a great space variation (stationary Kriging assumptions can be questionable for the whole region, for instance); 3) The thirteen (red diamond) samples of Quadrat II are considered false negative outliers due to their vertical axis closeness; 4)

Figure 31. The Moran location scatterplot of myGeoffice©

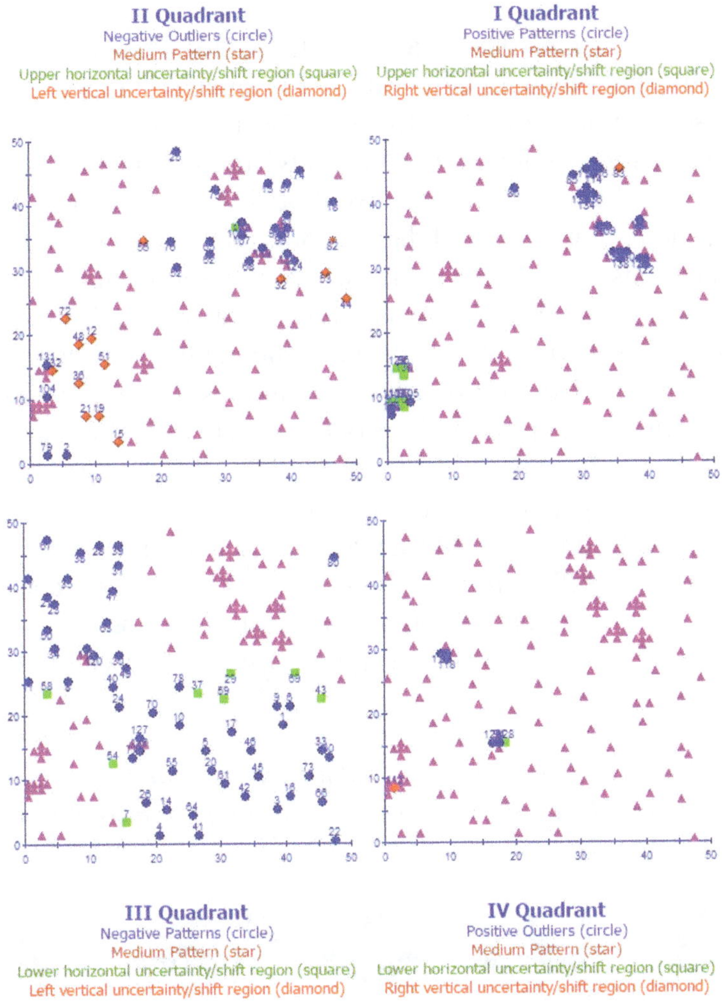

With the exception of sample 80, a negative and continuous spatial pattern can be found within Quadrant III (local variography for this zone may be a possibility for spatial interpolation); 5) Samples 7, 29, 37, 43, 54, 58, 59 and 69 were identified as transitional ones; 6) Two small positive outliers spots can easily be recognized in Quadrant IV (given their spatial location and by comparing carefully with Quadrant III, two small hills is a possibility for this variable behavior in space). These visual conclusions are based on the color code applied to each sample and its labeling.

MORAN VARIANCE SCATTER

One common way to test stationarity is the use of a variogram cloud. By selecting pairs of points with large variance at small distances, the mean and variance discontinuity will be reflected. The mean-variance computation in a sliding moving window context can also help in the finding of spatial heterogeneity. Still, the inability of conventional Moran scatterplot to measure heteroskedasticity among spatial data becomes real. This procedure reduces neighbor locations to a single mean, where variability among themselves and between the central site and its vicinity is totally lost. With regard to the cases presented in Figure 32, for instance, the neighbor average is close to thirty in spite of the lack of homogeneity amongst them. According to the conventional Moran scatterplot, the plot of I, II, III and IV within the Moran scatterplot will be the same thus not providing a solution to this problem. Case V stresses a negative outlier situation.

This section introduces the Moran variance scatterplot, a new ESDA tool based on the common 2D orthogonal chart. The basic idea is to test the second moment of the Moran scatterplot, i.e., the variance (the weighted square distance difference). Here, the x-axis (Figure 33, left) reflects the neighborhood

Figure 32. Layout of four-neighbor combinations yielding the same neighbor average. This spatial heteroskedasticity question is a problem not resolved by the conventional Moran scatterplot.

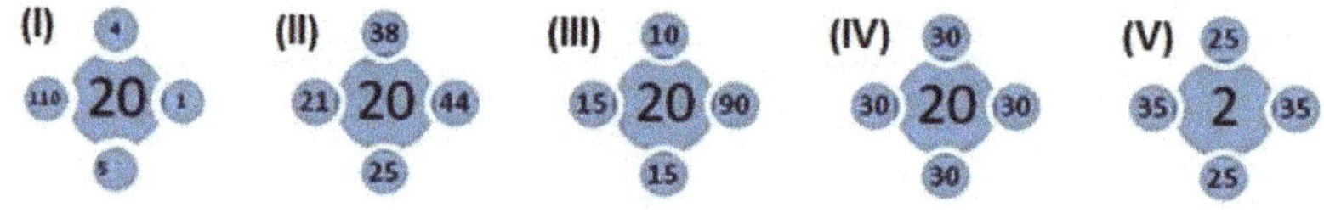

Figure 33. W_{ij} represents the binary contiguity or any other distance decay specification weight of i location regarding j neighbor, in which m equals the total number of observations considered for that particular vicinity, x_i is the value of the site itself and x_k and x_j correspond to two distinct neighbors of i location.

$$X\_axis_i = \frac{\sum_{j=1}^{m} w_{ij}(x_i - x_j)^2}{m} \qquad Y\_axis_i = \frac{\sum_{j=1}^{m} w_{ij} \sum_{k=1, j\neq k}^{m} (x_j - x_k)^2}{m(m-1)}$$

homogeneity/heterogeneity between the sample and its neighbors while the y-axis represents the neighborhood homogeneity/heterogeneity among neighbors (Figure 33, right).

For educational purposes, let us consider one generic central point with three neighbors whose W weights are a, b and c (Figure 34). The sum of them is one. If the binary contiguity is assumed for the W specification, those three weights are equal. If a distance decay contiguity is assessed then all weights are unequal (a>b>c). As expected, the final result of x_axis_i and y_axis_i is different, confirming the fuzzy concept of the W matrix.myGeoffice© presents the Moran variance scatterplot to address this variability issue, where the XX axis represents the mean variance between the central site and their neighbors (variable-neighbor) while the YY axis corresponds to the mean variance among all neighbors (neighbor-neighbor). In addition, the Moran variance scatterplot encloses its sample number to help researchers locate it in a 2D layout map. A low x-axis and y-axis indicate a stable pattern whereas a high one reveals an erratic region. However, if a low y-axis occurs with a high x-axis then an outlier emerges. Briefly, the Moran variance scatterplot emphasizes outliers (no positive/negative distinction is made) and stable/transition regions.

In order to categorize them, two cutoff values, one for each axis, should be setup by the user: The smaller this threshold becomes, the greater the number of outliers, high and medium variability regions can be considered. Probably, the number of observations classified as a stable region will be less. Under myGeoffice©, the default threshold for each axis is 30% of the respective highest variance found (although the researcher may modify this parameter). Theoretically, four categories can be detected (see Figure 36):

- If the neighborhood variance of each sample is greater than the product of the YY maximum variance by the selected percentage (cutoff 1)

Figure 34. Based on Figure 33, the computation of both axes for i site becomes as follows: $x\_axis_i=(c\times(i-j_1)^2+a\times(i-j_2)^2+b\times(i-j_3)^2)/3$; $y\_axis_i=(a\times(j_1-j_2)^2+b\times(j_1-j_3)^2+c\times(j_2-j_1)^2+a\times(j_3-j_2)^2+b\times(j_2-j_3)^2+c\times(j_3-j_1)^2)/6$

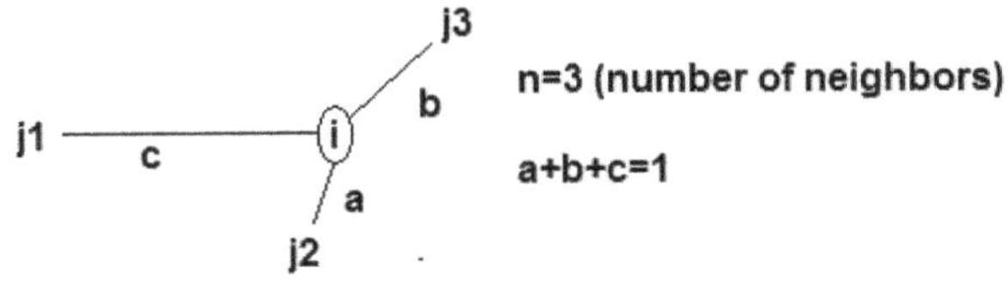

Figure 35. The Moran variance scatterplot corresponding to the five cases presented in Figure 32.

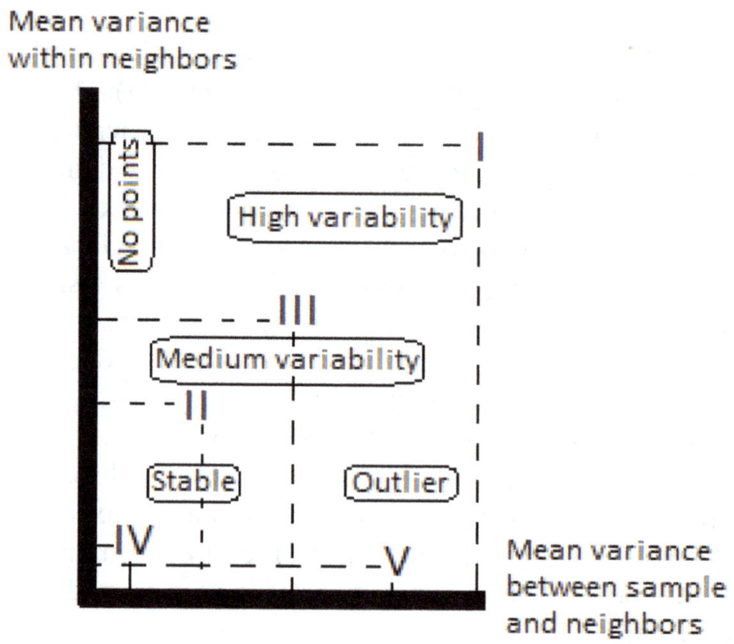

Figure 36. It should be emphasized that, if these highlighted outliers do not completely match those of the Moran location scatterplot, the cutoff value must be shifted to a lower or a higher one in an interactive process (in fact, a subset of the available samples can be classified differently according to different cutoffs).

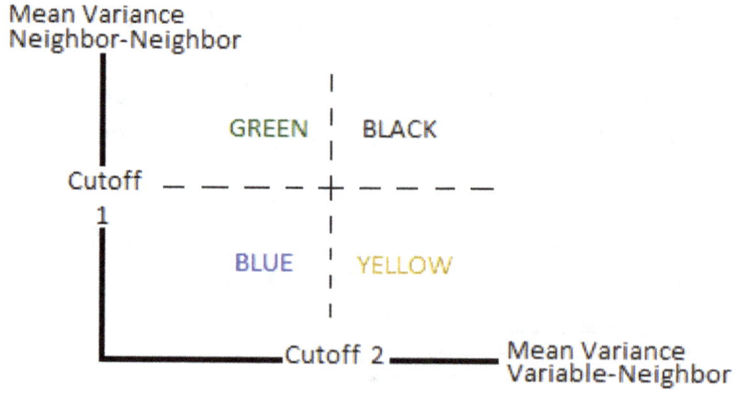

and the sample-neighborhood variance is greater than the product of the XX maximum variance by the same selected percentage (cutoff 2), that particular observation is classified as a 'High Variability' region (BLACK).

- If the neighborhood variance is greater than the product of the YY maximum variance by the selected percentage (cutoff 1) and the sample-neighborhood variance is less than the product of the XX maximum variance by the same selected percentage (cutoff 2), that particular observation is classified as a 'Medium Variability' region (GREEN).
- If the neighborhood variance is less than the product of the YY maximum variance by the selected percentage (cutoff 1) and the sample-neighborhood variance is greater than the product of the XX maximum variance by the same selected percentage (cutoff 2), that particular observation is categorized as an 'Outlier' (YELLOW).

Figure 37. The bottom-right illustration shows the Moran variance scatterplot (non-spatial view) for the present dataset

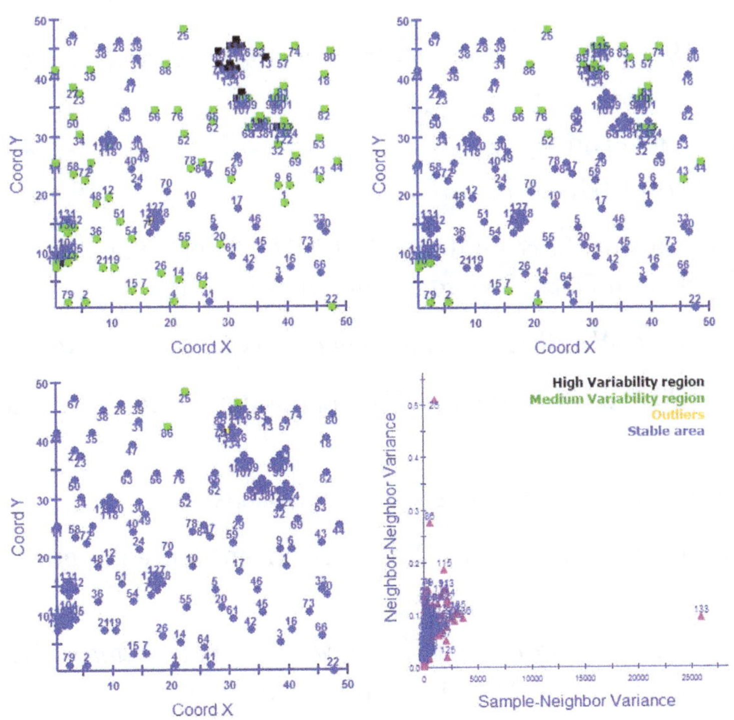

- The remaining samples are considered to be in the 'Stable' zone (BLUE).

Three spatial mappings with three cutoffs (5%, 15% and 30% of the maximum variance found for each axis) were setup for analysis (Figure 37, top-left, top-right and bottom-left, respectively). Confirming previous statements, the number of samples considered stable (blue dots) enlarges with an increase in the percentage considered. The two upper scatter maps reveal two different sub-regions of medium variability due to the transition between high and low spatial areas. No high variability field (black dots) was detected for 15% and 30% levels but, for a 5% cutoff, several clustered samples were classified as high variability regions. A positive outlier (sample 133 with the value of 58.32) was highlighted on the upper right section when a 30% level was setup. This also means that the categorization of some samples may change with different cutoffs. Further, the '2D-3D data posting' or the 'Cutoff layout map' options of myGeoffice© should be seen as complementary and not mutually exclusive within this ESDA framework.

REFERENCES

Anselin, L. (1998). *Spatial econometrics, methods & models.* Boston, MA: Kluwer.

Armstrong, M. (1998). *Basic linear geostatistics.* Springer-Verlag. doi:10.1007/978-3-642-58727-6

Banerjee, S., Carlin, B., & Gelfand, A. (2015). Hierarchical Modeling and Analysis for Spatial Data (2nd ed.). Chapman & Hall/CRC.

Clark, I. & Harper, W. (2000). *Practical geostatistics.* Ecosse North America.

Cressie, N. (1993, 2015). Statistics for spatial data. New York: John Wiley & Sons.

Ebdon, D. (1998). *Statistics in geography.* Blackwell.

Geary. (2015). In *Wikipedia.* Retrieved from: en.wikipedia.org/wiki/Geary%27s_C

Getis, A. (2008). *A history of the concept of spatial autocorrelation: a geographer's perspective.* The Ohio State University.

Getis, A., & Ord, J. (1992). *The analysis of spatial association by use of distance statistics in geographical analysis*. Ohio State University Press.

Goovaerts, P. (1997). *Geostatistics for natural resources evaluation*. Oxford University Press.

Griffith, D. A., & Layne, L. J. (1999). *A casebook for spatial statistical data analysis: a compilation of analyses of different thematic datasets*. Oxford University Press.

Isaaks, E., & Srivastava, M. (1989). *An introduction to applied geostatistics*. New York: Oxford University Press.

Matheron, G. (1965). *Les variables régionalisées et leur estimation*. Paris: Masson.

Negreiros, J. (2004). *SAKWeb - Spatial autocorrelation and Kriging web*. Ph.D. Thesis.

Negreiros, J. (2009). Spatial neighborhood matrix computation: inverse distance weighted versus binary contiguity. Geo-Spatial Crossroad GI_Forum, 154-157.

Negreiros, J., Costa, A., Painho, M., & Santos, J. (2007). Autocorrelation, autoregression and Kriging: appraisal for the spatial interpolation issue. *56th Conference of the International Statistic Institute*.

Negreiros, J., Painho, M., Costa, A., Cabral, P., & Aguilar, F. (2008). *The local confidence uncertainty plume of SAKWeb©*. In C. Brebia & E. Beriatos (Eds.), *Risk Analysis VI–Simulation and Hazard Mitigation* (pp. 81–90). WIT Press.

Negreiros, J., Painho, M., Oliveira, T., Aguilar, M., & Aguilar, F. (2008). *Synopsis of an Internet spatial (decision support) prototype*. WorldComp 08 – International Conference on eLearning, eBusiness and eGovernment, Lisbon, Portugal.

Smith, M. (2015). *STATSREF: Statistical analysis handbook - A web-based statistics resource*. Winchelsea, UK: The Winchelsea Press.

Soares, A. (2000). *Geoestatistica para as ciências da terra e do ambiente*. IST Press.

Wong, D., & Lee, J. (2005). *Statistical analysis of geographic information with ArcView® GIS and ArcGIS*. Hoboken, NJ: John Wiley & Sons.

Chapter 8
Deterministic Interpolation

ABSTRACT

The review of deterministic interpolation approaches of myGeoffice© is the main goal of this chapter. The first two sections focus on first and second order polynomial while sections three and four present multiquadratic techniques and their inverse. Inverse distance weight (IDW) and moving average interpolators are addressed in sections five and six. Nearest neighbor and triangulated irregular network (Voronoy and Delaunay) conclude this chapter.

FO POLYNOMIAL TREND

A polynomial trend surface, an inexact global basis model based on spatial coordinates, is used as a fitting regression procedure of a global surface for smoothing, filtering and data interpolation by separating the variable into two components: Large-scale variation (trend or regional features) and random error (non-systematic white noise fluctuation due to local features).

Developed by Whitten in 1957, the unknown b_i coefficients of the polynomial regression coordinates are found by solving a set of simultaneous non-linear equations of the cross-products sums of the (x,y) coordinates and their Z_i values. If the n terms presented under the A matrix (Figure 1) correspond to the coordinates' data points and the n terms of vector Z equal the correspondent variable values, b0 of vector B represents the constant term while the remaining b1 and b2 parameters are the regression coefficients of the linear polynomial (algebraically, this least square solution equals $B=(A'A)^{-1}(A'Z)$, where A' represents the transpose of A matrix). Thus, the

DOI: 10.4018/978-1-5225-3270-5.ch008

estimation of a particular (x_i,y_i) location equals $b0+b1xA_{xi}+b2xA_{yi}$, if a linear polynomial is considered, or $b0+b1xA_{Xi}+b2xA_{yi}+b3xA^2_{Xi}+b4xA_{xi}xA_{yi}+b5xA_{yi}^2$, if a quadratic one is contemplated.

It is this simplicity that makes this approach worth using and unlike Kriging, it avoids the estimation of the variogram weights. Yet, like any OLS approach, this method is highly susceptible to edge effects and rarely passes exactly through the original data points.

Analogous to the traditional regression analysis, the F global test can be used to check for a significant linear regression relationship between the present variable (Z vector) and the space coordinates (A matrix). According to a pre-defined level of confidence (95%, usually), if the computed F ratio (2.42 in this particular case) is smaller than the chosen F-table critical value (close to 3.03, with df1=3-1=2, df2=140-3=137), the present spatial regression does not account for the present dataset behavior (statistically speaking, the null hypothesis should be accepted, i.e., the estimator b1 and b2 of the linear regression cannot be considered different from zero). In other words, this linear trend surface does not explain the variation of the sample data.

SO POLYNOMIAL TREND

Generically, polynomial surface analysis is useful in two situations: 1) To detect physical or biological global trends attached to the spatial variable under study; 2) To overcome the lack of the intrinsic assumption of Kriging. In this case, the polynomial fits a plane on the data on a global scale, where Kriging is applied to the residuals to analyze the local scale structure (section "Kriging With Trend").

In particular, second order polynomials model spatial drifts with a reasonable amount of hills and holes across the sampling area. Since this approach respect OLS, again, this surface analysis is highly susceptible to

Figure 1. The linear (a constant dip in a single direction) spatial polynomial matrices used by myGeoffice© to compute the B vector parameter

$$\underset{Z}{\begin{vmatrix} Z1 \\ \dots \\ Zn \end{vmatrix}} = \underset{A}{\begin{vmatrix} 1 & X1 & Y1 \\ \dots & \dots & \dots \\ 1 & Xn & Yn \end{vmatrix}} \times \underset{B}{\begin{vmatrix} b0 \\ b1 \\ b2 \end{vmatrix}}$$

Figure 2. Green and pink points in the right map represent positive and negative residuals, respectively (the difference between the sample values and the linear trend estimation)

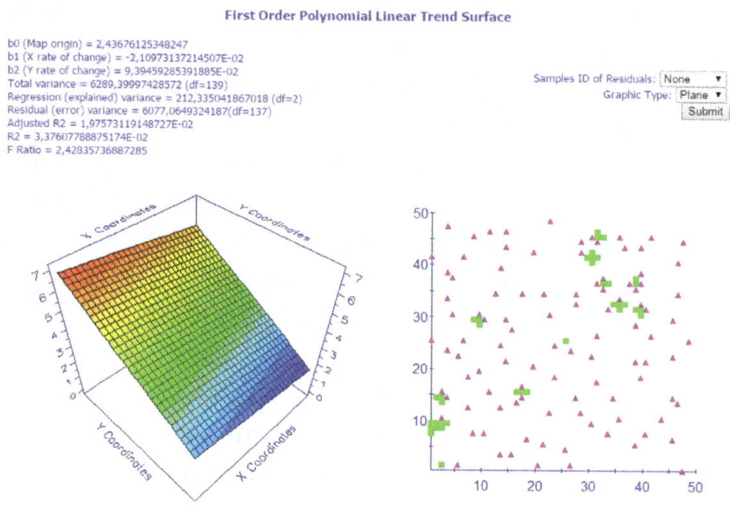

Figure 3. The quadratic (a bowl or dome shape) spatial polynomial matrices used to assess the B vector estimators under myGeoffice©

$$\overset{\textbf{Z}}{\begin{vmatrix} Z1 \\ \dots \\ Zn \end{vmatrix}} = \overset{\textbf{A}}{\begin{vmatrix} 1 & X1 & Y1 & X1{*}X1 & X1{*}Y1 & Y1{*}Y1 \\ \dots & \dots & \dots & \dots & \dots & \dots \\ 1 & Xn & Yn & Xn{*}Xn & Xn{*}Yn & Yn{*}Yn \end{vmatrix}} \times \overset{\textbf{B}}{\begin{vmatrix} b0 \\ b1 \\ b2 \\ b3 \\ b4 \\ b5 \end{vmatrix}}$$

edge effects and rarely passes exactly through the original data points (inexact interpolator). Another limitation is the lack of uncertainty assessment.

Consistent with Figure 4, the second order polynomial for the present dataset becomes $Z(X_i,Y_i)=7.59-0.18X_i-0.2Y_i-1.64X_iX_i+1.01X_iY_i+1.31Y_iY_i$, where $Z(X_i,Y_i)$ is the estimation value for any (X_i,Y_i) coordinate site. The adjusted-R^2 goodness-of-fit (the modified version of R^2 that has been adjusted by the number of predictors in the model) equals 10.33% (the explanatory power of this regression model). The degrees of freedom for this explained variation are 5 (number of independent predictors minus one) and 134 (the total number of samples, 140, minus the total number of independent predictors, 6). Inevitably, the total number of degrees becomes 139=134+5=140-1. Since the global F-test (regression_explained_ variance/error_unexplained_variance) is 9.13

(the critical F-table for a 95% level of confidence equals approximately 2.25), one can say that this second order polynomial explains a significant part of the variables' behavior in space. Statistically speaking, the null hypothesis should be rejected, i.e., the multiple regression coefficients should not be considered null. Moreover, this F index of 9.13 is higher than the linear F statistic, revealing (probably) a better fitness by the second order polynomial when compared to the first order.

To compare both polynomials, mathematically, a third F ratio ($df=df_{quadratic}-df_{linear}=5-2=3$) should be computed whose sum of squares (how much more variation is explained by the second order when compared with the first order) equals the difference between both of them (730.82-212.33=518.49). In this particular case, the new $F_{Quadratic\text{-}Linear}=\dfrac{518.49\ /\ 3}{5558.57\ /\ 134}$ equals 4.16, meaning that the extra coefficients of the quadratic term explains 2.039 times ($\sqrt{4.16}$) better than the first order.

Since small glaciers are particularly susceptible to the effects of climate change, they offer a powerful means of climate monitoring (Carrivick & Brewer, 2004). Glacier responses to climate change may be measured by a change in the Equilibrium Line Altitude (ELA), an important measurement that allows for the prediction of future glacier behavior. In respect of small

Figure 4. Negative values can also be achieved (check the right corner, for instance), thus depicting a serious distortion of reality

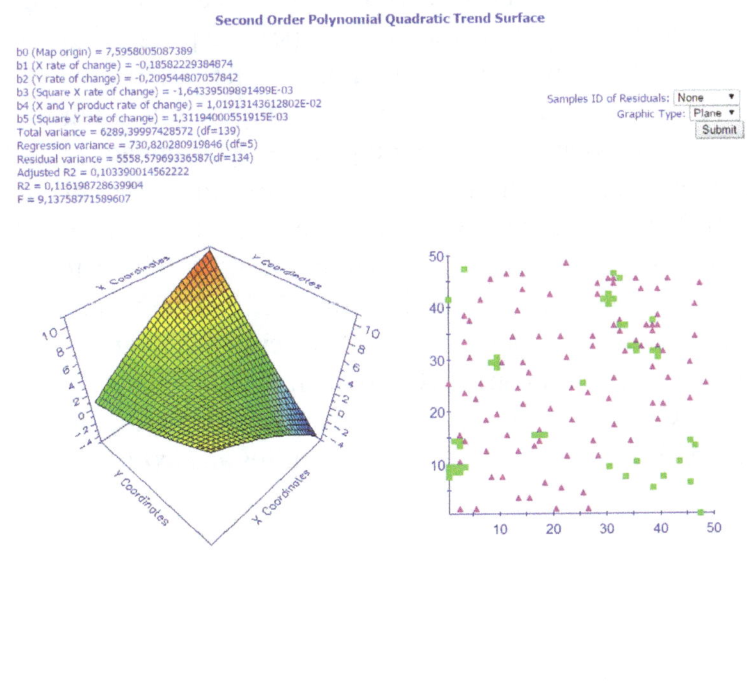

glaciers close to the glaciation level, ignoring the importance of local topographic components in mass balance estimates may lead to erroneous climatic reconstructions. For instance, of the 510 small valley and cirque glaciers digitized across northern Scandinavia, 284 were objectively deemed suitable for inferring an ELA. According to these authors, the glacier elevation, area, slope and aspect parameters were found to be the best predictors of the local ELA. These estimations improved from a 77% accuracy up to 94% when topographic parameters for every grid-cell (representing glacier surfaces) were computed (rather than using subjective measurements from topographic maps). With this aim in mind, a second-order polynomial trend surface was computed, achieving a better representation of the ELA variations across the study area (within the initial local measurement accuracy of ±100m). It is concluded that current subjective topographic map based analyses are unlikely to be sufficiently accurate for predicting the regional ELA of small, sensitive and marginal glaciers, unless GIS based spatial analyses are measured at a reasonable resolution (Carrivick & Brewer, 2004).

Another case concerns the variation partitioning by canonical correspondence analysis, where CCA was applied to analyze spatial variation in the species composition of a weed community for an area of farmland in southern Finland (Hyvönen, Holopainen & Tiainen, 2005). The farmland, covering 450ha, was sampled with a 60m×60m grid. Data on weed species were collected along with the following groups of explanatory terms: spatial variables (the regression terms of the second-order polynomial trend surface generated on the x and y coordinates: x, y, x^2, xy, y^2, x^2y and xy^2), farmer variables (nine farms), soil variables (four types and pH value), crop variables (barley, oats, sugar beet, potato and turnip rape) and physical variables (area of field, altitude, slope and aspect in four directions). In line with these three investigators, the main variation in species composition along the first and second CCA axes was caused by interplay between farmer and crop variables. These variables explained more of the variation than the soil or physical variables. Further, all the remaining variables were to some extent spatially structured. The spatial variables contributed 54.5% of the total variation with pure spatial variation accounting for 12.2%. The highest covariation with spatial variables was detected with farmer (33.7%) and crop variables (25.7%). According to Hyvönen, Holopainen and Tiainen (2005), variation partitioning by CCA is definitely recommended for studying the relationship between the spatial variation in weed communities and explanatory variables.

MULTIQUADRATIC

Radial Basis Functions (RBFs) are a series of exact interpolation techniques (thin-plate spline, splines with tension, completely regularized spline, inverse multiquadratic spline and multiquadratic function) that differ from Inverse Distance Weighting (IDW), another exact interpolator, in the following respect: RBFs can predict values above the maximum and below the minimum measured observations. Normally, it produces good results for varying surfaces, where elevation is slight. Still, this technique is inappropriate when there are large changes in surface values within a short horizontal distance and/or when the researcher suspects that the available data sample is prone to error or uncertainty (ESRI, 2005).

The first choice for the multiquadric option in myGeoffice© was based on Aguilar, Aguilar, Aguera and Carvajal (2005) work. According to these authors, spatial interpolation by this function proved to be more accurate than Logarithmic and Multilog functions, including Natural Cubic and Thin-Plate Splines. For the shape or smoothing factor, the best value mostly for this parameter is close to zero. Theoretically, the larger this parameter is, the rounder the peeks and the smoother the contour lines become. Still, there is no rule for the best shape factor although major GIS software sets up its decision based on cross-error minimization.

By using Figure 5 for demonstration purposes to interpolate x0, the multiquadratic form uses the following mathematical expression: Z(x0)=b_1x sqrt(d_1xd_1+cxc)+$b_2$xsqrt(d_2xd_2+cxc)+$b_3$xsqrt(d_3xd_3+cxc)+$b_4$xsqrt(d_4xd_4+cxc), where sqrt() means square root, c denotes the shape or smooth factor while d_1, d_2, d_3 and d4 are the four considered distances between the four available neighbors and the estimation point. Therefore, b_1, b_2, b_3 and b_4 weights may be estimated by the following four equation system:

Z(x1)=b1xsqrt(d1,1xd1,1+cxc)+b2xsqrt(d1,2xd1,2+cxc)+b3xsqrt(d1,3xd1,3+cxc)+b4xsqrt(d1,4xd1,4+cxc)

Z(x2)=b1xsqrt(d2,1xd2,1+cxc)+b2xsqrt(d2,2xd2,2+cxc)+b3xsqrt(d2,3xd2,3+cxc)+b4xsqrt(d2,4xd2,4+cxc)

Z(x3)=b1xsqrt(d3,1xd3,1+cxc)+b2xsqrt(d3,2xd3,2+cxc)+b3xsqrt(d3,3xd3,3+cxc)+b4xsqrt(d3,4xd3,4+cxc)

Figure 5. Under myGeoffice© (analogous to the inverse quadratic, moving average and inverse distance deterministic interpolators), the pre-defined default neighborhood boundary is the distance at which the Moran I reaches its high of value (according to different lags) and where the maximum possible active lag distance equals the initial variogram range

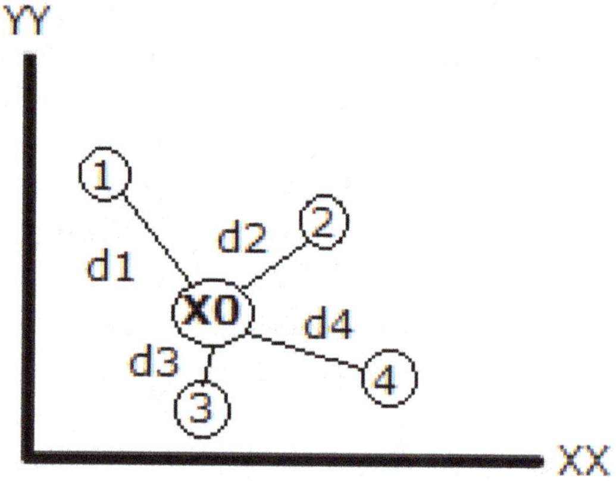

Z(x4)=b1xsqrt(d4,1xd4,1+cxc)+b2xsqrt(d4,2xd4,2+cxc)+b3xsqrt(d4,3xd 4,3+cxc)+b4xsqrt(d4,4xd4,4+cxc)

Logically, Z(x1), Z(x2), Z(x3) and Z(x4) equal the four neighbor observations values. Generically, $d_{a,b}$ represents the distance between site a and site b. In terms of matrices, these four equations will be transformed as [Z]=[B]x[DC], where D and C respectively equals the distance among neighbors and the smoothing factor. When using the Gauss-Jordan approach to compute each [B] weight, it becomes compulsory to assess the inverse square matrix of [DC] in order to multiply it by the [Z] vector.

An interesting study by Šiljeg, Lozić and Šiljeg (2015) whose aim was choosing the most appropriate interpolators for geomorphologic purposes was carried out by comparing seven deterministic methods. This research used a set of elevation data gathered by aero-photogrammetry and stereo-restitution. The accuracy of interpolation methods was tested by analyzing eight statistical parameters, which were obtained by methods of cross-validation, split sampling and jackknifing. Apart from analyzing the parameters, the research also compared the interpolation methods by visual means, through graphic representation of data (two and three dimension). This study also tested the effect of power, number of neighbors, distance and sectors on the quality of output data. According to these three authors, the multiquadratic radial basis

Figure 6. Based on the present dataset and for a two smooth input factor, the total sum error component (true sample minus estimation) equals 259.098523 (Positive sum error=101.07609; Negative sum error=158.022433)

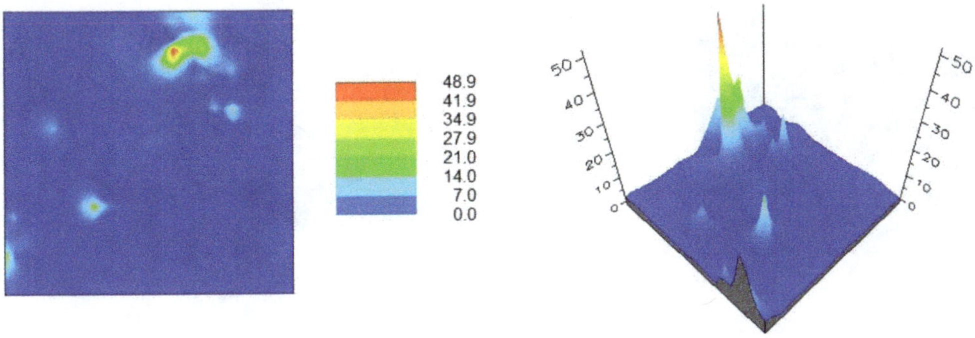

Figure 7. For a one smooth input factor, the total sum error component equals 199.482443 (Positive sum error=84.887432; Negative sum error=114.595011)

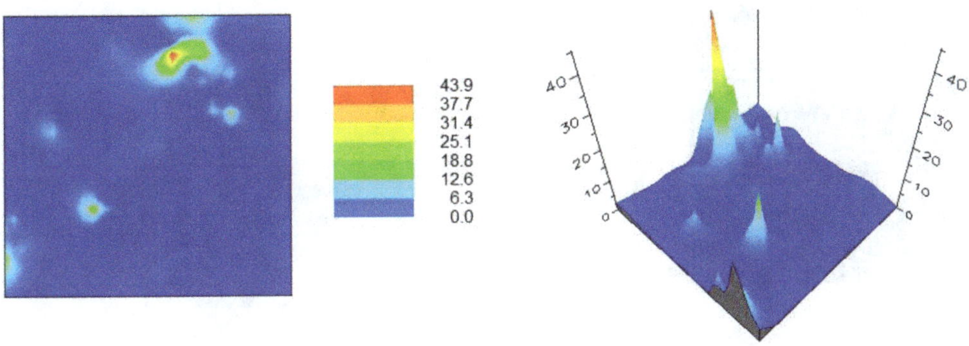

Figure 8. For a 0.5 smooth input factor, the total sum error component equals 186.671419 (Positive sum error=83.911315; Negative sum error=102.760104) and is, statistically, the best result

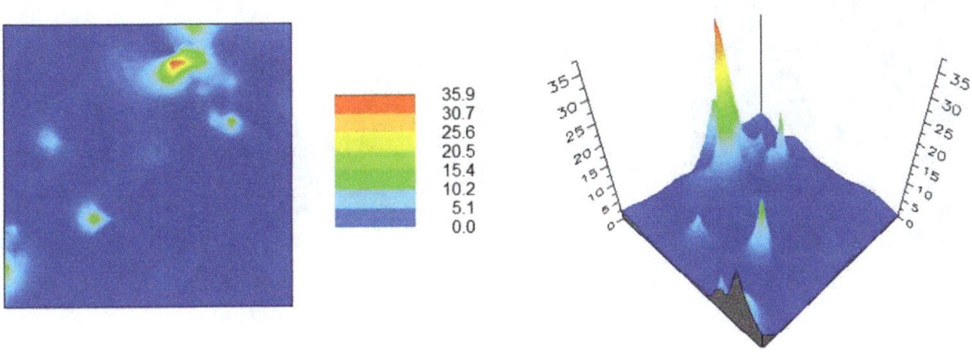

Figure 9. For a 0.1 smooth input factor, the total sum error component equals 188.272326 (Positive sum error=88.272884; Negative sum error=99.999442)

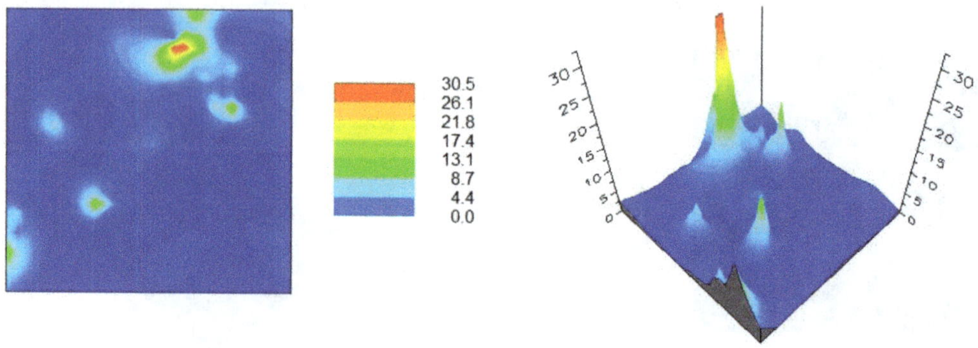

Figure 10. For a 0.01 smooth input factor, the total sum error component equals 193.269178 (Positive sum error=93.164512; Negative sum error=100.104666)

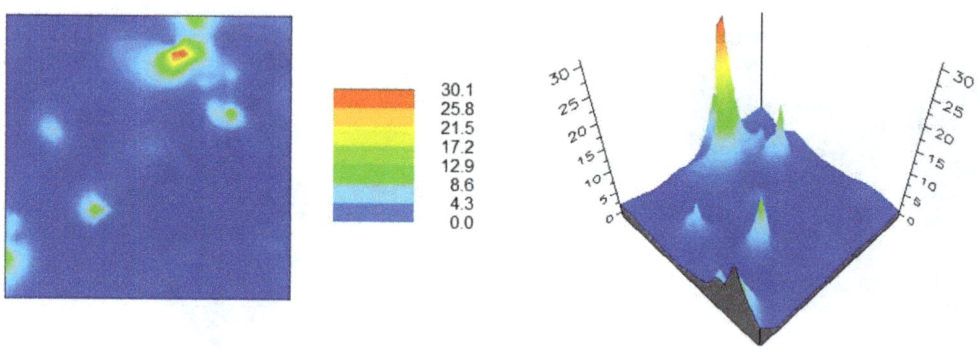

Figure 11. For a zero smooth input factor, the total sum error component equals 394.714256 (Positive sum error=370.479823; Negative sum error=24.234433)

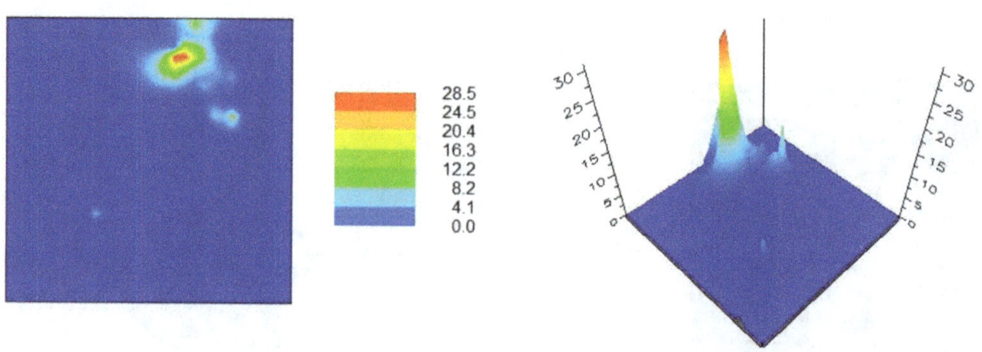

function proved to be the best deterministic method of interpolation for all the relevant parameters.

Another report by Chaplot, Darboux, Bourennane, Leguédois, Silvera and Phachomphon (2006) partially confirms the previous findings. One of the most important scientific challenges of digital elevation modeling is the development of numerical representations of large areas with a high resolution. Although there have been many studies on the accuracy of interpolation techniques for the generation of digital elevation models (DEMs) in relation to landform types and data quantity or density, there is still a need to evaluate the performance of these techniques on natural landscapes of differing morphologies and over a large range of scales. To perform such an evaluation, these authors investigated a total of six sites, three in the mountainous region of northern Laos and three in the more gentle landscape of western France, with various surface areas ranging from micro-plots, hillslopes and catchments. The techniques used for the interpolation of point height data with density ranges from 4 to 109 points/km^2 include inverse distance weighting (IDW), ordinary Kriging (OK), universal Kriging (UK), multiquadratic radial basis function (MRBF) and regularized spline with tension (RST). The study sites exhibited coefficients of variation (CV) in altitude between 12% and 78% and isotropic to anisotropic spatial structures with strengths from weak (with a nugget/sill ratio of 0.8) to strong (0.01). Irrespective of the spatial scales, variability and spatial structure of altitude, few differences existed between the interpolators although MRBF performed slightly better. However, at lower sampling densities, Kriging yielded the best estimations for landscapes with a strong spatial structure, low CV and low anisotropy. RST yielded the best estimations for landscapes with a low CV and weak spatial structure. Under conditions of a high CV, strong spatial structure and strong anisotropy, IDW performed slightly better than the remaining ones. These results indicate that the accuracy of interpolation techniques for DEM generation should be tested not only in relation to landform types and data density but also to their applicability to multi-scales (Chaplot, Darboux, Bourennane, Leguédois, Silvera & Phachomphon, 2006).

INVERSE MULTIQUADRATIC

The major difference between inverse multiquadratic and multiquadratic RBF is the internal computation formula (all remaining concepts are equivalent). Under myGeoffice©, the inverse multiquadratic applies the inverse square root of the sum of the squares of two parameters: The c shape factor and the d distance between the vicinity samples and the estimation site, i.e., 1/ sqrt(dxd+cxc).

Figure 12. For a 0.1 smooth input factor, the total sum error equals 320.241608 (Positive sum error=267.916494; Negative sum error=52.325114). Parallel to multiquadratic and other available options, myGeoffice© also allows countered mapping of the final output.

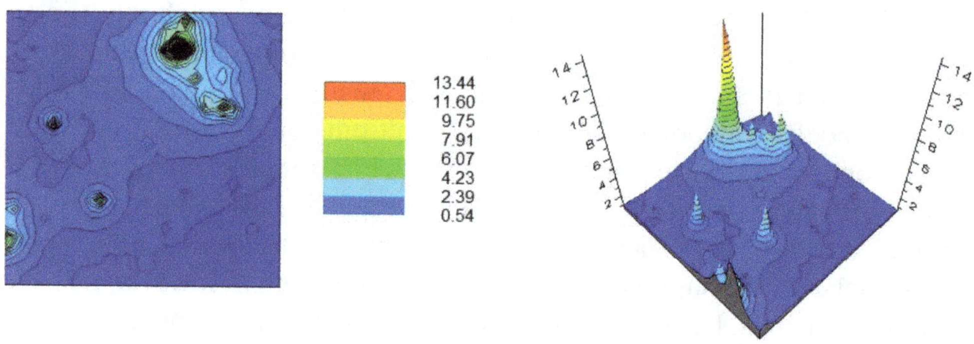

Figure 13. For a 1 smooth input factor, the total sum error equals 181.547246 (Positive sum error=84.479876; Negative sum error=97.06737) and is, statistically, the best result.

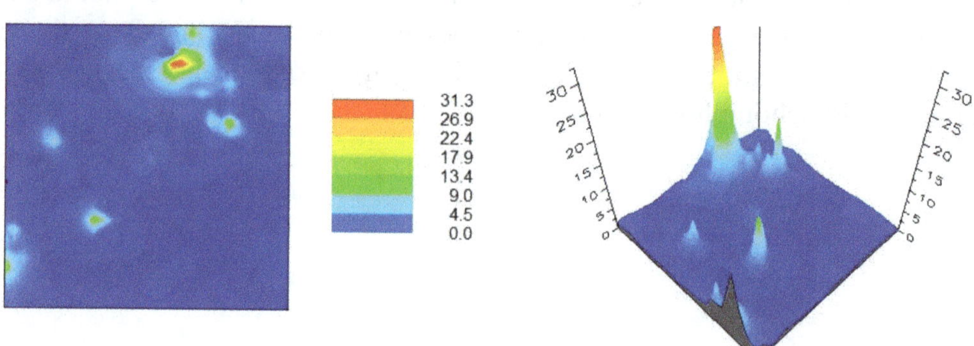

Figure 14. For a 1.5 smooth input factor, the total sum error equals 186.51925 (Positive sum error=80.552492; Negative sum error=105.966758)

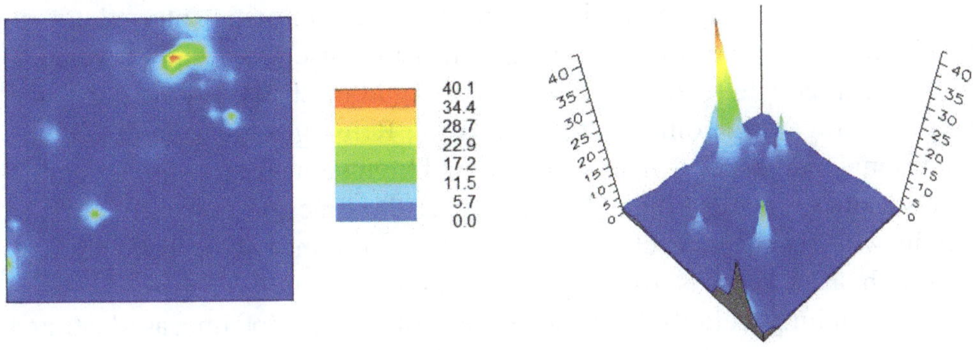

Figure 15. For a 2 smooth input factor, the total sum error equals 211.164348 (Positive sum error= 87.415036; Negative sum error=123.749312)

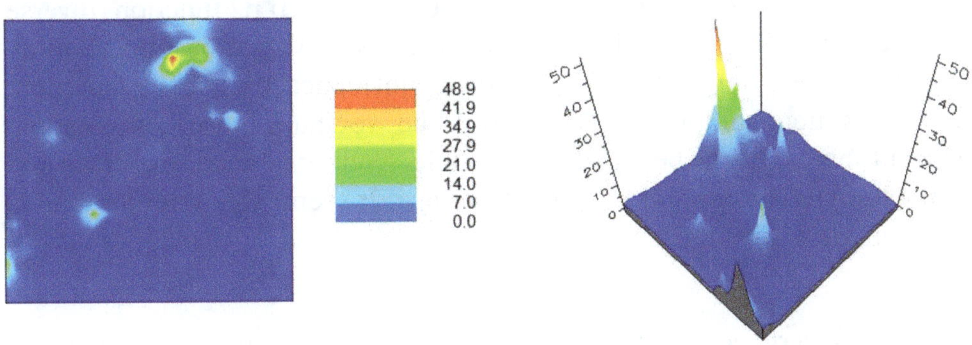

Figure 16. For a shape or smoothing factor of 3, the total sum error equals 266.267983 (Positive sum error=102.547866; Negative sum error=163.720117)

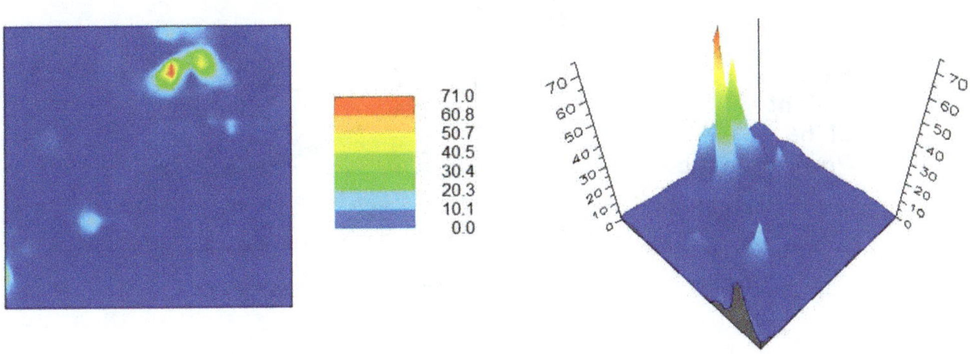

INVERSE DISTANCE

Inverse Distance Weighted (IDW), an exact deterministic interpolator, is a weighted average of the surrounding measures used to predict a value at an unmeasured site (Negreiros, Painho & Aguilar, 2008). Like all distance weight averages, its common concern is the Kriging regression effect (the conditional bias issue): Under-estimation of high values and over-estimation of low values. As per Figure 17, the higher the function power, the higher weight will be given to closer samples in accordance with Tobler's Law (where the best guess is given by the closest observations).

When vicinity weights do not decrease with distance (p0 line, as illustrated in Figure 17), all considered neighbors hold the same weight. Hence, the final estimation becomes the simple average. For a vicinity sample, which falls apart from the estimation site at d1 distance, for instance, its contribution weight will be close to 0.3=1-0.7=30%, if p1 function (inverse distance and myGeoffice© choice) is chosen or 0.75=1-0.25=75%, if p2 function (inverse square distance) is preferred. As expected, as the p weight becomes higher, only the closest neighbor will principally influence the final prediction. Further, as more samples are included in the weighted linear combination, the resulting estimates become less variable, leading to fewer extremes and a smoothing effect around the global average. It seems that interpolation is synonymous with smoothing.

Figure 17. Generic power neighbors weights for an inverse distance interpolation approach

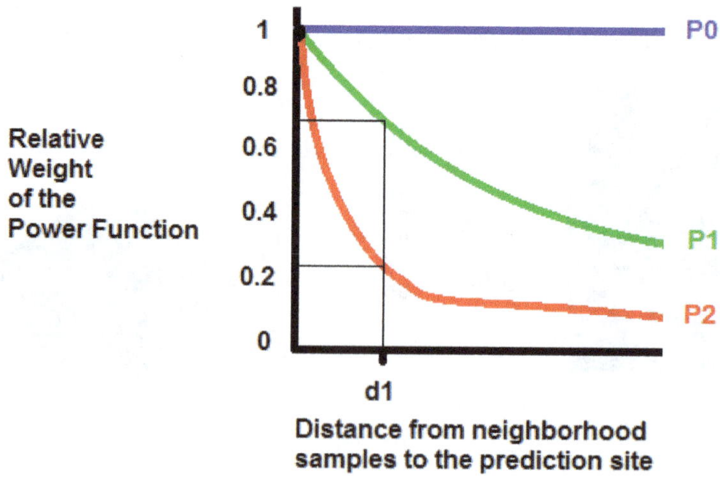

In respect of programming, the resulting map first sets a regular grid over the area while the estimated value at each node is calculated. Yet, in some of those grid nodes, there are already samples in it, implying an infinity value since the distance between the sample and the estimation point equals zero. Therefore, a special case for this computation should be made (for most software, the observed values are copied over, forcing this technique to be an exact interpolator). Another exception regards the moon hole-effect due to the lack of samples within its neighborhood. In this situation, myGeoffice© assigns the global sample average. An additional dilemma concerns the uniform population distribution assumption within areas and on how to choose the best neighborhood weight. For the latter, the common statistical search possibility for the optimal IDW is based on the cross-validation procedure. By using the original observations and their predictions, the best weight parameter is the one that holds the lowest root-mean-square prediction error.

The purpose of site specific farming entails fine scale detection of field parameters that affect harvests whose purpose is to select areas that improve the profitability of the field-scale cropping system. According to Wu, Hung & Patton (2013), only limited technologies are currently available to evaluate spatial variability in soil properties on a fine scale. Therefore, information is typically generated by collecting discrete samples and utilizing spatial interpolators to estimate data for the unsampled locations. In their analysis, pH soil samples were collected from a 12.15ha agricultural field in northwest Missouri using two grid-sampling regimes: 0.11ha with 110 samples and

Figure 18. For a search radius of 5×0.25=1.25 units (multiplicative factor=0.25), the total sum error equals 335.0955 (Positive sum error=286.0178; Negative sum error=49.0777). Note that 5 units was the default boundary distance computed by myGeoffice© for this particular dataset (at which the Moran I reaches its high value).

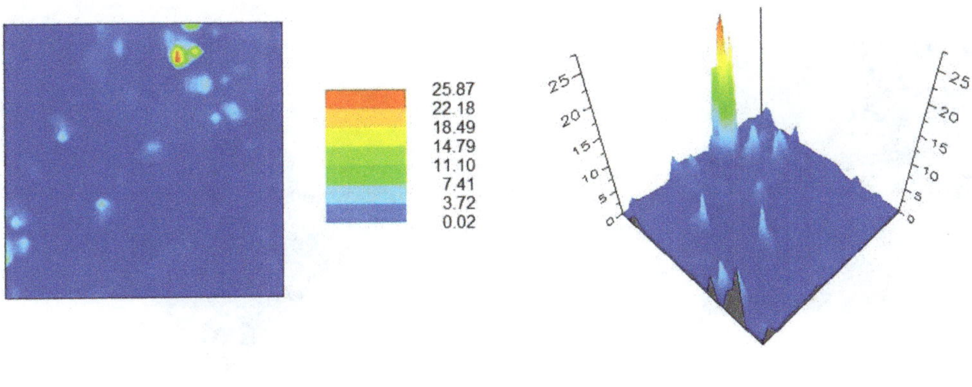

Figure 19. For a search radius of 5×0.5=2.5 units (multiplicative factor=0.5), the total sum error equals 353.9721 (Positive sum error=306.2299; Negative sum error=47.7422)

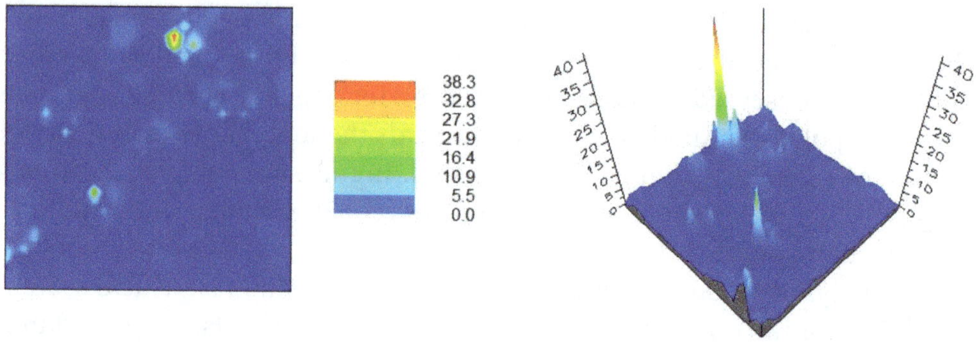

Figure 20. For a search radius of 5×1=5 units (multiplicative factor=1), the total sum error equals 476.2418 (Positive sum error=430.3884; Negative sum error=45.8534)

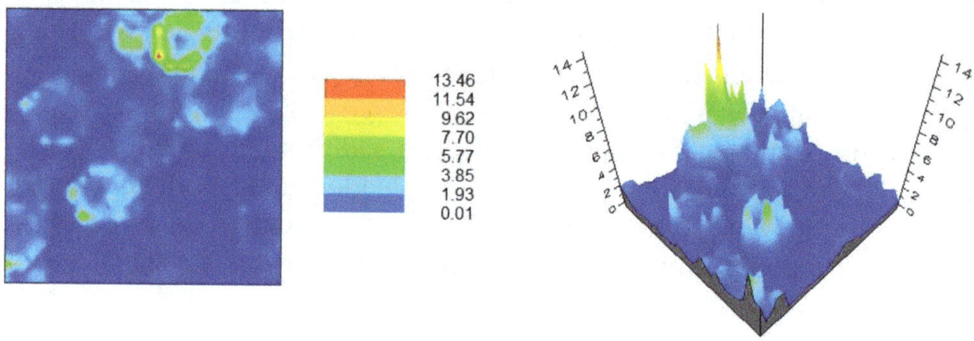

Figure 21. For a search radius of 5×1.75=8.75 units (multiplicative factor=1.75), the total sum error equals 523.6602 (Positive sum error=498.9958; Negative sum error=24.6643). Under myGeoffice©, other seven search radius multiplicative factors can be generated: 4, 3, 2.5, 2, 1.5, 1.25 and 0.5

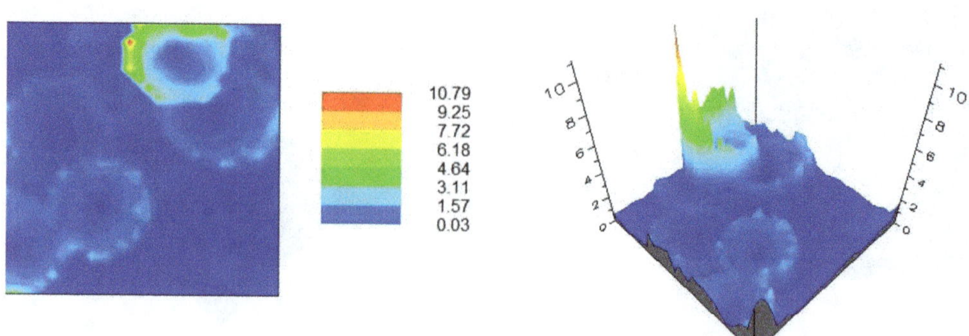

0.98ha with 12 samples. Three spatial interpolators (inverse distance weighted, spline and Kriging) were tested to evaluate the effects of interpolation. In addition to quantitative validation evaluations, results were also assessed by 2D and 3D visualization. Following Wu, Hung and Patton (2013), although each assessment approach provided useful information, the inverse distance weighted overall estimated pH soil values better than the remaining approaches.

Yet, a further contradictory study of the previous results is stated by Di Piazza, Conti, Viola, Eccel, and Noto (2015). An exhaustive comparison among different spatial interpolation algorithms was carried out to derive annual and monthly air temperature maps for Sicily (Italy). Deterministic, data-driven and geostatistical algorithms were used and, in some cases, the elevation information and other physiographic variables were added to improve the performance of interpolators and the reconstruction of the air temperature field. The dataset came from 84 stations (spread around the island of Sicily). In a standard procedure, these interpolators only used a subset of the available initial dataset while the remaining one was used to validate the results in terms of the accuracy and bias of the estimates. According to them, cross-validation results indicate that univariate methods (which neglect information from physiographic variables) significantly result in the largest errors while performances improved when such parameters were taken into account. The best results at the annual scale have been obtained using ordinary Kriging of residuals from linear regression and the artificial neural network algorithm (Di Piazza, Conti, Viola, Eccel & Noto, 2015).

MOVING AVERAGE

The moving average (MA) methodology computes the estimation of grid nodes by averaging the available data within the grid nodes. If there are no neighbor samples on the square/ellipse search, the grid node can be considered blank. In myGeoffice©, the sample mean is instead assigned (to avoid this last radical solution, myGeoffice© tries to use a wider neighborhood cutoff distance first). Some other MA features can be enhanced such as:

- An MA with a narrow search window emphasizes best spatial variation.
- The recommended number of neighbors is close to six although the layout map depends on data clustering.
- Anisotropy can be partially resolved with the quadrant/octant search.

Figure 22. For a search radius of 5×0.5=2.5 units (multiplicative factor=0.5), the total sum error equals 244.0075 (Positive sum error=109.4304; Negative sum error=134.5770)

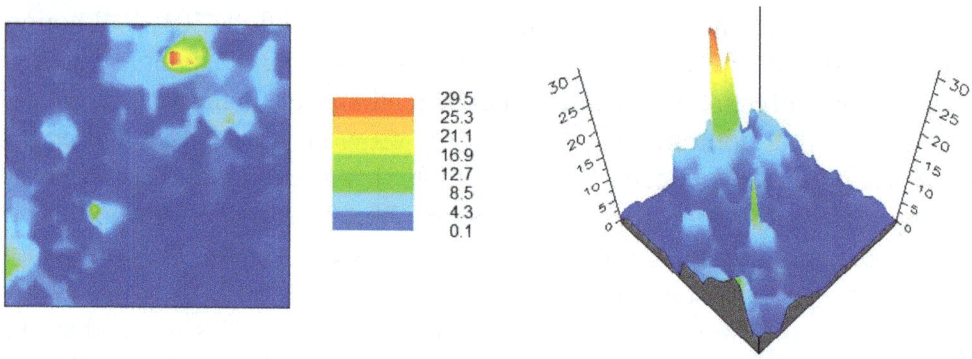

Figure 23. For a search radius of 5×1=5 units (multiplicative factor=1), the total sum error equals 377.2408 (Positive sum error=164.5716; Negative sum error=212.6691). Notice that the countered mapping option was setup to "on".

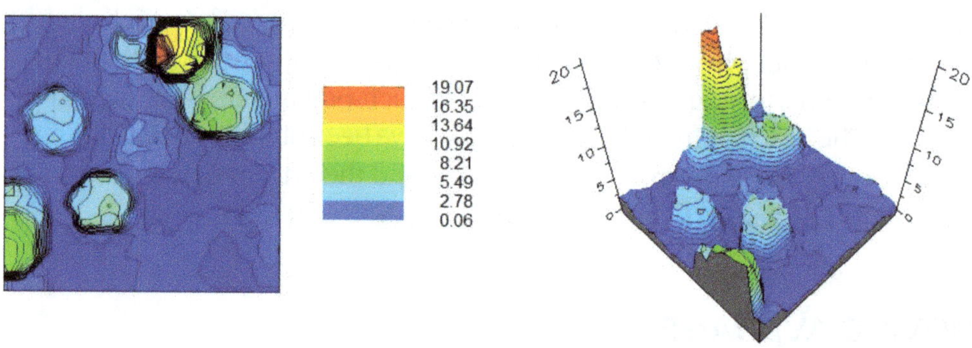

Figure 24. For a search radius of 5×2=10 units (multiplicative factor=2), the total sum error equals 467.3176 (Positive sum error=234.5591; Negative sum error=232.7585).

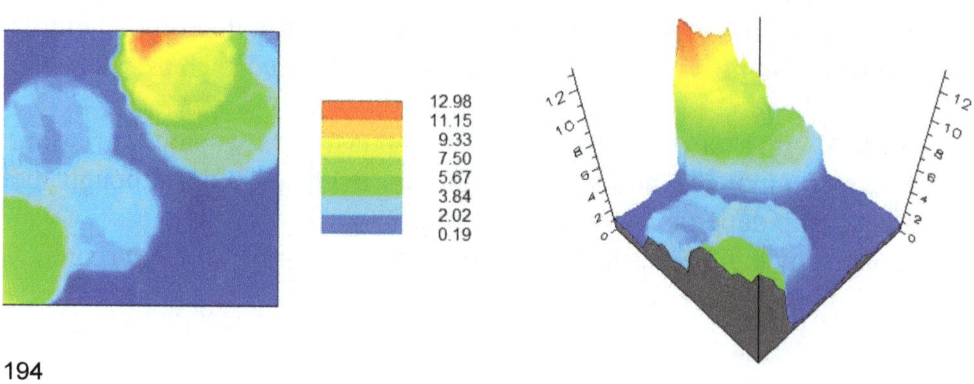

The central issue of these last four interpolators is the lack of a sustainable decision-making approach regarding their input required parameters, which lead to wide different outcomes and raising more questions than answers. If the aim is to reach the optimal prediction (interpolation and reality completely matches), Clark & Harper (2000) raised eleven brilliant questions and quite difficult (particularly with deterministic approaches) to respond: 1) What function of distance should we use in any given application? 1/d, 1/d^2, 1/$\sqrt{d}$ ...; 2) How do we handle different continuity in different directions (anisotropy)? In fishing, the offshore may have different level of predictability to the longshore direction; 3) How many samples should we include in the estimation? Under myGeoffice©, 8 was chosen as the default value but it could be different; 4) How do we compensate for irregular spaced highly clustered sampling? It is almost inevitable in mining that a geologist will schedule more sampling in the rich areas than in the poor ones; 5) How far should we go to include samples in our estimation process? Rainfall is a pretty continuous phenomenon over oceans but not in mountains; 6) Should we honor the sample values? Based on the precise and accurate sampling assumption, the answer for this question appears to be affirmative; 7) How reliable is the estimate when we have it? After all, classical statistics offers a set of methods to compute confidence intervals for the mean and the variance. 8) Why is the final map too smooth? Geostatistical simulation is one way to avoid this issue; 9) What happens if our sample data is not Normal? Prior data standardization will be the first choice. Rank uniform Kriging the last one; 10) What happens if there is a strong trend in the values? Universal Kriging (myGeoffice© option) is an alternative to trend removal (by a first or second order polynomial, for instance) before variography is applied on residuals; 11) How do estimate average values over areas or volumes? Frequently, sampling of points holds higher values than block ones.

NEAREST NEIGHBOR

The Nearest Neighbor (also known as proximal interpolation) gridding approach assigns the estimation value of the nearest sample to each grid node, a methodology equivalent to the Voronoi diagram. Compared with other major interpolation techniques, this methodology is less affected by the smooth effect and it can be effective for filling data, if grid maps present large holes due to the lack of available observations (Negreiros, Painho, Aguilar & Aguilar, 2010).

Figure 25. The total sum error equals 270.68 (Positive sum error=103.62; Negative sum error=167.06)

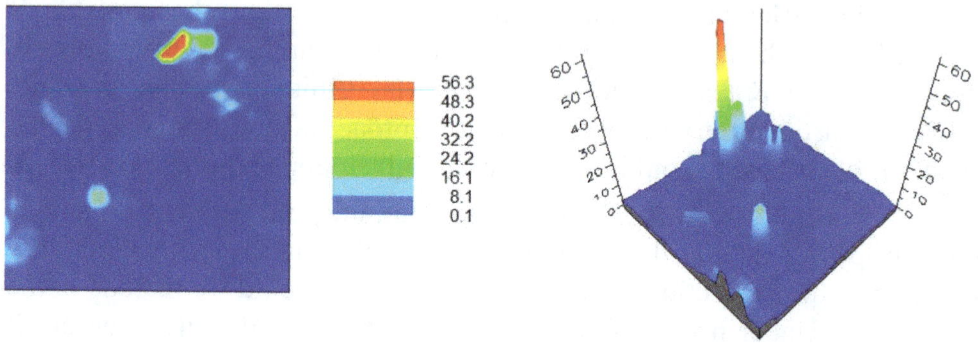

An example of the use of nearest neighbor regards the slope maps obtained in 1997 with the declivity maps generated by Bertotti, Filho, Pelegrina, Boas Gomes, Costa Toledo, and Piasecki (2015) in Brazil. The input data from the testing areas include topographic maps and Shuttle Radar Topography Mission (SRTM) images and stored in Spring® databases. The test regions were built with well-defined isolines concerning the horizontal spacing between the contour lines. The interpolation procedures analyzed were weighted mean, simple mean, nearest neighbor, linear and quintic surface with break lines. These researchers used all these methods to draw rectangular grids with 1m x 1m spatial resolution. Out of these approaches, the TIN grid with Delaunay was selected, which allowed the generation slope image with a spatial resolution of 1m. They concluded that the results were on average 10% higher for regions classified as plan relief and over 100% higher for areas of corrugated and mountainous relief. The input data from topographic maps and SRTM images with spatial resolution of 90m were satisfactory.

Precipitation is a key hydrological variable analyzed before being supplied as an input to rainfall-runoff models. The difficulties to quantify the amount of rain in order to truly represent the event that occurs in the watershed primarily come from the high precipitation spatial variability. For instance, Silva, Monteiro, and Negreiros (2010) have already experienced this thinking at Cape Verde, Africa. For Viana (2011), one way to quantify the amount of rainfall in a given area for use in concentrated rainfall-runoff models (from a network of rain gauges) can be based on the inverse square distance, nearest neighbor, natural neighbor and linear triangulation. The interpolator most appropriate to the situations found on his precipitation series study was selected by using the cross-validation technique, which the natural neighbor method showed better performance. As well, the average rainfall obtained by this natural interpolator was statistically more representative for most of the scales analyzed (Viana, 2011).

VORONOI AND DELAUNAY

With these exact interpolators, the sample points are connected by lines to form triangles, where variations are assumed only at borders. With Voronoi or Thiessen vector polygons, the best information about an unvisited site becomes the nearest data prediction point, a weighted linear combination approach, where all weights are given to the closest observation. Therefore, polygon output depends on the sample layout, a point-in-polygon computation problem resolved by the semi-line algorithm (the pair line linkage of neighboring points is bisected at right angles through the polygon boundary). Hence, its size varies inversely with point density. On the other hand, the Delaunay triangulation corresponds to the dual graph of the Voronoi diagram, where triangulations maximize the minimum angle of all angles of the involved triangles (Wikipedia-Delaunay, 2015).

A key issue on the construction of the Voronoi diagrams concerns the computing efficiency of these polylines/polygons issue. Jiatian, Qili, Fuli, Huilan, and Yan (2015) proposed a new decomposition and combination algorithm, where the main procedure transforms polylines/polygons into discrete feature points. According to its cross-connection features, vicinity point's pairs were constructed whose cross parts of Voronoi diagrams could be substituted for the equal distance boundaries of each polyline/polygon. Stated by these authors, it can be concluded that the time complexity of the proposed algorithm is $O(n^2)$ at high precision (the algorithm performance is directly proportional to the square of the size of the input dataset).

Mota, Takano, and Taco (2014) described the use of Voronoi on a commuter rail system (Brasilia, Brazil). Their technique comes from a previously defined track line but considered the stations area coverage. These investigators first used point density to represent the maximum concentration activity area. Afterwards, they also employed trip generating rate weights for point density. The final product was a T-Trunk-Feeder (TTF) framework that allows an integrated transportation planning but considered other existing transit systems (Mota, Takano & Taco, 2014). After the stations were classified, according to their degree of activity density and its importance for integration, the TTF became the representation of the commuter rail network while the Trunk-Distribution stands for the transport network (responsible for the passenger's distribution in central business district). The proposed model features 69Km extension with an estimated total travel time of 62 minutes and 17 stations being integrated into two undergrounds: Four unified into the LRT stations and three combined into the main bus stops of Brasilia.

An additional example is given by Varouchakis and Hristopulos (2013). In sparsely monitored basins, accurate mapping of the spatial variability of the groundwater level requires the interpolation of scattered data. Their aim was to present a comparison of deterministic interpolators (inverse distance weight (IDW) and minimum curvature (MC)) with stochastic ones (ordinary Kriging (OK), universal Kriging (UK) and Kriging with Delaunay

Figure 26. myGeoffice© Java Applet layout produced by Safari® for the present dataset counterparts (IDW and MC). DeK yields the most accurate cross-validation estimate by presenting the lowest value. According to Varouchakis and Hristopulos (2013), OK and UK lead to a smooth isolevel contours while DeK and IDW generate more edges.

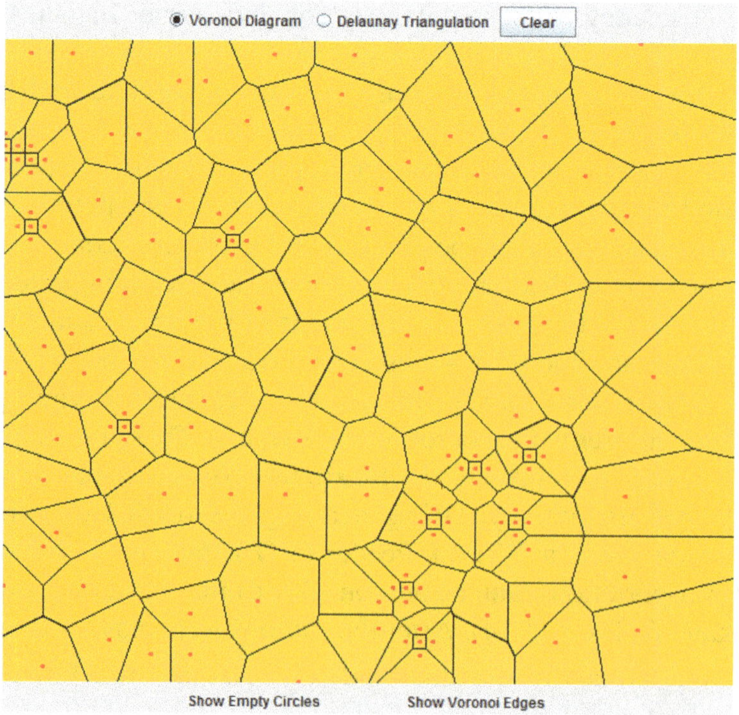

Figure 27. myGeoffice© Java Applet layout produced by Safari® for the present dataset

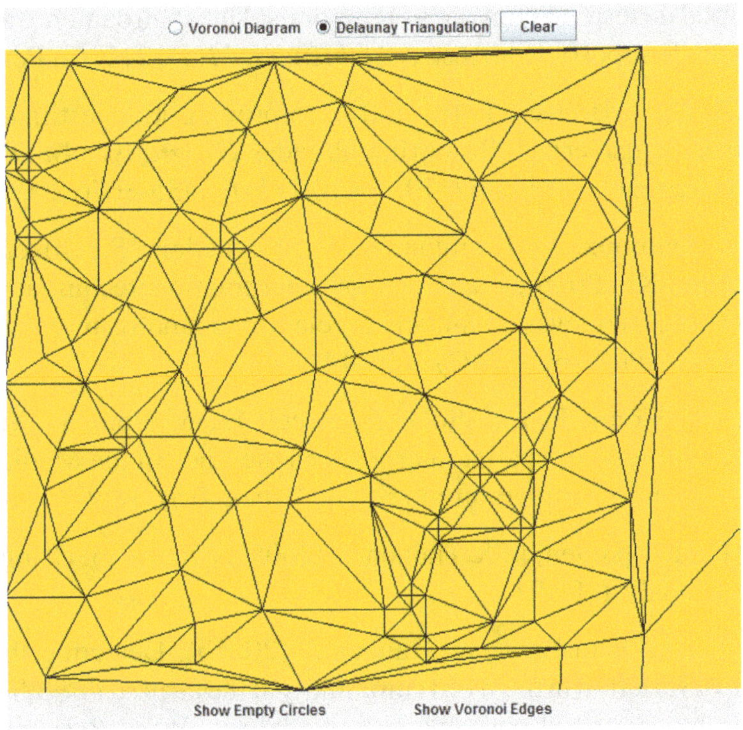

triangulation (DeK)). The study area was the Mires Basin of Mesara Valley in Crete (Greece) because the sparsely sampled basin has limited groundwater resources, which are vital for the island's economy. As well, spatial variations of this groundwater level are essential for developing management and monitoring strategies. The Spartan variogram was applied for the first time to hydrological data and was shown to be optimal with respect to stochastic interpolation of this dataset. In brief, the three stochastic methods (OK, DeK and UK) perform overall better than the deterministic

REFERENCES

Aguilar, F., Aguilar, M., Aguera, F., & Carvajal, F. (2005, July 1). Effects of terrain morphology, sampling density and interpolation methods on grid DEM accuracy. *Photogrammetric Engineering and Remote Sensing*, *71*(7), 805–816. doi:10.14358/PERS.71.7.805

Bertotti, L. G., Filho, M. C., Pelegrina, M. A., Boas Gomes, M. V., Costa Toledo, B. H., & Piasecki, J. S. (2015). Avaliação de técnicas de geoprocessamento no modelo digital de terreno com dados temporais: Um abalroamento abalizado na inclinação da superfície terrestre. *Ambiência, 11*(1), 215. (in Portuguese)

Carrivick, J. L., & Brewer, T. R. (2004). Improving local estimations and regional trends of glacier equilibrium line altitudes. *Geografiska Annaler. Series A. Physical Geography, 86*(1), 67–79. doi:10.1111/j.0435-3676.2004.00214.x

Chaplot, V., Darboux, F., Bourennane, H., Leguédois, S., Silvera, N., & Phachomphon, K. (2006). Accuracy of interpolation techniques for the derivation of digital elevation models in relation to landform types and data density. *Geomorphology, 77*(1/2), 126–141.

Di Piazza, A., Conti, F. L., Viola, F., Eccel, E., & Noto, L. V. (2015). Comparative analysis of spatial interpolation methods in the Mediterranean area: Application to Temperature in Sicily. *Water, 7*(5), 1866.

Hyvonen, T., Holopainen, J., & Tiainen, J. (2005a, February). Using ArcGIS geostatistical analyst. ESRI.

Hyvönen, T., Holopainen, J., & Tiainen, J. (2005b). Detecting the spatial component of variation in the weed community at the farm scale with variation partitioning by canonical correspondence analysis. *Weed Research, 45*(1), 48–56. doi:10.1111/j.1365-3180.2004.00430.x

Tian, L., Qili, Y., Li, L., Huilan, L., & Yan, L. (1545). Decomposition combined line/area Voronoi diagram generation algorithm. *Geomatics & Information Science of Wuhan University, 40*(11). doi:10.13203/j.whugis201418

Mota, D. R., Takano, M., & Taco, P. G. (2014). A method using GIS integrated Voronoi diagrams for commuter rail station identification: a case study from Brasilia (Brazil). *Procedia - Social And Behavioral Sciences, 162*, 477-486.

Negreiros, J., Painho, M., & Aguilar, F. (2008). Principles of deterministic spatial interpolators. *Tekhne – Polytechnical Studies Review, 6*(9), 308-318.

Negreiros, J., Painho, M., Aguilar, F., & Aguilar, M. (2010). Geographical information systems principles of ordinary Kriging interpolator. *Journal of Applied Sciences, 10*, 852-867.

Šiljeg, A., Lozić, S., & Šiljeg, S. (2015). The accuracy of deterministic models of interpolation in the process of generating a digital terrain model – The example of the Vrana Lake nature park. *Tehnicki Vjesnik / Technical Gazette, 22*(4), 853-863.

Silva, J., Monteiro, P., & Negreiros, J. (2010). Spatial modulation of rainfall in Santiago, Cape Verde, with geostatistical analyst®. *Proceedings of the International Association for the Scientific Knowledge – InterTIC International Conference*, 272-276.

Varouchakis, E., & Hristopulos, D. (2013). Comparison of stochastic and deterministic methods for mapping groundwater level spatial variability in sparsely monitored basins. *Environmental Monitoring and Assessment, 185*(1), 1–19. doi:10.1007/s10661-012-2527-y PMID:22311559

Viana, S. (2011). *Estimativa de precipitação pontual em diferentes escalas para uso em modelo concentrado chuva-vazão*. Ipswich, MA: RCAAP. (in Portuguese)

Delaunay. (2015). In *Wikipedia*. Retrieved from: en.wikipedia.org/wiki/Delaunay_triangulation

Wu, Y., Hung, M., & Patton, J. (2013). Assessment and visualization of spatial interpolation of soil pH values in farmland. *Precision Agriculture, 14*(6), 565–585. doi:10.1007/s11119-013-9316-7

Chapter 9
Stochastic Interpolation

ABSTRACT

Spatial analysis includes an expanding array of methods which address different spatial issues, ranging from remote sensing to spatial error uncertainty. Each of these methods focuses on geographically raw data correlated by statistical methods. In general, spatial interpolation and stochastic Kriging, in particular, will be addressed in this chapter. Ordinary Kriging (OK) foundations are presented in the first section which encompasses eight sub-sections (in accordance with the eight myGeoffice© options). Section two introduces Kriging with Trend (KT but sometimes known as Universal Kriging) including five sub-sections: Geocomputation of KT, estimation mapping, the cross-validation procedure, validation using an extra dataset and KT versus OK comparison. Finally, Indicator Kriging (IK) is explored in section three together with nine sub-sections: First and second cutoff definition, first and second probabilistic interpolation maps, construction of the conditional cumulative distribution function, entropy of Shannon, E-type spatial estimation (including misclassification risks and economic classification), morphologic geostatistics and probabilistic interval mapping.

ORDINARY KRIGING

The difference between inverse distance weighted (IDW) and Ordinary Kriging (OK) derives from the mathematical process that minimizes the minimum squared error variance of the estimation. The OK equations system can be obtained from the following two algebraic expressions stressed in Figure 1.

DOI: 10.4018/978-1-5225-3270-5.ch009

Figure 1. w_j are the weights to be assigned to the ith observation and Ψ represents the Lagrange multiplier or slack value required for forcing total weights to equal one when the partial derivatives system is resolved (set to minimize the estimation variance). $\gamma(x_i,x_j)$ is the variogram value between sample i and j while $\gamma(x_i,x_0)$ equals the variogram value between i sample and x_0 site.

$$\sum_{j=1}^{n} w_j \gamma(x_i, x_j) - \psi = \gamma(x_i, x_0), \; i = 1 \ldots n \quad \text{and} \quad \sum_{j=1}^{n} w_j = 1$$

Analogous to Thiessen and B-Splines, Kriging is an exact interpolator in the sense that the sample and estimation are equal. This happens because the variogram value between k sample and x_0 estimation, $\gamma(x_k,x_0)$, equals zero since the estimated site is the sample itself. Therefore, its Kriging variance equals zero. By resolving the Kriging matrix system presented in Figure 2, the w_k weight is one while all other weights (including the Lagrangean dummy parameter) become zero (the k row of matrix B changes to zero, if the nugget-effect is considered zero).

Under this elegant matrix layout, each interpolated value is calculated as the sum of weighted known points whose weights are calculated from the (n+1) simultaneous linear equations set: $A \times W = B$ or $W = A^{-1} \times B$. The statistical distance (it incorporates anisotropic distance, direction, spatial autocorrelation and clustering information) between n samples and distances from each sample to the grid estimation point is used to compute the model variance reproduced on A (among samples) and B matrices (between each sample and the estimated location). While A^{-1} underlies the declustering factor, B represents the structural distance between the estimation and all samples. In addition, the product of A^{-1} by B adjusts the raw inverse statistical distance weights of B matrix to account for possible redundancies between samples (Isaaks & Srivastava, 1989). As expected, if no spatial autocorrelation is found among the available samples, the Kriging estimator equals the sample average.

Figure 2. The Kriging matrix system
(Lima & Negreiros, 2011)

$$A = \begin{vmatrix} Cov(x1,y1) & \ldots & Cov(x1,yn) & 1 \\ \ldots & \ldots & \ldots & \ldots \\ Cov(xn,y1) & & Cov(xn,yn) & 1 \\ 1 & \ldots & 1 & 0 \end{vmatrix} \quad B = \begin{vmatrix} Cov(x1,x0) \\ \ldots \\ Cov(xn,x0) \\ 1 \end{vmatrix} \quad W = \begin{vmatrix} w1 \\ \ldots \\ wn \\ \Psi \end{vmatrix}$$

With a zero nugget-effect, the diagonal of A matrix equals zero while Kriging weights decrease as the dataset location gets farther from the estimated one. Hence, major software computation uses the covariogram model to setup A and B matrices to avoid the diagonal zeros of A. After all weights are calculated, the estimation of the regular grid surface can be determined with a great deal of calculation. In fact, the computation intensity to solve OK weights is proportional to the cube of the number of observations. Regarding the uncertainty layout, this is closely related to OK variance and given by Figure 3.

f the sum of the weights is one and the average estimation error equals zero then the Kriging error variance (minimum estimation variance) for a general estimated x_0 point becomes as revealed in Figure 4. Notice that no measurement error is included here. Critically, OK variance is not sensitive to local error for two major reasons: (1) It is based on the same global variogram; (2) Distances among locations are the only relevant factor. Hence, OK variance is mainly a geometric dependent measure. OK variance is too much of a spatial operation.

For theoretical reference, some other uncertain measures should be highlighted. Due to economic land evaluation, both linear programming and sensibility analysis are possibilities. Burrough (1991) presented a gradual deterministic response for pH levels in crops, illustrating the impact of soil

Figure 3. C_{00} equals the variance of the estimated point value (as the variable becomes more erratic, this term increases in magnitude), C_{ij} is the covariance between i^{th} and j^{th} sample (as samples get closer together, a clustering issue, the average covariance increases), w_i and w_j denote the OK weights while C_{i0} represents the covariance between the ith sample and the unknown value being estimated (uncertainty decreases as samples are closer to the value being estimated).

$$\sigma_e^2 = C_{00} + \sum_{i=1}^{n}\sum_{j=1}^{n} w_i w_j C_{ij} - 2\sum_{i=1}^{n} w_i C_{i0} = -\gamma_{00} - \sum_{i=1}^{n}\sum_{j=1}^{n} w_i w_j \gamma_{ij} + 2\sum_{i=1}^{n} w_i \gamma_{i0}$$

Figure 4. n equals the number of neighbors intervening in the interpolation process

$$\sigma_{OK}^2 = \sum_{i=1}^{n} w_i \gamma(x_i, x_0) + \psi$$

acidity on crop growth: (1) No crops, if PH>=7; (2) Normal growth, if PH<=5; (3) (7-PH)/2 of crops, if 5<PH<7. Seven years later, Burrough and McDonnell (1998) presented the Kenyan annual soil erosion simulation using the Universal Soil Loss Equation for rainfall-rainoff impact. This yielded an annual soil loss rate of 9 $\pm$ 6cm in 40 years: R_e(297 $\pm$ 72)×S_e(0.1 $\pm$ 0.05)×S_l(2.13 $\pm$ 0.045)×S(1.169 $\pm$ 0.122)×C(0.63 $\pm$ 0.15)×P(0.5 $\pm$ 0.1), where R_e, S_e, S_l, S, C and P represent, respectively, rainfall erosion, soil erodibility, slope length, slope, cultivation and practice.

On the basis of the assumption of null dependency among random samples, classical statistics present other options to expose how reliable our estimates are. Based on the best unbiased estimator for the true population standard deviation, it is possible to approach uncertainty in three different ways (Negreiros, Painho, Aguilar & Aguilar, 2008):

1. Computation of the best standard deviation confidence interval based on the true unbiased population assumption: $\sqrt{\frac{(n-1)\sigma^{*2}}{\chi^2 lower}} < \sigma < \sqrt{\frac{(n-1)\sigma^{*2}}{\chi^2 upper}}$, where is the estimated sample variance while $\chi^2 lower$ and $\chi^2 upper$ equals the lowest and upper X^2 table value for a certain confidence level whose degrees of freedom (df) equals the total number of samples minus one.
2. Assessment of the mean confidence interval using the T-Student distribution: $\overline{x} + t_{df,Lower}\frac{\sigma^*}{\sqrt{n}} < \mu < x + t_{df,Upper}\frac{\sigma^*}{\sqrt{n}}$, where σ^* is the dataset standard deviation, $t_{df,Lower}$ and $t_{df,Upper}$ equal respectively the lowest and upper T-table values for a particular confidence level (df=n-1).
3. Estimation of the probability that is likely to be above or below any specific threshold. Thus, each cutoff must be mathematically transformed (Z=(cutoff-sample_mean)/standard_deviation) into the Normal standard table distribution for probability calculation purposes.

According to Clark and Harper (2000), these approaches may help users determine how much of the total agriculture production would provide higher yields than a certain crop cutoff or assist when examining toxicity levels of pollution thresholds. Using a coal seam simulated dataset in Southern Africa (n=96, mean=24.62, sd=2.44), both authors concluded that for a 90% level of confidence, the standard deviation value lies between 2.19 and 2.79, the mean varies between 24.2 and 25.04 while the probability of a sample being greater than 28.5 is 5.74%, for instance.

OK Geocomputation

Kriging is a surface estimation method that minimizes the variance of the error, the mean error becomes close to zero while predictions are a result of a weighed linear combination of the original samples. Under myGeoffice©, four models will be estimated: (A) Non-exact Kriging version (due to the inclusion of the measurement error within Kriging matrix system); (B) Model 1 - Discontinuity Kriging version (inclusion of the nugget-effect for all lag distances with a zero variogram value for a zero lag distance); (C) Model 2 - Continuity Kriging version (no nugget-effect is considered); (D) Model 3 - Nested Kriging version (nugget-effect plus two nested structures for estimating distances below and above the shortest sampling interval, the distance at which no information is provided by the spatial dataset layout for the variogram geocomputation). Confirming Clark and Harper's (2000) suspicions, instead of a nugget-effect of C_0, use a nugget-effect of zero and an extra Spherical component with a very short range of influence and a sill of C_0. Following this principle with myGeoffice©, two slightly differences can be found in this version: 1) The first variogram component follows the second one (which means it may be Exponential, Spherical or Gaussian); 2) The sill of the first nested structure equals the value of the second variogram at SSI (the shortest sampling interval among all samples, a distance provided by the nearest neighborhood analysis). In doing so, the contiguity of the overall variogram is guaranteed.

If the researcher wants to add residual variation due to instrument measurement error or for spatial sources of variation at distances shorter than

Figure 5. Variogram behavior of scenarios (B), (C) and (D)

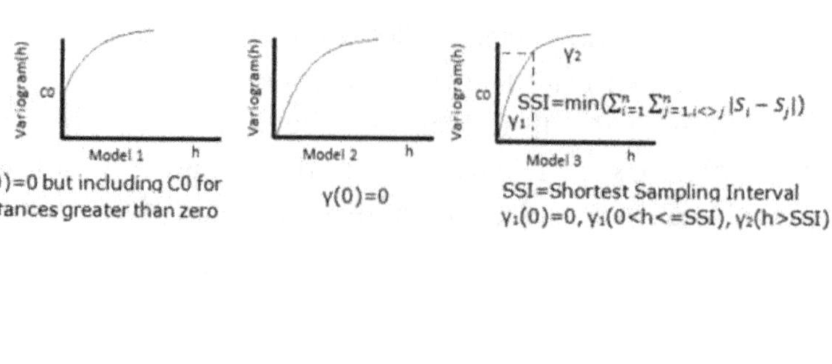

the sampling interval, it is possible to include this aspect within myGeoffice© whose nugget-effect percentage ranges between 0% and 30% (see Figure 6).

OK Default

Theoretically, the variogram sill should equal the global variance of the dataset. However, this frequently does not always happen. This discrepancy may be explained by the inadequacies of the variogram model for small separation distances (Isaaks & Srivastava, 1989). After all, we are using a result of our conceptual random function model, which is based on an averaging of all other possible outcomes of the random function (outcomes that have no counterpart in reality). Therefore, a sill rescaled operation is needed to reflect the global variance (it should be emphasized that this operation only affects the Kriging variance map and not the Kriging estimation). Under myGeoffice©, all Kriging variances are rescaled by the ratio between the estimated global variance (whose weights are estimated by the nearest neighborhood analysis) and the variogram sill (see Figures 7 and 8). In this particular case, this ratio equals 0.493 (basically, the OK variance is downsized by half).

Figure 6. For academic purposes, a 12% of measurement error was considered in this example

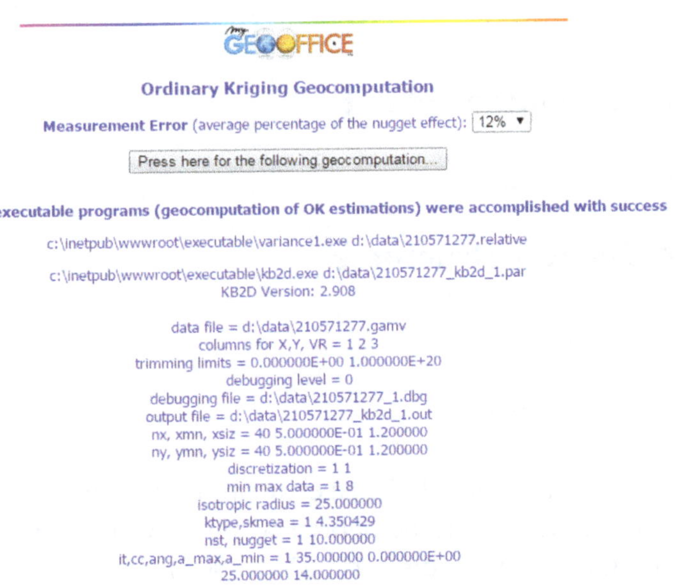

Figure 7. Estimated map based on Model 1

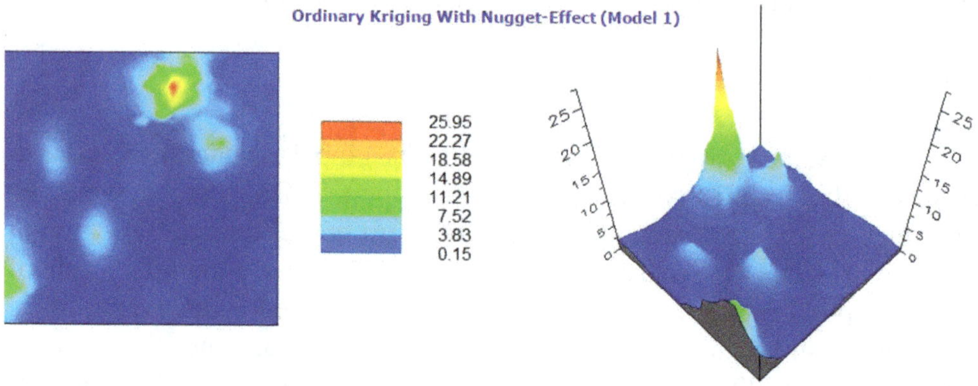

Figure 8. Mapping of the Ordinary Kriging variance for Model 1

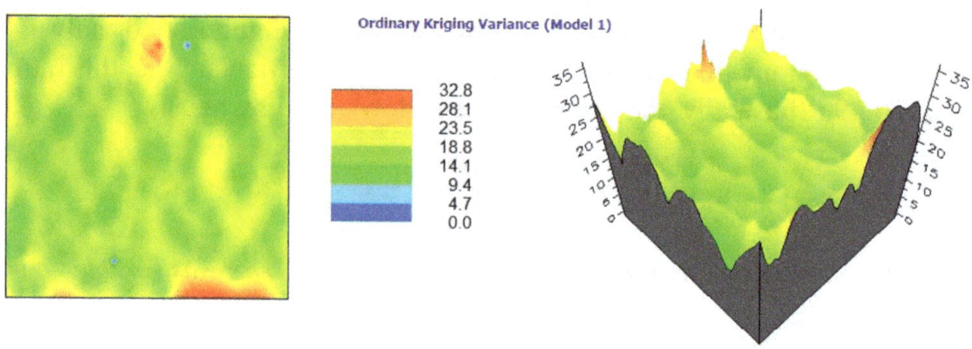

OK With No Nugget-Effect

See Figures 9 and 10.

OK With Nested Structures

See Figures 11 and 12.

OK Cross-Validation

The consistency checking of biased estimations can be tested by temporarily dropping any sample from the dataset while its value is re-estimated using the remaining ones. According to ESRI (2005), when repeating this cross-

Figure 9. Isotropic nugget-effect and no octant search are considered within myGeoffice©

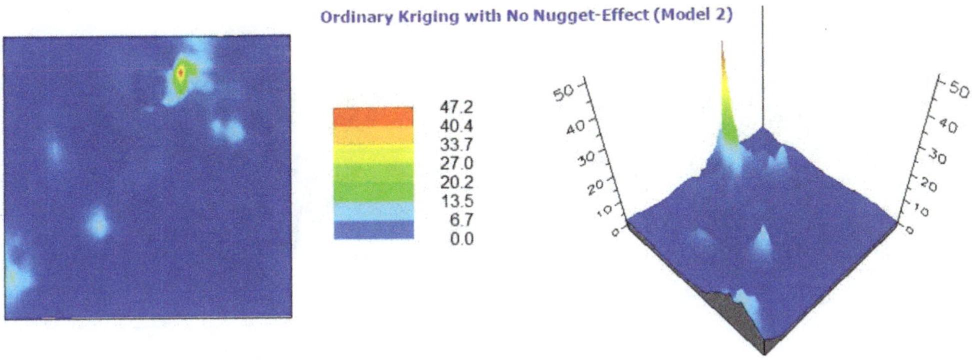

Figure 10. Under myGeoffice©, the considered number of samples for each estimated block ranges between 1 and 8

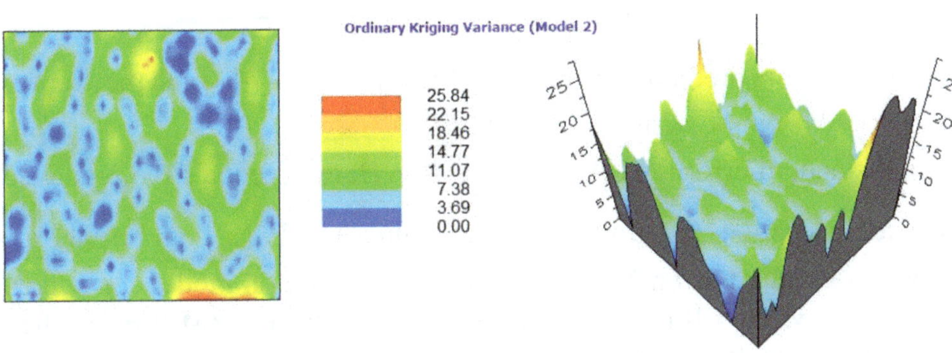

Figure 11. Estimated map based on Model 3

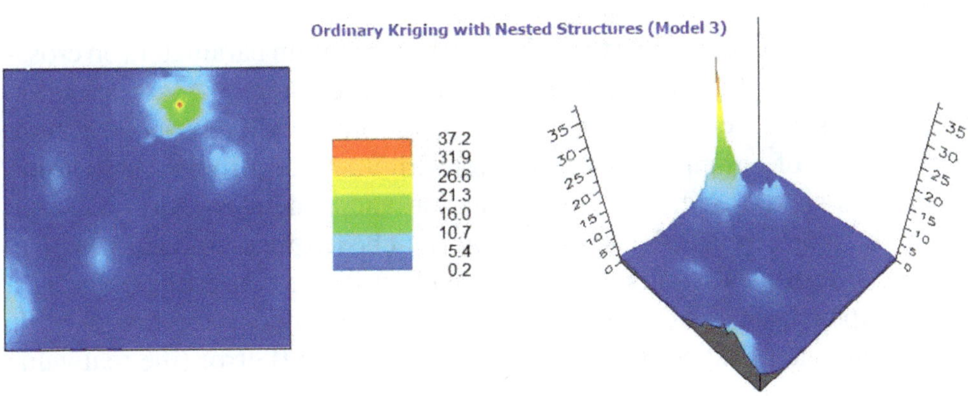

Figure 12. Mapping of the Ordinary Kriging variance for Model 3

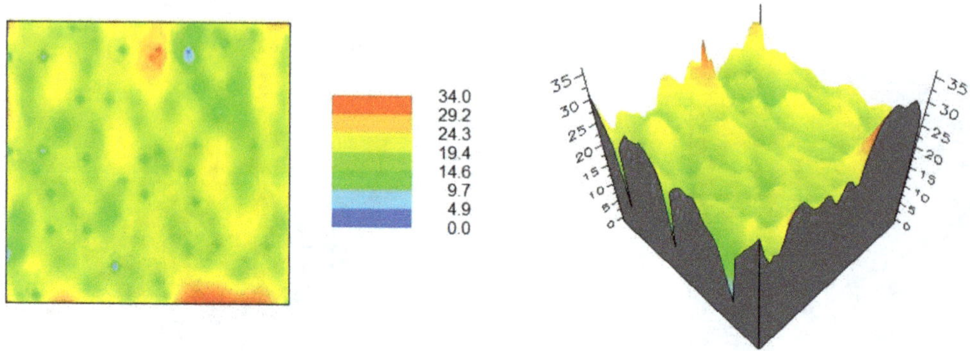

validation procedure (estimation minus real) for all observations, it is possible to obtain the experimental mean prediction error (MPE=$\frac{\sum\left(x_i^* - x_i\right)}{n}$), the root-mean-square prediction error (RMSPE=$\sqrt{\frac{\sum\left(x_i^* - x_i\right)^2}{n}}$) and the average Kriging standard error (AKSE=$\sqrt{\frac{\sum\sigma^*\left(x_i\right)}{n}}$). Here x_i^* equals the predicted value for i location, x_i is the observed value and $\tilde{A}^*\left(x_i\right)$ represents the prediction standard error. Naturally, the cross-validation model considered the best one whose MPE is close to zero with the lowest RMSPE. Further, if the average standard error is close to the root-mean-square prediction error, the user is correctly assessing his/her prediction variability. However, if the average standard error is greater or less than the root-mean-square prediction error, the user is over or under estimating variability, respectively (ESRI, 2005).

Yet, it can be hard to judge the influence of variogram parameters on cross-validation errors. If the original dataset is spatially clustered, cross-validation pursues the same trend, leading to a situation where residuals become more representative of certain regions. Moreover, Kriging suffers from the smoothing effect (less spread), i.e., the overestimation and underestimation of small and high values, respectively. Based on the chosen threshold, in practical situations, residuals results can be misleading in the sense of obtaining a higher Alpha, false positive or type I error (the real value is less than the OK estimation) and/or higher Beta, false negative or type II error (the real value is greater than the OK estimation). Thus, a good cross-validation statistical model does not imply a good Kriging model.

The QQ plot comparison layout between the true and the estimations (the closer the slope of this 45° line, the better) or the residual map (myGeoffice© option) to check location bias are common graphic alternatives to cross-validation indexes (myGeoffice© also plots the scattergram of the original samples values against their predictions for outliers detection).

Some comments can be made about Figures 13-16: The total number of false positives and false negatives decreases with threshold increase; Globally, model 3 (nested structures) performs slightly better with a cutoff of one (less Alpha and Beta sum errors) and ten (along with model one); Following this criteria, Model 2 (no nugget-effect) behaves best with the five and twenty thresholds; The spatial layout of false positive errors (green squares) in quadrat III and IV reveals a similar cluster pattern for the two highest cutoffs.

Figure 13. The three cross-validation indexes outcomes computed by myGeoffice© for the present dataset are as follows: MPE (Model 1)=0,28; MPE (Model 2)=0,17; MPE (Model 3)=0,3; RMSPE (Model 1)=5,81; RMSPE (Model 2)=6,46; RMSPE (Model 3)=5,78; AKSE (Model 1)=4,53; AKSE (Model 2)=3,15; AKSE (Model 3)=4,74. Although model 2 presents the smallest mean prediction error, its prediction variability is considered under estimated.

Kriging Cross-validation (Part I)

Sample_id	Coord_x	Coord_y	Sample value	Cross-validation Error (Model 1)	Cross-validation Error (Model 2)	Cross-validation Error (Model 3)	Estimation Variance (Model 1)	Estimation Variance (Model 2)	Estimation Variance (Model 3)
1	39,5	18,5	0,06	,48689980	,21673180	,52683880	21,730080	11,265570	23,732120
2	5,5	1,5	0,06	3,6368090	2,4247730	3,8507890	28,754630	18,979980	30,581600
3	38,5	5,5	0,08	,64623250	,29964260	,69446440	26,138660	15,695770	28,024860
4	20,5	1,5	0,09	,29999550	-,19170320E-01	,35288820	26,374320	17,049120	28,108210
5	27,5	14,5	0,09	,54799940	,20998360	,58853730	22,572210	12,541040	24,443200
136	31,5	41,5	22,75	-9,9983720	-17,380700	-9,4225120	16,535120	5,8875160	18,646890
133	29,5	41,5	58,32	-47,182220	-48,299590	-47,134680	16,045260	4,8818520	18,241320

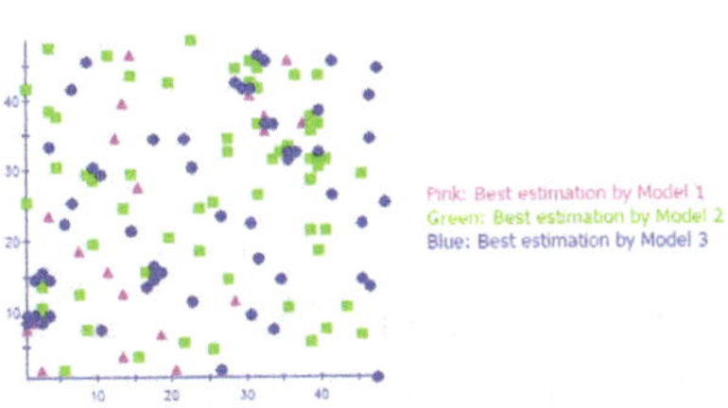

Minimum Sample Value: 0,06 **Maximum Sample Value:** 58,32

Chosen Model: OK Default (Model 1)
OK Default (Model 1)
Threshold Value: OK Without Nugget-effect (Model 2)
OK With Two Nested Structures (Model 3)

Figure 14. Cross-validation of Model 1 for a cutoff of one (True Positive: 86, False Positive: 29, True Negative: 18, False Negative: 7), five (True Positive: 34, False Positive: 22, True Negative: 79, False Negative: 5), ten (True Positive: 7, False Positive: 7, True Negative: 119, False Negative: 7) and twenty (True Positive: 0, False Positive: 3, True Negative: 133, False Negative: 4).

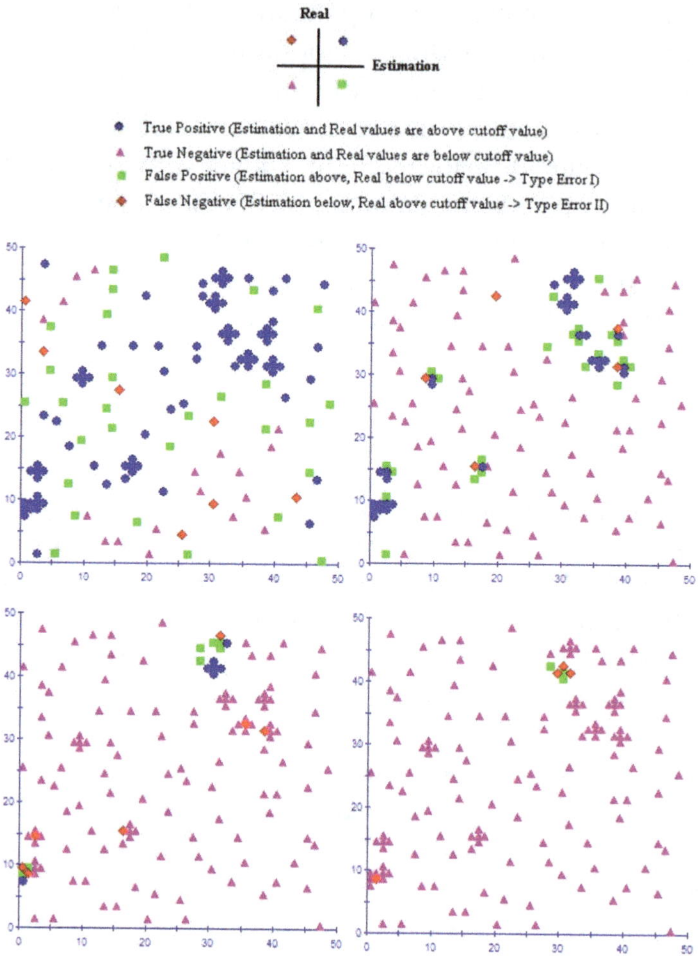

OK Differences

With Model 1, observations are considered precise and accurate although sudden jumps at the variogram's origin may emerge. This means forcing the nugget-effect to zero at zero lag distance. Another possibility is to consider $\gamma(0)=0$ in a continuous mode (Model 2). Concerning the micro-

Figure 15. Cross-validation of Model 2 for a cutoff of one (True Positive: 86, False Positive: 30, True Negative: 17, False Negative: 7), five (True Positive: 32, False Positive: 18, True Negative: 83, False Negative: 7), ten (True Positive: 5, False Positive: 11, True Negative: 115, False Negative: 9) and twenty (True Positive: 1, False Positive: 3, True Negative: 133, False Negative: 3).

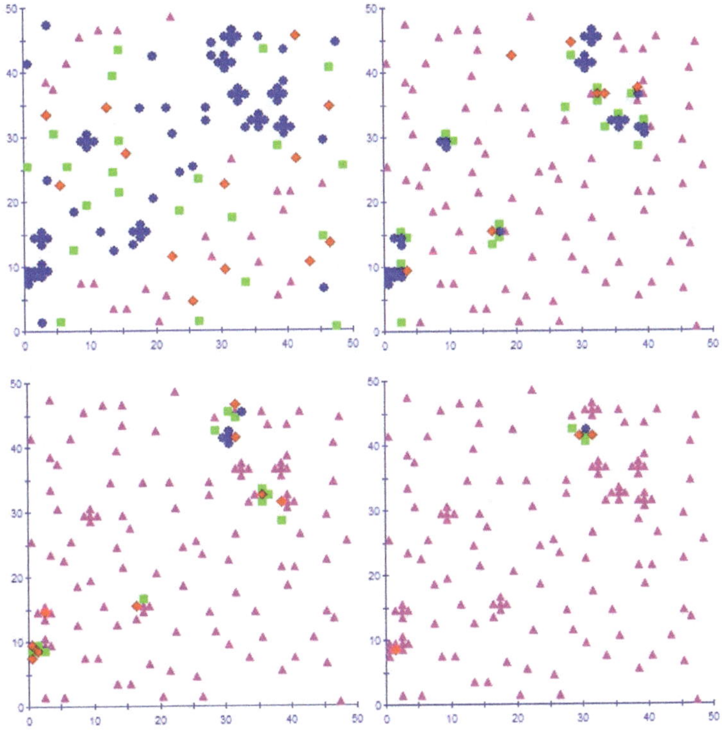

scale component of Model 3, the nugget-effect is divided into two nested factors: Lag distance less than the shortest sampling interval (SSI) and lag distance between the shortest sampling interval and the variogram default range. Therefore, the first micro-scale range equals the SSI lag, its nugget-effect is zero and its total sill matches the extrapolated value of the second variogram structure (i.e., the given value of the second variogram structure at SSI lag distance becomes the first variogram sill parameter). For Cressie (1993), this micro-vision reflects different variability processes on different scales: Large-scale deterministic variation, smooth small-scale variation, micro-scale variation and a zero-mean white-noise process.

Figure 16. Cross-validation of Model 3 for a cutoff of one (True Positive: 80, False Positive: 22, True Negative: 25, False Negative: 13), five (True Positive: 33, False Positive: 22, True Negative: 79, False Negative: 6), ten (True Positive: 7, False Positive: 7, True Negative: 119, False Negative: 7) and twenty (True Positive: 0, False Positive: 3, True Negative: 133, False Negative: 4).

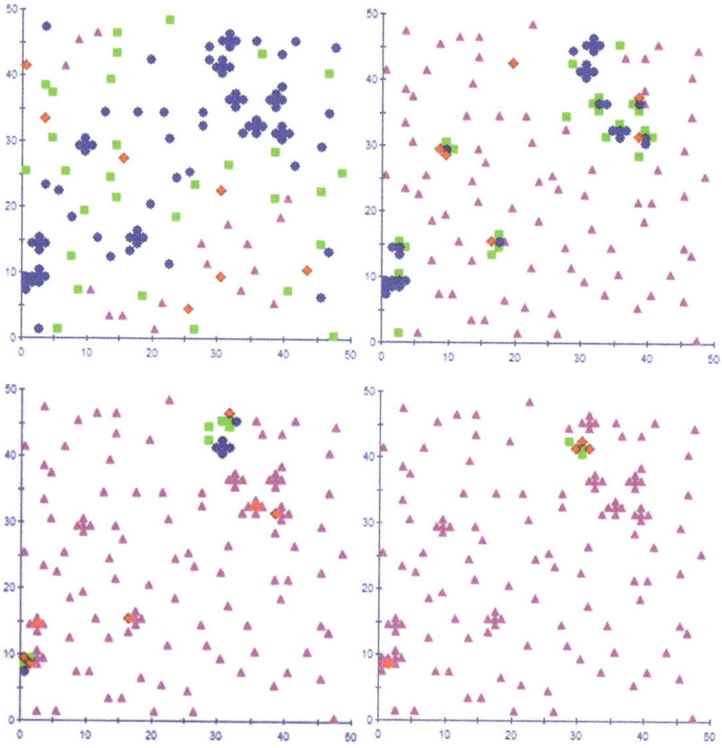

As a rule of thumb, the smaller the difference between the nugget-effect (C_0) and the total sill, the bigger the difference between models 1 and 2. In particular, the greater the nugget-effect, the more likely the estimations are close to the global average (the Kriging estimates will become overly smoothed, an undesirable situation when the aim is to outline areas characterized by high and low values) (see Figure 17).

OK Validation

See Figures 18 and 19.

Figure 17. It is noteworthy that the main disparities among the three nugget-effect models occur in the high sub-regions (the sum difference between these three interpolators are 572.4, -85.9 and -658.3, respectively, which means model 3 globally interpolates higher extreme values than models 1 and 2).

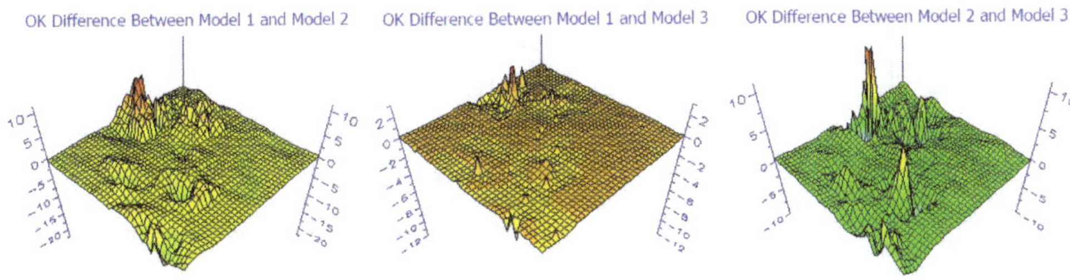

Figure 18. Between the curved parentheses, the user may find the difference between the sample value and the three interpolators

Validation with an Extra Dataset (Ordinary Kriging)

Sample ID	Coord X	Coord Y	Sample Value	OK Default	OK Without Nugget-effect	OK With Two Nested Structures
1	33	33	3,9	4,7970310 (-0,897031)	2,3785270 (1,521473)	5,0194530 (-1,119453)
2	44	44	3,1	2,2017520 (0,898248)	2,0467640 (1,053236)	2,2267360 (0,873264)
3	11	11	0,4	,61503910 (-0,2150391)	,18729130 (0,2127087)	,70649680 (-0,3064968)
4	22	22	1,5	2,2287560 (-0,728756)	1,9584130 (-0,458413)	2,2831150 (-0,783115)
5	22	11	1,6	1,5982590 (1,74099999999999E-03)	,13274740 (1,4672526)	1,8023540 (-0,202354)
6	44	11	1,9	1,7610020 (0,138998)	2,3720920 (-0,472092)	1,8960140 (3,98599999999982E-03)
7	44	33	1,8	2,7059850 (-0,905985)	,92288180 (0,8771182)	2,9569960 (-1,156996)
8	11	25	4,1	3,2244790 (0,875521)	3,2292990 (0,870701)	3,2702310 (0,829769)
9	33	44	10,5	11,597990 (-1,09799)	11,328460 (-0,82846)	11,573080 (-1,07308)
10	20	20	3,6	2,6142300 (0,98577)	2,1204510 (1,479549)	2,7610590 (0,838941)

OK With Measurement Error

If measurement error is given, the C_{me} factor should be included in the variogram of the $B(x_i,x_0)$ matrix (see Figure 20). Computationally, the B Kriging matrix follows the same rules unless the estimation site coincides with the samples locations, leading to a non-exact version of the Kriging interpolator. The measurement error must be setup (instead of zero as in the exact version) leading to a smoother prediction of the Kriging estimations. It should be perceived that all other OK estimations are the same whether

Figure 19. Globally, model 1 (OK Default) works best

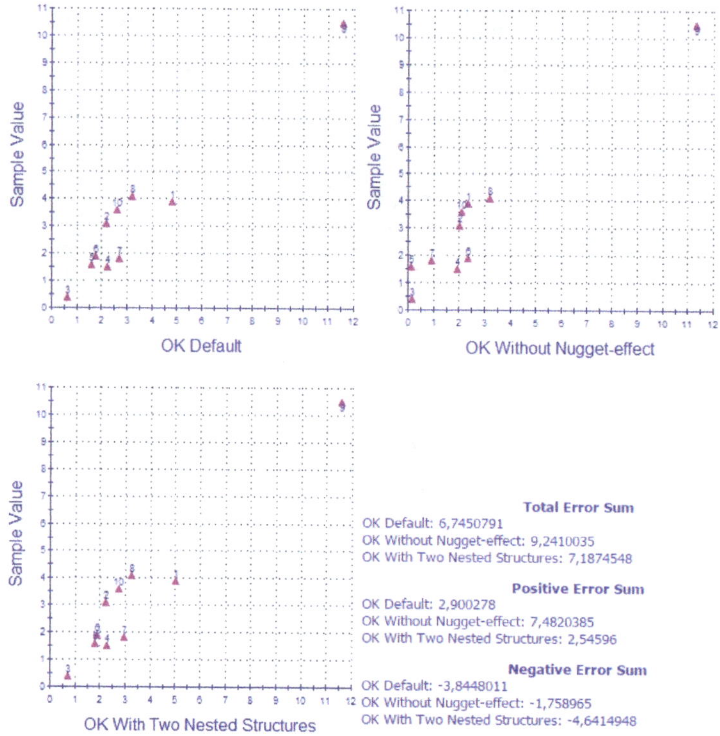

measurement error is considered or not. Therefore, instead of the final OK map being laid out, only the available samples are re-estimated in myGeoffice©.

KRIGING WITH TREND

One of the Kriging assumptions is that the data being estimated is stationary. This means that as the interpolator moves from one region to the next one, the average value of the dispersed sites should be relatively constant. However, this assumption is difficult to make with many spatial datasets. Further, when there is a significant spatial trend in the data such as a sloping surface or a localized flat region (often due to a prior known physical cause), this assumption is also violated.

In such cases, the stationary condition can be temporarily imposed on the data by using a drift term. This drift is a coordinate polynomial function that models the average value of the scatter sites, where residuals become the

Figure 20. If the variogram model does not have a nugget-effect (C_0=0), this option will produce the noiseless exact version

Ordinary Kriging with Measurement Error

Sample ID	X Coordinate	Y Coordinate	Sample Value	Estimation with 12% of measurement error
1	39,5	18,5	0,06	0,040467
2	5,5	1,5	0,06	0,278111
3	38,5	5,5	0,08	-0,021242
4	20,5	1,5	0,09	0,045666
5	27,5	14,5	0,09	0,131
6	40,5	21,5	0,1	0,099248
7	15,5	3,5	0,1	0,033616
8	6,5	25,5	0,11	0,268699
112	1,5	8,5	20,35	19,299496
135	30,5	42,5	21,08	20,760353
136	31,5	41,5	22,75	22,122744
133	29,5	41,5	58,32	54,352294

difference between the drift (myGeoffice© only covers the first and second order) and the real samples (see Figure 21). If this polynomial deterministic model fits the original spatial data well, the residuals should be stationary and Kriging can be performed on them (instead of the original samples). After all, if Kriging interpolated residuals are computed for the whole region, they will be added back to the polynomial model in order to compute the final spatial estimations. The use of a deterministic drift model before Kriging is often called Universal Kriging (UK) or Kriging with Trend (KT) since the residual component is usually modeled as a stationary random function with zero mean and C(h) covariance. According to Clark and Harper (2000), Kriging with Trend generally requires an increase in the search radius (unless the user is using a dense dataset) so as to obtain enough equations to satisfy all of the conditions.

KT Geocomputation

See Figure 22.

Figure 21. After the residuals have been assessed (based on the polynomial deterministic trend), Kriging analysis is performed on those residual terms

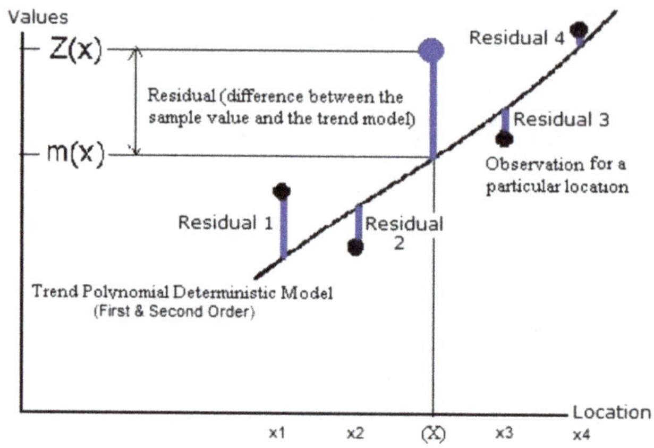

KT Estimation

See Figures 23 and 24.

KT Cross-Validation

Like Ordinary Kriging (OK), cross-validation (a technique for performance estimation of a predictive model) may also be applied with Kriging with Trend (see Figures 25, 26 and Table 1). As stated by Clark and Harper (2000), if there is little correlation between actual and predicted values, Kriging will not assist in filling in our knowledge gaps. Further, the scatterplot of the real samples' values against each estimation (the positive, negative and total sum) is also computed by myGeoffice© for possible comparison among the different models, including the previous OK versions. Theoretically, the aim is to achieve a small average error close to zero with a constant variance behavior.

KT Validation

See Figure 27.

KT vs. OK

See Figure 28.

Figure 22. The nine drift possibilities (top) offered by myGeoffice© towards Universal Kriging

Figure 23. Interpolation maps using the following nine trend models (from top to bottom): Linear drift in X, Linear drift in Y, Linear drift in X and Y, Quadratic drift in X, Quadratic drift in Y, Quadratic drift in X and Y, Linear drift in X plus quadratic in X and Y, Linear drift in X and Y plus cross-quadratic drift in XY and Linear drift in X and Y plus quadratic in X and Y plus cross-quadratic drift in XY.

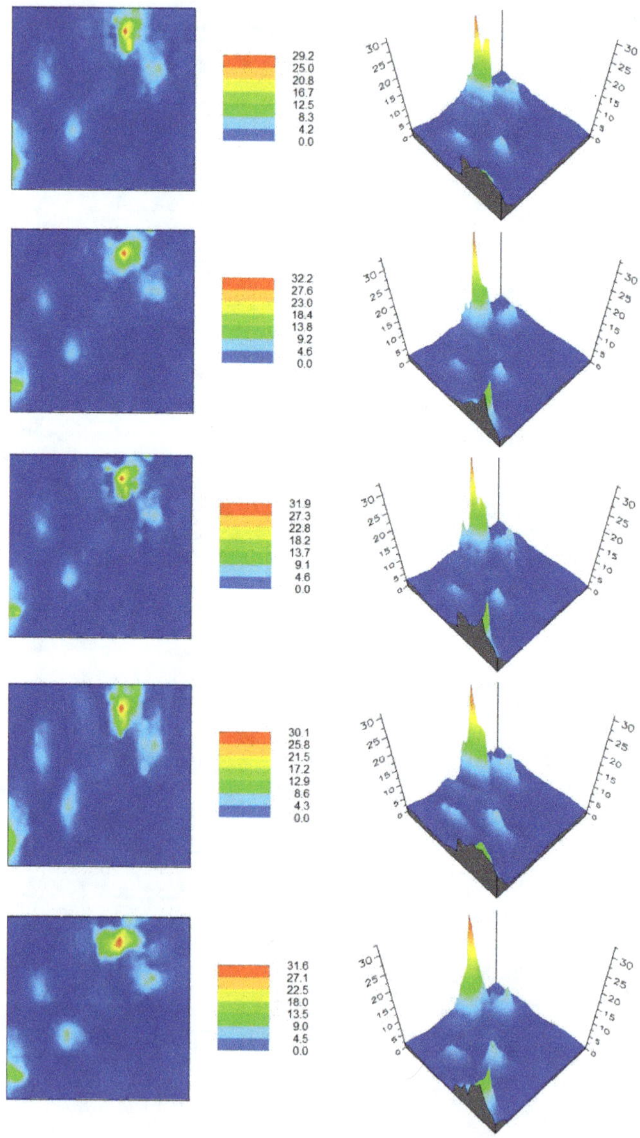

Figure 24. Interpolation maps using the following nine trend models (from top to bottom): Linear drift in X, Linear drift in Y, Linear drift in X and Y, Quadratic drift in X, Quadratic drift in Y, Quadratic drift in X and Y, Linear drift in X plus quadratic in X and Y, Linear drift in X and Y plus cross-quadratic drift in XY and Linear drift in X and Y plus quadratic in X and Y plus cross-quadratic drift in XY.

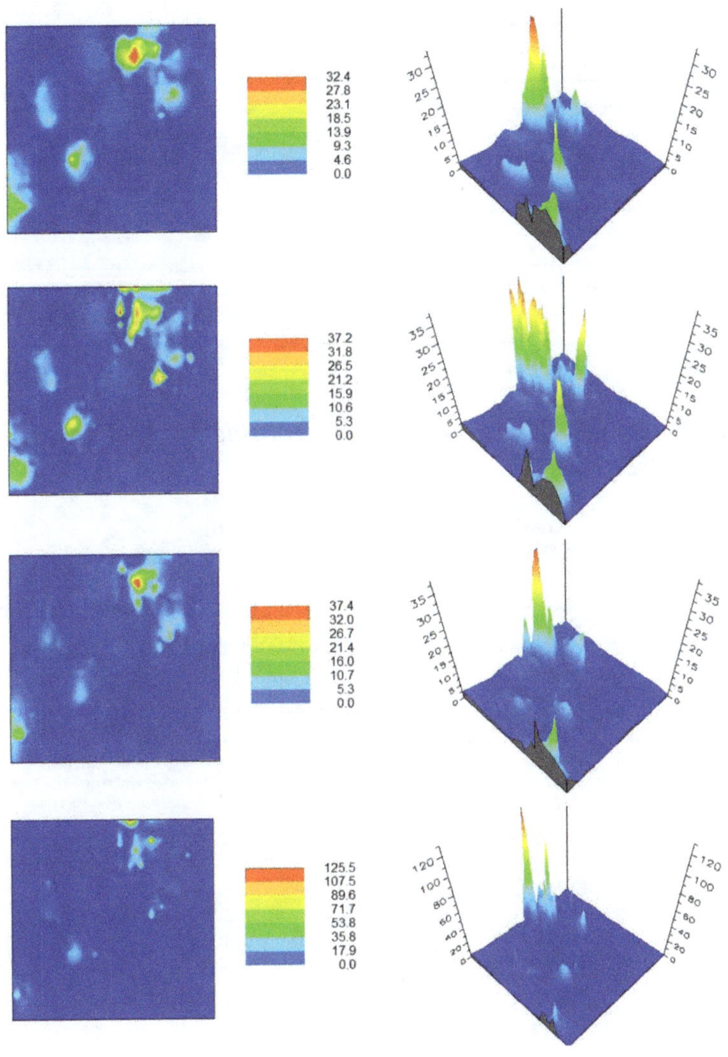

Figure 25. The cross-validation errors using the linear drift in X by myGeoffice©. Keep in mind that the 'Cross-validation Error' column equals the difference between the estimation and the real sample. For reference, the Mean Prediction Error, the Root Mean Square Prediction Error and the Average Kriging Standard Error results are 0.25, 6.18 and 4.61, respectively.

Cross-validation of Kriging with Trend (Part I)

Sample_id	Coord_x	Coord_y	Sample value	Cross-validation Error	Estimation Variance
1	39,5	18,5	0,06	,51241140	21,747310
2	5,5	1,5	0,06	2,2393250	29,853470
3	38,5	5,5	0,08	,64334550	26,142850
4	20,5	1,5	0,09	,29602750	26,384200
125	16,5	15,5	19,44	-15,834730	16,291050
112	1,5	8,5	20,35	-11,393480	14,514140
135	30,5	42,5	21,08	-2,4635300	14,715030
136	31,5	41,5	22,75	-16,862330	17,346900
133	29,5	41,5	58,32	-48,257090	16,367990

Figure 26. Like with OK cross-validation, the considered cutoff of 2 was used for illustration purposes

Cross-validation of Kriging with Trend (Part II)

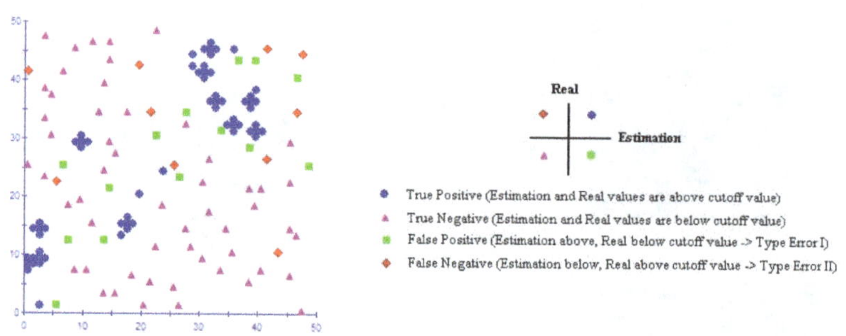

Number of Samples
True Positive: 62
False Positive: 14
True Negative: 54
False Negative: 10

Indicator Kriging

Indicator Kriging is a geostatistical model of probabilistic mapping and is based on the local conditional cumulative distribution function (ccdf) whose aim is to evaluate the probability of any variable at any x location being lower/higher than a particular cutoff. The essence of this indicator approach is binomial

Table 1. According to these cross-validation results, it appears that KT with a linear drift in Y presents the best spatial interpolation for the present dataset

Kriging Model	Positive Sum Error	Negative Sum Error	Total Sum Error
Linear drift in X	245	209	454
Linear drift in Y	221	204	426
Linear drift in X and Y	240	227	468
Quadratic drift in X	281	201	483
Quadratic drift in Y	263	196	459
Quadratic drift in X and Y	319	218	538
Linear drift in X and Y plus quadratic drift in X and Y	360	328	688
Linear drift in X and Y plus quadratic drift in XY	305	283	589
Linear drift in X and Y plus quadratic drift in X and Y plus cross-quadratic drift in XY	489	767	1256

Figure 27. For reference, the values between the curved parentheses of the above table equal the difference between the sample value of the extra dataset and the estimated (X drift) Universal Kriging version. Thus, it is possible to build a table comparing the different order of the KT versions for the same extra dataset.

Validation with an Extra Dataset (KT)

Sample ID	Coord X	Coord Y	Sample Value	Kriging with Trend
1	33	33	3,9	3,6618460 (0,238154)
2	44	44	3,1	2,2073810 (0,892619)
3	11	11	0,4	,57255910 (-0,1725591)
4	22	22	1,5	2,4845320 (-0,984532)
5	22	11	1,6	1,4936870 (0,106313)
6	44	11	1,9	1,7485540 (0,151446)
7	44	33	1,8	2,5894270 (-0,789427)
8	11	25	4,1	3,1284220 (0,971578)
9	33	44	10,5	12,968850 (-2,46885)
10	20	20	3,6	3,2622240 (0,337776)

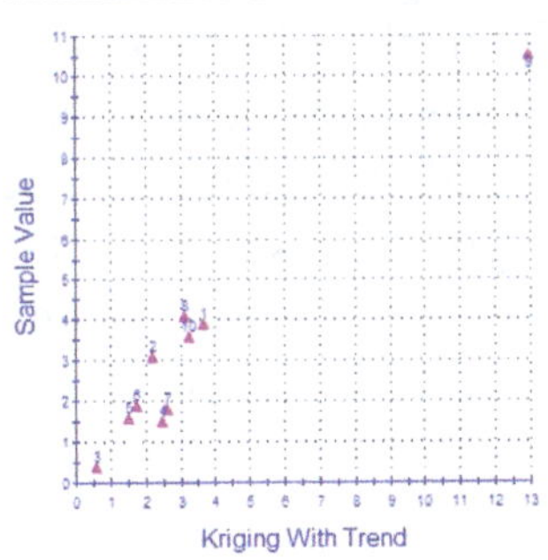

Total Error Sum: 7,1132541
Positive Error Sum: 2,697886
Negative Error Sum: -4,4153681

Figure 28. Notice that the different concepts considered in these three maps equals OK minus KT estimations (Y drift). Positive peaks illustrate that OK versions estimate higher values than KT. The computations in the bottom-right refer to the average, maximum and minimum estimations for each four interpolators.

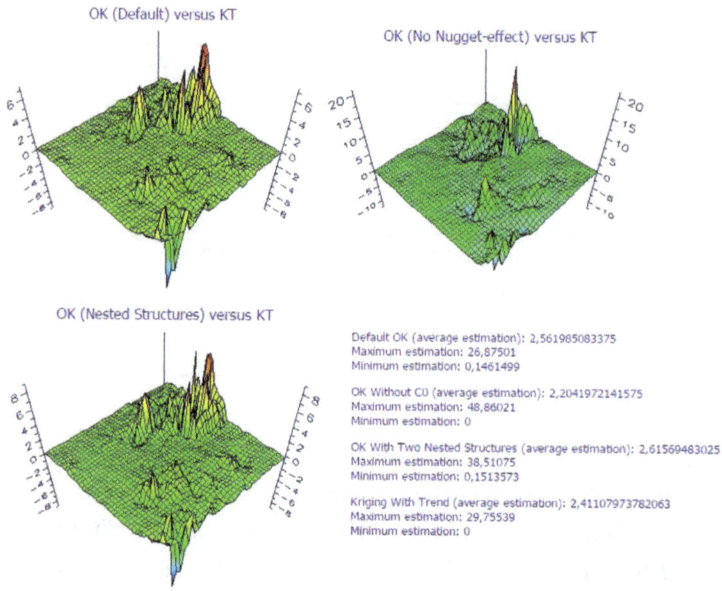

data coding into either 1s or 0s, depending on the variable's relationship to a particular threshold, z_k. For a given z(x) observation value, the indicator variable I(x) of myGeoffice© is decoded as follows: 1, if $z(x)<z_k$, and 0, otherwise. This is a non-linear transformation of the input data into binary form. Values, which are much greater than a given cutoff, z_k, will receive the same indicator value as those values, which are only slightly greater. Theoretically, two advantages of using IK are the following: 1) It provides an estimate of the local cumulative distribution function rather than a single estimate of the attribute itself; 2) It copes with skewed data distributions and outliers since this indicator approach transforms the effects of these outliers into 0s and 1s. Regarding the latter advantage, the indicator transformation of data is thus an effective way of limiting outliers' influence. Based on a set of transformed values, Indicator Kriging gives a result between zero and one for each estimated point, the probability of being greater or smaller than the considered threshold. To solve this probability issue, the following steps must be performed:

- Transform the original data into indicators by using a pre-selected cutoff (z_k).
- Set up the estimated indicator variogram, $\gamma(h,z_k)$.
- Use the indicator value and the indicator variogram of the previous steps in an OK system to solve the indicators' weights at each location of the study site. The obtained indicator predictions should be interpreted as the probability of exceeding that particular threshold (or not, depending on the coding system used).
- Repeat the first three steps for several different z_k thresholds to provide discretized estimates of the local ccdf for each site. Nevertheless, some smoothing or adjustment of the estimated ccdf's may still be needed to preserve the relationship order required for a valid distribution function. IK does not always guarantee, for instance, the following relationship order: $I_{Q2}(x)<=I_{Q3}(x)$, where Q_2 and Q_3 represent the second and third quartile. In order to resolve this matter, the simple average between each pair of values should be computed, e.g., according to the following rule: If $I_{Q1}(x) < I_{Q2}(x)$ and $Q_1>Q_2$ then $I_{Q1}(x) = I_{Q2}(x) = \frac{I_{Q1}(x) + I_{Q2}(x)}{2}$.

Under this framework, if several N thresholds (Z_K, for k=1,2,…N) are set up, then N indicator variograms and N IK systems are needed (heavy computation is expected). A good starting point for selecting the Z_k cutoffs is to use the 0.10, 0.25, 0.50, 0.75 and 0.90 quantile values. Afterwards, the user might add some other cutoffs to achieve a good representation of the distribution through a set of distinct indicator variograms.

Median IK (MIK) allows simplifying this multiple variography needed to build each ccdf by using the common median variogram for all thresholds (myGeoffice© approach). As expected, MIK is a quite fast procedure because it requires only one variogram (of the median) and a single IK system to be solved at each location. However, according to Goovaerts (1997), this solution does not allow different shapes or anisotropy patterns for the several indicator variograms.

Knowledge of this ccdf for any particular location allows a direct way to measure uncertainty about the unknown. The probability that any location is between 15 and 29 units, for instance, can be assessed as the probability difference between the ccdf values for those thresholds (see Figure 29). Hence, it is perfectly reasonable that R location holds a probability of 70%

Figure 29. For illustrative purposes, this example reveals the ccdf for a particular location based on seven thresholds. The probability of that location being greater than 90 units is 50%, while the probability of being less than 2 units is close to zero.

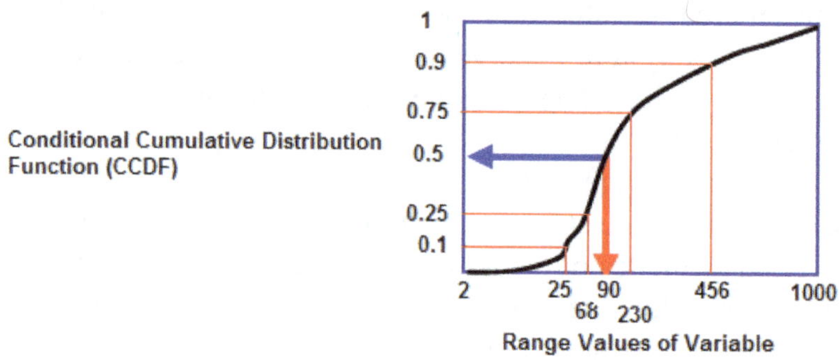

of being between 15 and 29 while T location holds a probability of 46% for the same range.

The following nine sub-sections covers myGeoffice©'s general perspective towards uncertainty and Indicator Kriging (IK), in particular. The two IK thresholds setups (including their variography) are covered in the first two sub-sections. Thereafter, both resulting interpolations are presented while sub-section CCDF Model illustrates the capability of myGeoffice© to compute the ccdf for any particular sample of the dataset in a graphical mode. The entropy of Shannon and E-type estimations (including health and remediation costs) is discussed in sub-sections "Entropy of Shannon" and "E-Type Estimation", respectively. Morphologic geostatistics and probability intervals conclude this chapter.

IK First Cutoff

Any continuous variable may be converted into a binary form (zeros and ones) by choosing some cutoff. Under myGeoffice©, if sample values are below the chosen threshold, they are decoded as one. Otherwise, they become zero. As previously stated, this procedure leads to interpolations between zero and one that can be interpreted as probabilities exceeding (or being below) the chosen threshold. In other words, instead of assuming a Normal distribution at each estimated location, Indicator Kriging (IK) builds the ccdf at each point based on the behavior and correlation structure of the indicator by transforming data near their neighbors.

To achieve this aim, IK requires a series of thresholds (two under myGeoffice©) within the range of the dataset (see Figure 30). The median threshold is then used to numerically build the ccdf of the estimation point. Given its neighborhood of transformed data and the chosen IK cutoff correlation structure, IK estimates the probability of the estimation being lower than any threshold. After the ccdf has been built, it may be used to create probability and risk maps. As expected, a spatial correlation model (variogram) must be available for each of the IK cutoffs (see Figure 31). The definition of a threshold becomes, hence, the present step (under myGeoffice©, this cutoff varies between 120% of the minimum observation value and 80% of the maximum one to avoid samples lacking the required variography).

IK Second Cutoff

See Figures 32 and 33.

Figure 30. Pink triangles and green squares represent samples which are below and above the chosen cutoff, respectively

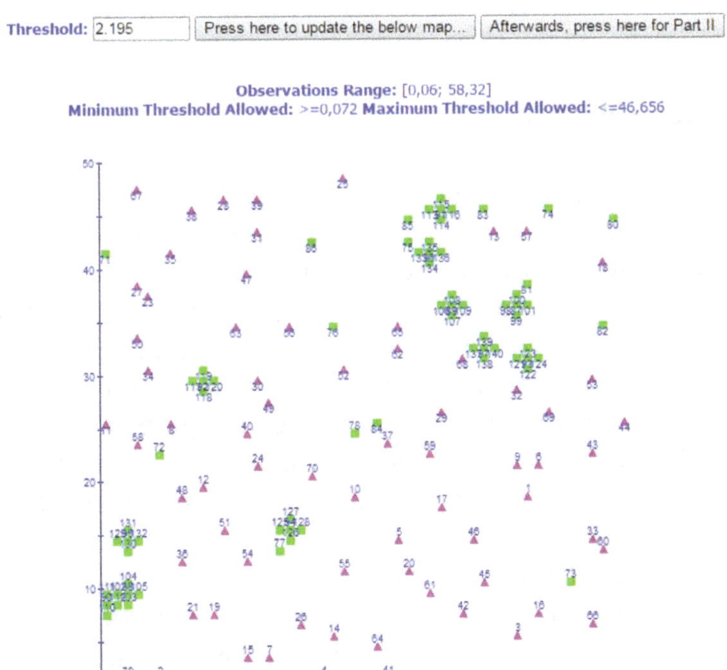

Figure 31. Variography for the first cutoff. It is vital to emphasize that the Entropy of Shannon and Sequential Simulation/Conditional Indicator option of myGeoffice© requires that one of the chosen cutoffs equals the median value (2.195 for this particular dataset).

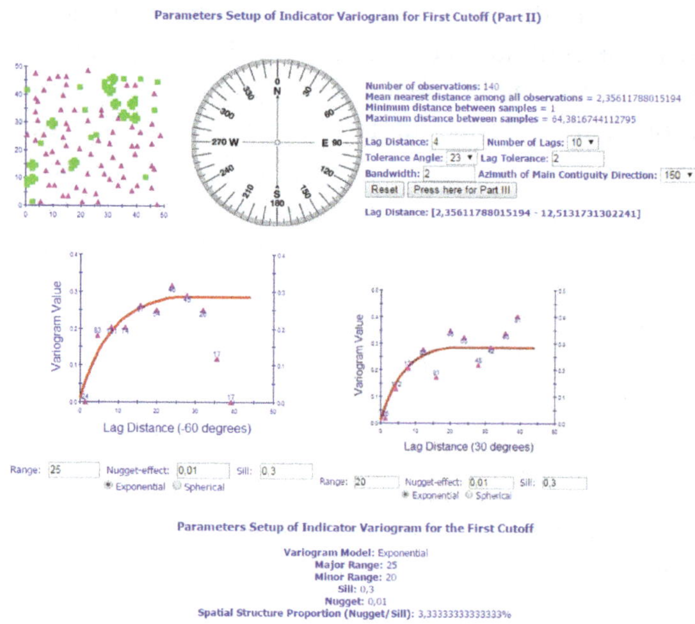

IK First Interpolation

The result of the interpolation map of IK reveals the probability of a variable exceeding or being below the chosen K threshold by transforming the continuous variable into an indicator (binary) one. It assumes an unknown mean while the phenomenon is considered continuous in space. This non-linear and non-parametric approach takes into account the structure of each indicator, produces an estimated variance, does not require prior modeling of the theoretical distribution and does not require global stationarity (only local). Statistically, IK provides a least square estimate of the ccdf at any cutoff. Thus, the ccdf for any location can be built by employing the Indicator Kriging system for several cutoffs.

Regarding the drawbacks of this non-parametric method, five issues should be highlighted: (1) There is a loss of information as it does not distinguish among observations if they are both below or above the threshold; (2) Many variograms have to be set up (according to the levels considered); (3) It is

Figure 32. It is important to stress that this second cutoff should be greater than the first one if the user intends to run the Probabilistic Interval option (see sub-section "Probabilistic Interval")

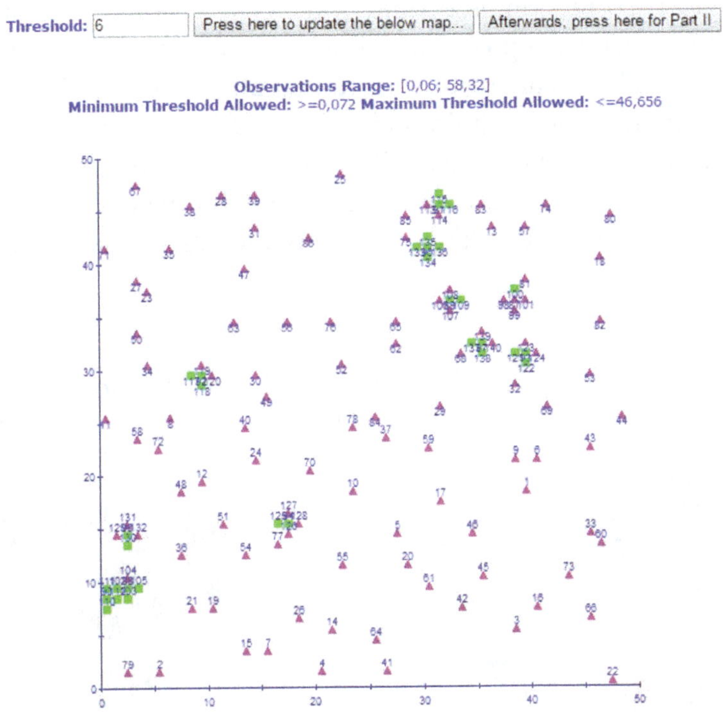

Figure 33. Based on a lag distance of 3 units (lag tolerance=bandwidth=1.5), the indicator exponential variogram for the second cutoff is as follows: Major range=10, minor range=6, sill=0.2 and nugget-effect=0.1

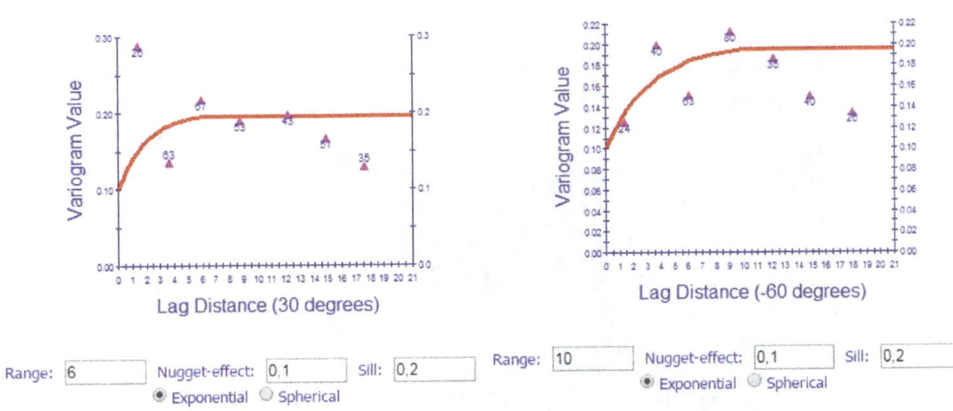

possible to get estimates greater than 1 and below 0; (4) With extreme values, the variogram may correspond to a pure nugget-effect; (5) Concerning a series of several cutoffs, order relation correction must be taken into account with a posteriori correction procedure of the ccdf (a non-decreasing one).

This ability to assess a probability risk map is thus the main function of IK (see Figures 34, 35 and 36). Observe that all negative estimations were setup to zero (particularly within regions with a lack of samples) or one if IK estimations were greater than 100%. Based on this methodology, the user may also set up the lowest and highest probability confidence levels for the chosen threshold by calculating the spatial locations that fall within any specific range.

Figure 34. The below layout should be interpreted as the estimation probability of being less than or equal to the first chosen threshold (2.195 in this particular case)

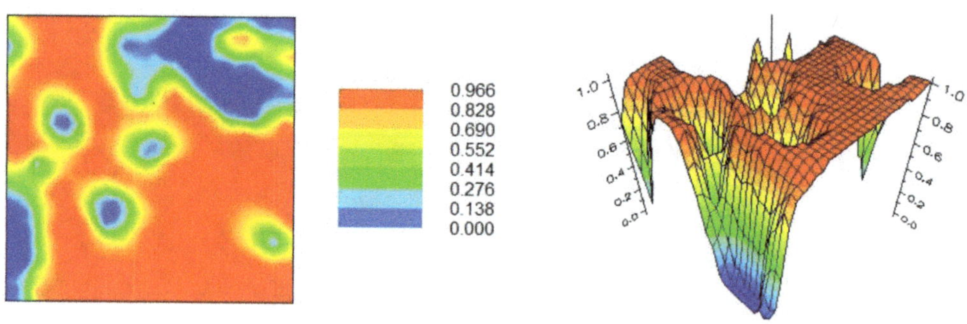

Figure 35. In a complementary view, this layout should be interpreted as the estimation probability of being greater for the same previous cutoff

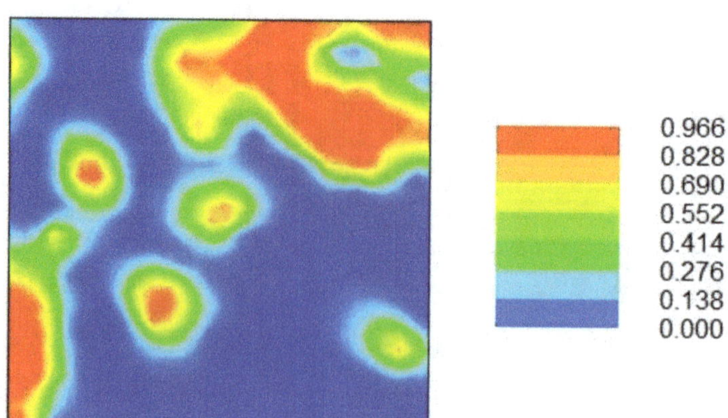

Figure 36. This map shows the probability of any spatial location being less than 2.195 is between 40% and 80%

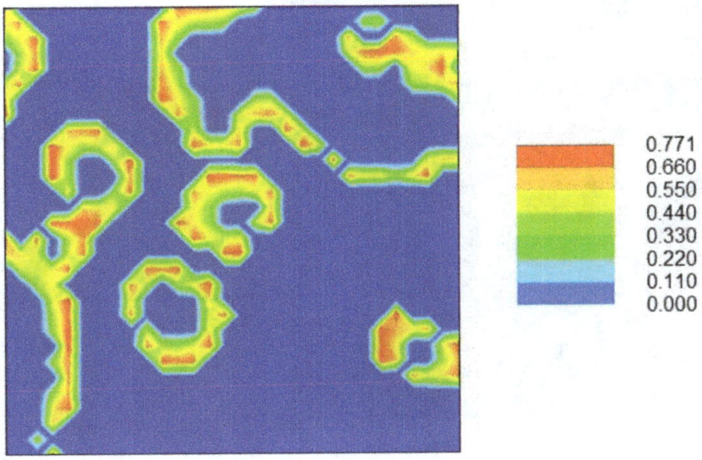

IK Second Interpolation

See Figures 37, 38 and 39.

CCDF Model

Indicator Kriging (IK) provides the ccdf at different known locations to assess uncertainty about any unknown location. The aim of this option is to plot the probability model for different observations using MIK (Median Indicator Kriging).

Figure 37. The below layout should be interpreted as the estimation probability of being less than or equal to the second chosen threshold (6 in this particular case)

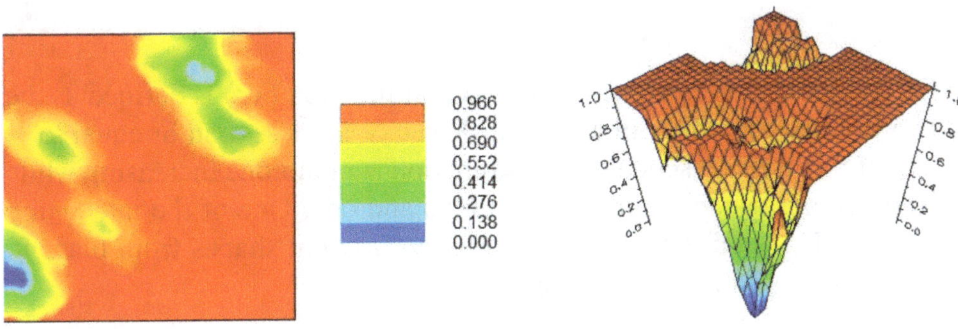

Figure 38. This layout equals the estimation probability of being greater than 6

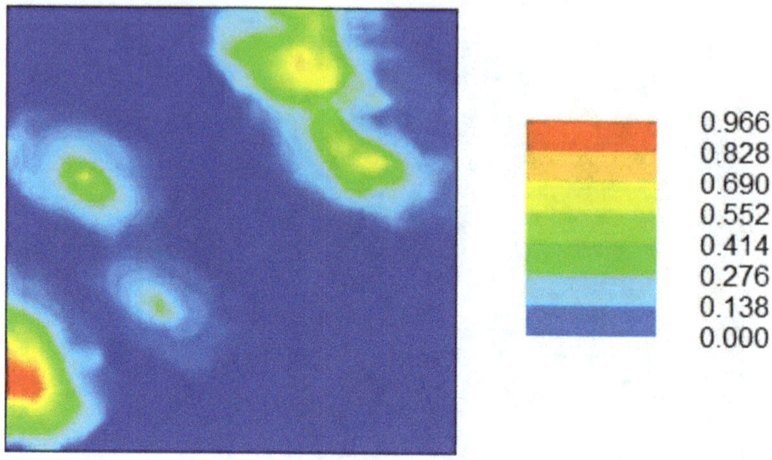

Figure 39. The below map illustrates the probability of any spatial location being less than 6 is between 60% and 95%

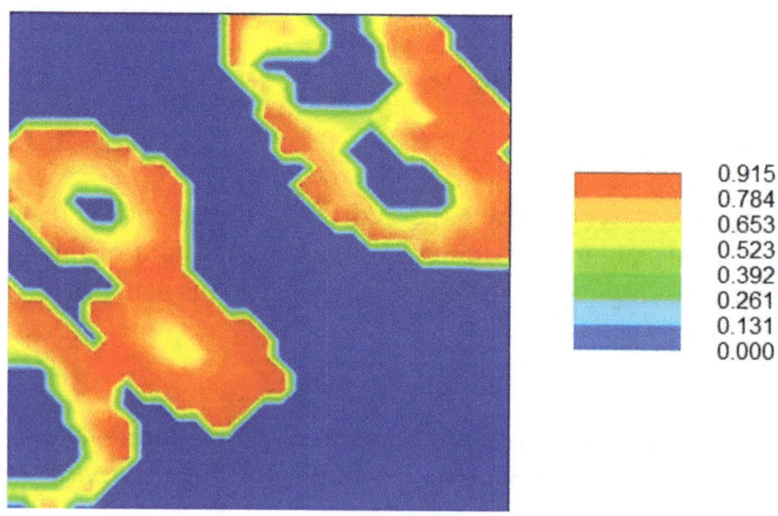

For reference, the cumulative distribution function F(x) is defined as the probability that all values in a random X vector are less than or equal to x, i.e., Prob(X=<x). Consequently, this function is monotone increasing while the vertical axis range varies between zero and one. One of the interesting features of IK is that the ccdf can take very different shapes from place to place (working as a local uncertainty measure).

Keep in mind that the nine chosen cutoffs divide the present dataset into ten non-overlapping class ranges whose sum of probabilities equals 100%. Therefore, it is possible to compute the probability of the final estimate being within any specific interval. For instance, the estimation probability of this observed site varying between 0.92 and 7.53 (which corresponds to the difference of both cutoffs) is close to 66%. Similarly, the probability of the site being higher than 9.42 is close to 30% (see Figures 40-43).

Entropy of Shannon

Indicator Kriging (IK) allows for the possibility of estimating the ccdf for any particular spatial location. By defining L cutoffs (nine, under myGeoffice©, which means ten non-overlapping classes), it is possible to deduce the probability of uncertainty for each class (differences between the ccdf of any adjacent cutoffs).

Using this information, the Shannon entropy (a disorder measure connected to the spatial organization of an attribute) provides inferences about the interpolation local uncertainty of unknown sites (see Figure 44). Given L sets of IK probabilities for any u location ($P_i(u)$, where i=1...L), the Shannon entropy equals $-\Sigma(P_i(u) \times LN(P_i(u)))$, where LN() denotes the Neperian logarithm,

Figure 40. By choosing one of the Sample IDs, the researcher may generate a graphical cumulative distribution function based on the same MIK with nine consecutive cutoffs

Sample ID	Coordinate X	Coordinate Y	Sample Value
1	39,5	18,5	0,06
2	5,5	1,5	0,06
3	38,5	5,5	0,08
4	20,5	1,5	0,09
5	27,5	14,5	0,09
...			
135	30,5	42,5	21,08
136	31,5	41,5	22,75
133	29,5	41,5	58,32

Choose a 'Sample ID' and press here

Figure 41. The below graph confirms this perspective by using nine cutoffs and the same MIK variogram (a practical shortcut to avoid each IK variography for each of the nine cutoffs) for the following case: Sample ID - 116; Coord X - 32,5; Coord Y - 45,5; Value - 10,2

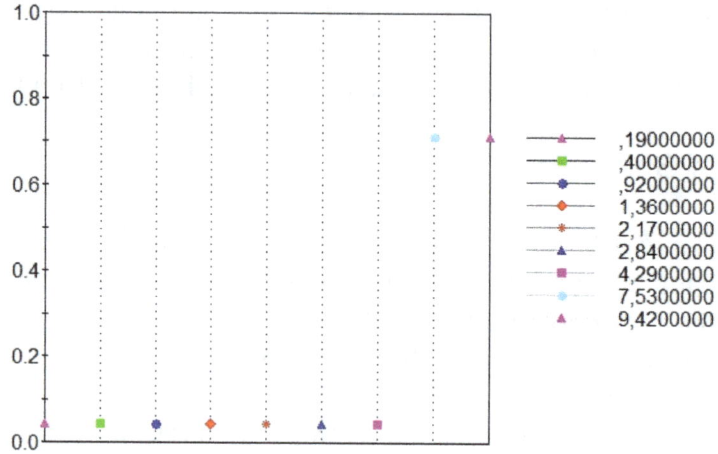

Figure 42. CCDF of sample 30: (a) The estimation probability of this sample varying between 0.92 and 7.53 is close to 97%; (b) The probability of being higher than 2.84 is close to 4%

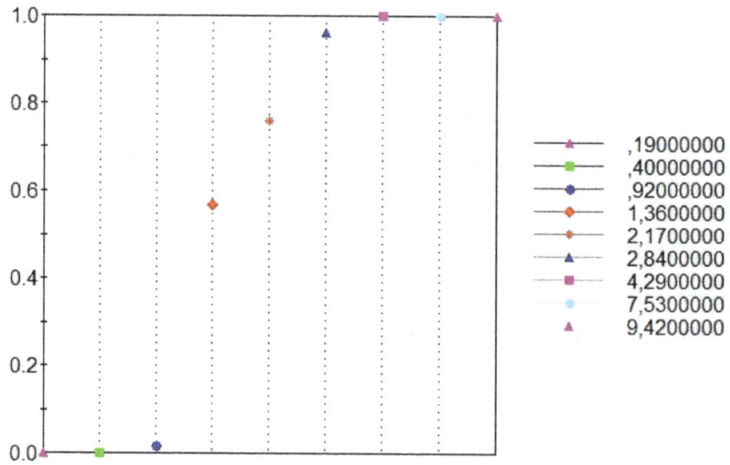

Figure 43. CCDF of sample 60: (a) The estimation probability of a variation between 0.92 and 7.53 is close to 23%; (b) The probability of the variation being higher than 0.92 is close to 75%

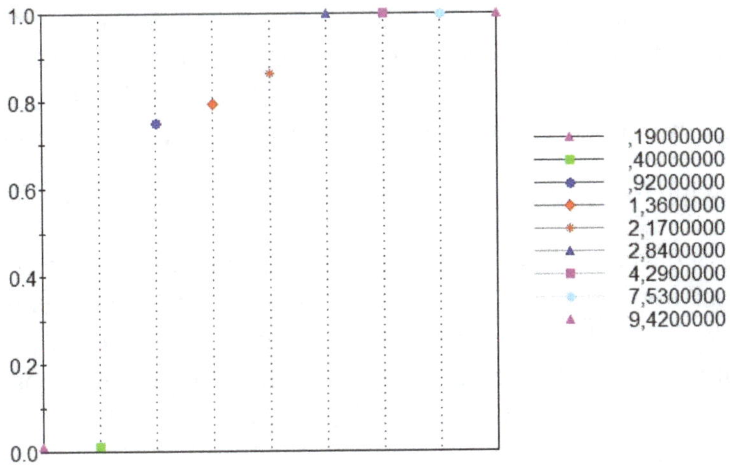

Figure 44. Ordinary Kriging variance is data-independent whereas this local uncertainty measurement is data-dependent

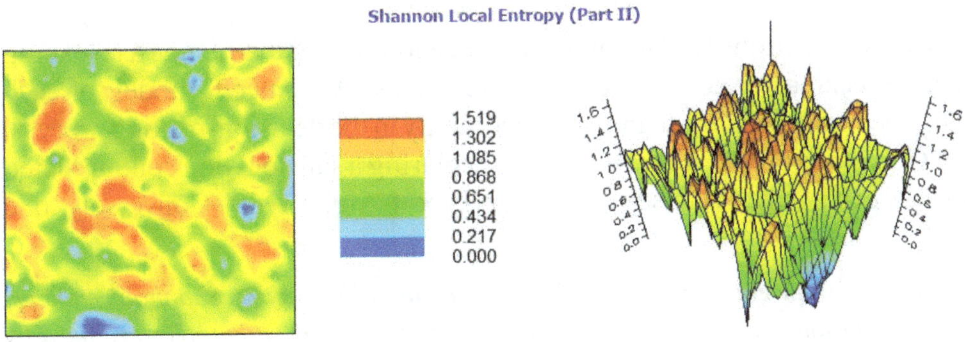

$P_i()$ represents the IK probability estimation for each class and $-\Sigma()$ is the negative sum of all probability classes. As expected, the red color mapping signifies a high lack of estimation confidence (low trust), which means the researcher should consider further sampling measurements for these sub-areas. Dark blue denotes low instability (high certainty) for spatial estimation (the smaller the entropy at any location, the greater the certainty associated with that location). Bear in mind, that under myGeoffice©, the Shannon entropy is not standardized and the MIK (the indicator variogram of the median cutoff value) is used for the nine thresholds due to geocomputation simplicity.

E-Type Estimation

E-type is considered an optimal estimator under the least square criterion whose estimation depends on data values. According to Goovaerts (1997), the advantage of E-type estimation lies in the availability of an uncertainty model, which is much more informative than the Kriging variance. Mathematically, the expected value of the ccdf at any x0 location is based on the following formula: $z^{*}_{(x0)} = \sum_{k=1}^{k+1} z^{'}_{k} \times \left[F^{*}_{z}\left(z_{k}, x0\right) - F^{*}_{z}\left(z_{k-1}, x0\right)\right]$, where $z^{'}_{k}$ is the mean of class $[z_{k-1}, z_{k}]$, which depends on the within-class interpolation model (under myGeoffice©, it is the average of both thresholds that limits each class), z_k (k=1…k) represents the k consecutive thresholds that discretizes the range of variation of all z-values (consistent with previous options, nine cutoffs were considered within myGeoffice©) and $z^{*}_{(x0)}$ stands for the expected value of the ccdf. $F^{*}_{z}\left(z_{k}, x0\right)$ and $F^{*}_{z}\left(z_{k-1}, x0\right)$ equals the probability at x0 location of being less than the z_k and z_{k-1} cutoff, respectively.

Misclassification Risks

Many investigations lead to important decisions being made such as cleaning hazardous areas or correcting soil deficiencies. With a ccdf model for each site, it becomes possible to include uncertainty in the decision-making process. For Goovaerts (1997), there is a risk of declaring a safe location contaminated for a specific cutoff (false positive or Type I Alpha risk). Conversely, one

Figure 45. Analogous to the 'CCDF Model' and 'Entropy of Shannon' choices, the MIK parameters (the indicator variogram of the median cutoff) were used by myGeoffice© to calculate these E-type maps

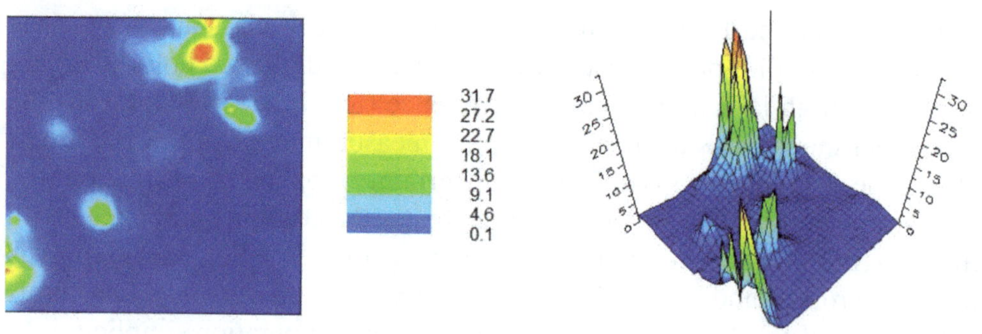

might declare a contaminated location safe (false negative or Type II Beta risk). These two misclassification risks can be assessed using the ccdf model with the following guidelines: Risk Alpha = Prob(z(x0)<=Cutoff | $z^*_{(x0)}$ >Cutoff) and Risk Beta = Prob(z(x0)>Cutoff | $z^*_{(x0)}$ <=Cutoff). Thus, for any particular x0 location, the size of misclassification risk depends on the ccdf model, i.e., this type of misclassification risk depends on the E-type estimation previously assessed.

According to the previous Alpha and Beta statistical risk definitions and based on Figure 46 and Figure 47, some inferences can be stated. Although the red areas of the Alpha risk map illustrate contaminated regions (E-type estimation is greater than the given 2.195 cutoff), there is a strong probability of this being incorrect (Alpha risk). The yellow and green spatial patterns indicate some suspicions about our hazardous classification (30%-70%). As expected, these locations are larger between high and low transition sub-areas (close to the borders of the cutoff regions). On the contrary, the dark blue color may indicate two notations: (1) The E-type estimation is smaller or equal to the chosen cutoff (myGeoffice© sets up a zero value for these locations and, therefore, these sites should not be analyzed under this risk context); (2) The associated probability of being higher than 2.195 is close to zero.

For the Beta risk map, this same dark blue color indicates the sub-regions whose E-type estimation is greater than the chosen cutoff or where the associated probability is lower than 2.195 and close to zero. However, a red pattern indicates that those sites were classified as clean (E-type estimation is smaller than the given cutoff) but with a major certainty (80%-100%) of them being considered hazardous (Beta risk). Yellow and green colors represent clean sections with some probability of misclassification (curiously, the Beta risk is not defined whereas the Alpha risk is).

Looking at the classification map below, the dark blue color represents regions below or equal 2.195 (where Beta risk interacts) while the strong red reveals areas considered above this same threshold (Alpha risk).

Economic Classification

Another uncertainty approach consists of evaluating the economic impact of two possible decisions by using the concept of an economical loss function. Consider, for example, the estimation of a toxic concentration. Underestimation of that concentration (negative estimation error) may cause ill health and even death, leading to insurance claims and lawsuits. Conversely, overestimation

Figure 46. Alpha risk of wrongly declaring a location as hazardous (overestimation, left image) and Beta risk of wrongly declaring a location as safe (underestimation, right image) for a 2.195 units cutoff. Consistent with previous sub-sections, myGeoffice© provides two cutoff options for Alpha and Beta misclassification.

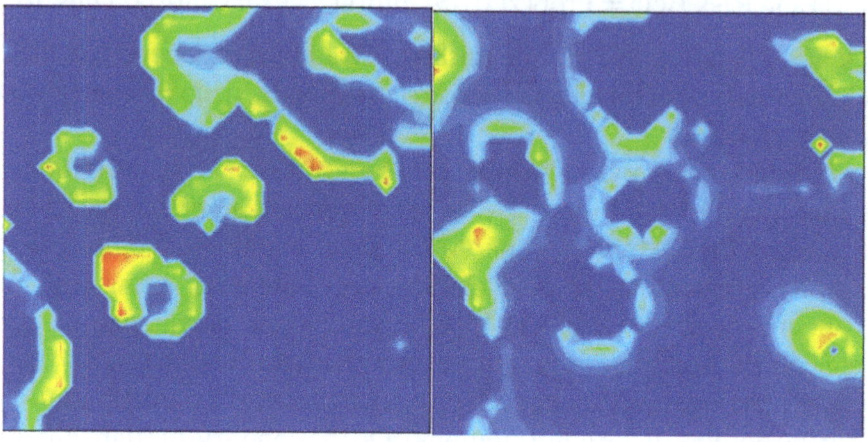

Figure 47. Note that the yellow, green and light blue bands fit the border lines between both risk areas

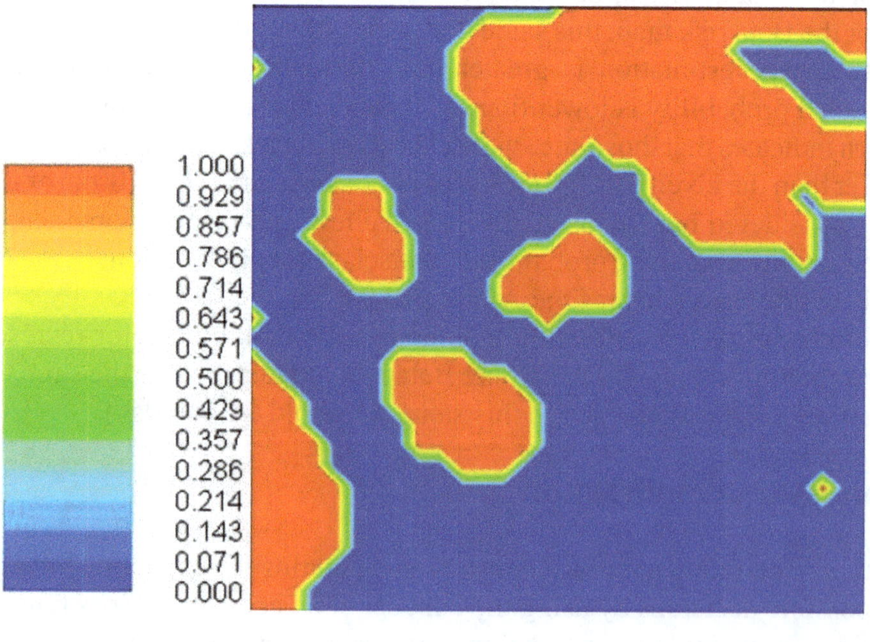

Figure 48. Alpha risk of wrongly declaring a location as hazardous (overestimation, left image) and Beta risk of wrongly declaring a location as safe (underestimation, right image) for a cutoff of six units

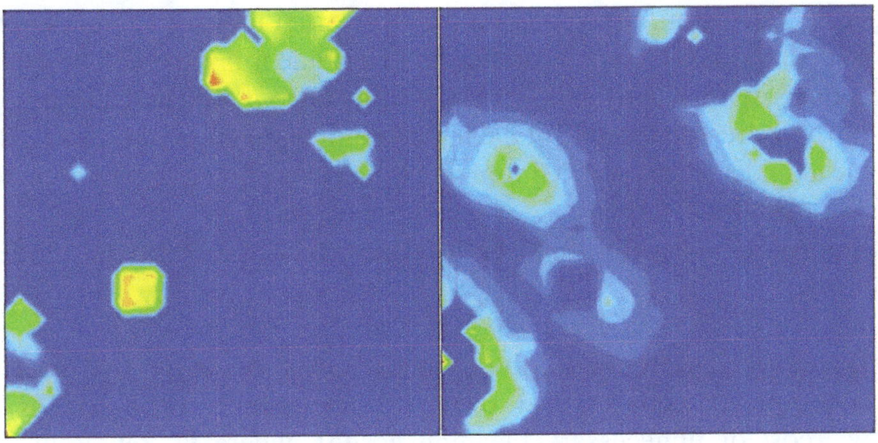

Figure 49. Misclassification map for a cutoff of six units

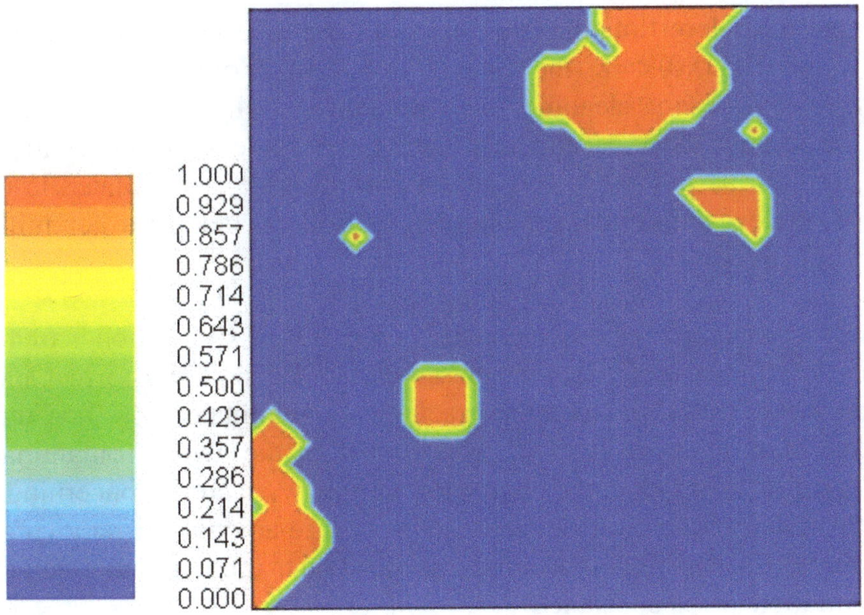

of a toxic concentration (positive estimation error) may result in costly and unnecessary cleaning (Goovaerts, 1997). When using a damage cost function, any x0 site estimate should be chosen so as to minimize the loss from making an incorrect decision. Since the real z(x0) is often unknown, the idea is to use the ccdf model (E-Type estimate) to determine this expected loss based on two types of misclassification: (1) Cleaning uncontaminated areas: Alpha risk due to overestimation; (2) Leaving contaminated areas uncleaned: Beta risk due to underestimation. Minimizing both costs of misclassification thus becomes the aim.

Using a particular z_c threshold, both situations can be respectively modeled as follows:

- Due to overestimation, the financial loss of each site declared contaminated equals 0 if $z(x0)>z_c$ (the site is indeed polluted) or else w_1 (cost of unnecessary cleaning a site), where w_1 represents, for instance, the constant cost of overestimating the toxic concentration. For Goovaerts (1997), this constant remediation cost is calculated in units of money such as dollars.
- Due to underestimation, the financial loss of each site declared safe equals 0 if $z(x0)<=z_c$ (the site is indeed clean) or otherwise $w_2 \times [z(x0)-z_c]$ (cost of not cleaning the contaminated site), where w_2 equals the relative cost of underestimating the same toxic concentration. For Goovaerts (1997), this health cost variable is calculated in units of money and concentration such as dollars and parts per million, respectively.

In a further graphical cost analysis (Figure 50), if x0 location is found to be safe, Z(x0)<=Cutoff, then the classification decision was correct and no loss occurred. But, if the location is actually contaminated (Z(x0)>Cutoff), the cost of misclassification is proportional to the actual contamination (Z(x0)-Cutoff) and directly connected to the w_2 factor. Concerning the complementary scenario (the cleaning procedure amounts of removing the upper layer of soil), this cost becomes independent of the actual contamination concentration (w_1 is constant).

Figure 50. Mapping layout of the four types of decisions, given a particular cutoff

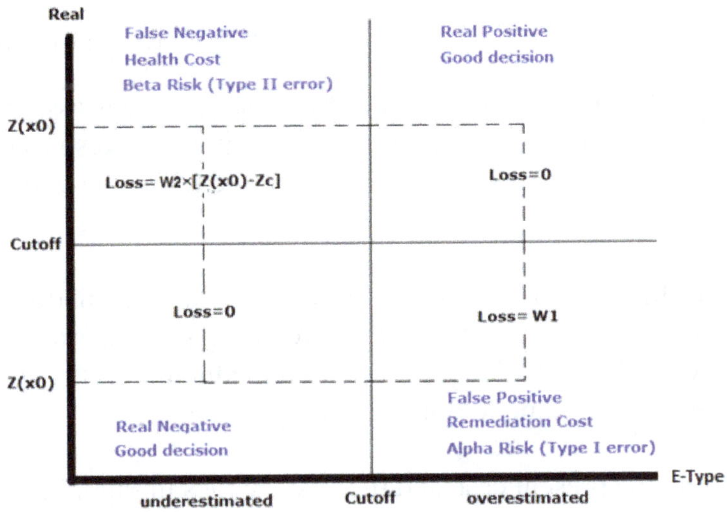

From the computational perspective, the optimal financial loss estimation is the value that minimizes both the over and under estimation loss costs. This expected loss assessment follows three steps: (1) The uncertainty about the unknown z(x0) value is first modeled using the ccdf of the E-type estimator; (2) From this model, both health and contamination estimation costs can be inferred based on the sum of the local probabilities of each ten sub-classes (under myGeoffice© it is multiplied by each loss function cost). Thus, regarding the health costs, if the mean value of each sub-class $z_k^{'}$ is greater than the chosen z_c cutoff, this proportional cost depends on w_2, i.e.,

$$\text{HealthTotalCost}\left(x_0\right) \approx \sum_{k=1}^{10} w_2 \times \left[z_k^{'} - z_c\right] \times \left[F_z^*\left(z_k, x0\right) - F_z^*\left(z_{k-1}, x0\right)\right],$$ where

$F_z^*\left(z_k, x0\right)$ equals the probability at x0 location being less than the z_k cutoff, $z_k^{'}$ is the mean of the class $[z_{k-1}, z_k]$ (under myGeoffice©, it is the average of both thresholds that limits each sub-class) while k=1…10 represents the 10 consecutive classes that discretize the range of variation for all z-values (see Figure 51). As expected, $F_z^*\left(z_k, x0\right) - F_z^*\left(z_{k-1}, x0\right)$ indicates the local probability for the value at x0 location of being between the sub-class limited by (k-1) cutoff and K cutoff (Figure 51). Regarding contamination costs, if the mean of each sub-class $z_k^{'}$ is less than the chosen z_c cutoff, the remediation cost always equals w_1, that is, $\text{RemediationTotalCost}\left(x_0\right) \approx \sum_{k=1}^{10} w_1 \times \left[F_z^*\left(z_k, x0\right) - F_z^*\left(z_{k-1}, x0\right)\right]$. Otherwise, the health remediation becomes zero. (3) At last, the site is assigned to a category. Any x0 location is declared safe or contaminated so as to minimize the resulting expected loss based on the following two principles:

- If the health loss cost is greater or equal to the remediation loss cost, x0 site is classified as contaminated (red regions, under myGeoffice©).
- If the health loss cost is less than the remediation loss cost, x0 site is classified as safe (blue regions under myGeoffice©).

Considering the present case and the cost charges in Figure 52, the range of health expenditures becomes [0 – 107.25]. Analogous to the remediation costs, this range equals [0 - 7]. Any x0 location is then declared safe or contaminated (bottom map) so as to minimize the resulting expected loss based on the previous two principles. For reference and based on the present unit amounts, the total health and remediation cost for the whole region amounts to 3718.99 and 5967.79 units of money, respectively.

Figure 51. Illustration of a ccdf divided into 10 consecutive sub-classes by 9 cutoffs and, where the red area represents the probability of the E-type estimate being between zc_6 and zc_7

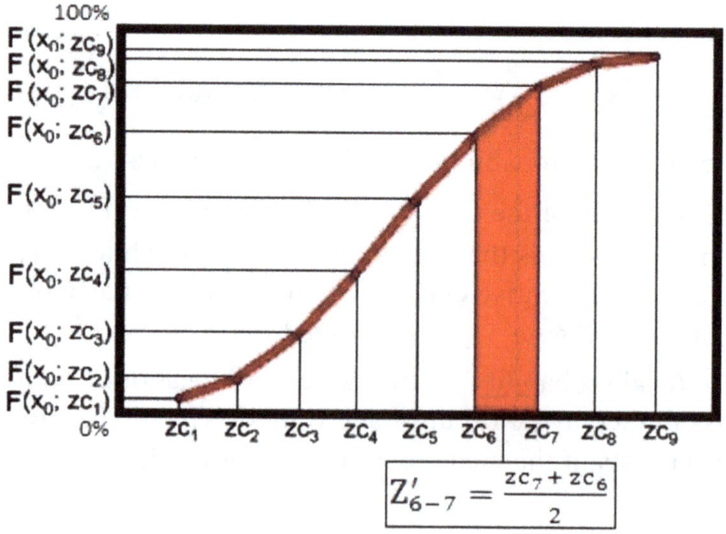

Figure 52. For Goovaerts (1997), the dichotomy between the uncertainty assessment and cost estimation clearly confirms that there is no best estimate for all situations. Rather, the effectiveness of an estimate depends on its intended use

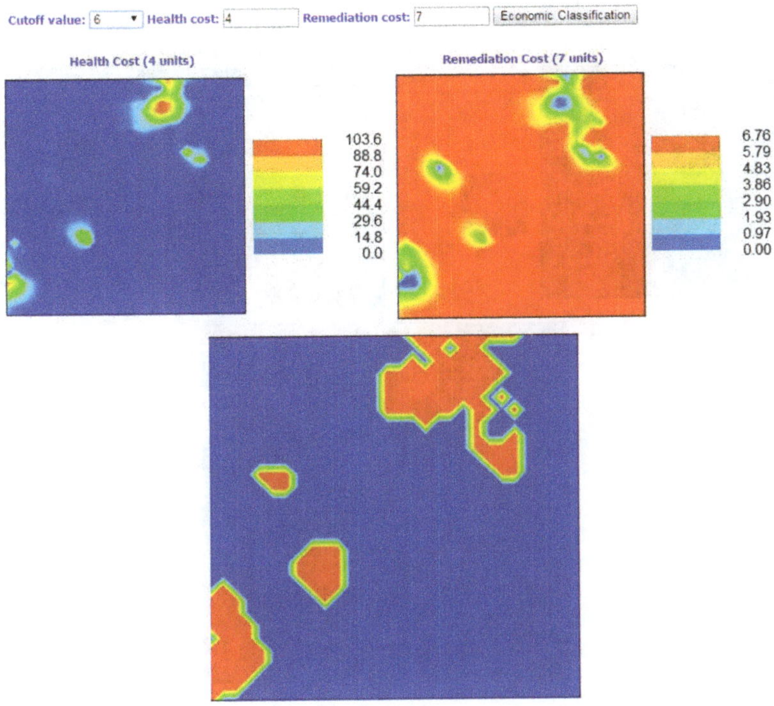

Morphologic Geostatistics

Morphologic Kriging (MK) of a two-phase structure is a classification allocation procedure that transforms continuous attributes into two categorical ones (plus an optional uncertainty region). Unlike continuous attributes, categorical ones such as rock types or land uses, cannot be estimated as a mere linear combination of neighboring data (Negreiros, Costa & Painho, 2010).

A straightforward technique of estimation or classification would consist in assigning unknown locations to the category (a geographical range entity that is distinguished from its neighborhood) of the nearest datum using the Thiessen or polygonal method. Another type of approach consists in allocating the site to a category with the largest conditional probability of occurrence. Because conditional probabilities of any location are likely to be large if the corresponding marginal probability is already large, most of the sites tend to be allocated to the most frequent category (Goovaerts, 1997).

Figure 53. For a 2.195 threshold and based on the previous health and remediation costs, the range of health expenditures is [0 – 122.46] while the remediation fee range equals [0 - 7]. The total health (blue area of the bottom map) and remediation costs (red area of the bottom map) totals 5911.45 and 4418.08 units of money, respectively.

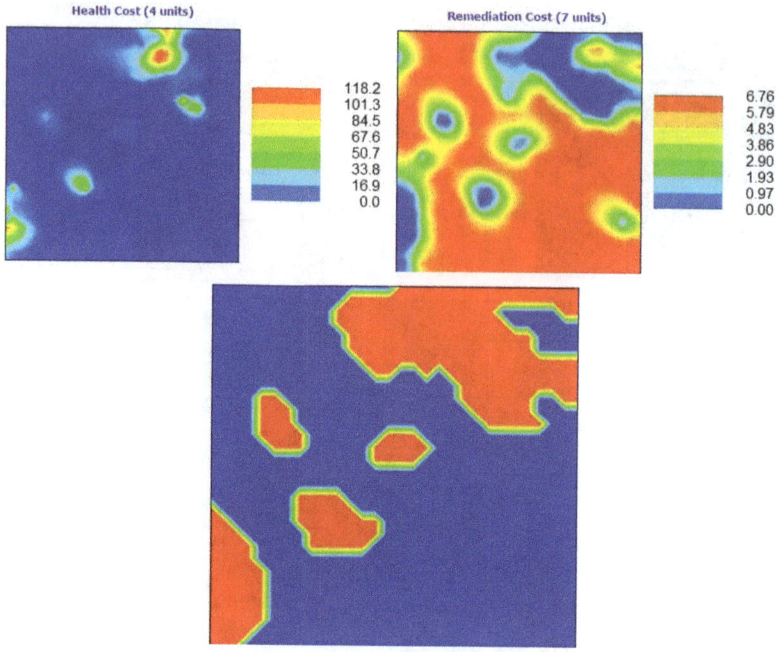

In practical terms and analogous to the Indicator Kriging (IK) approach, any spatial location is classified as one if that site is considered to be part of a certain X category or zero otherwise (the location belongs to the complement of X body, i.e., Xc). Although morphologic variography is similar to IK, its interpretation is quite different in the sense that there is no relationship order between X and Xc. Any location is classified according to its probability value (greater or smaller than the chosen cutoff level). Thus and following MK, the variogram range gives the amplitude average of X and Xc classifications while the nugget-effect presence illustrates the transition frequency between X and Xc categories at a narrow scale.

As previously stated, the purpose of MK is to transform those belonging probabilities into a binary map, where X and Xc complementary bodies are reproduced (see Figure 54). Basically, the Soares (2000) approach (and adopted by myGeoffice©) consists of allocating any spatial location to the category with the largest conditional probability of occurrence under the reproduction constraints of the global proportions. This algorithm proceeds as follows:

- Choose one of the IK thresholds of myGeoffice© (to be used for dual category morphologic mapping) and an uncertainty percentage for the transition area between X and Xc.
- From largest to smallest, the 900 (the total estimation number of myGeoffice© grid maps) IK probability estimations (recall that this IK mapping generates values between 0 and 1) should be sorted in an internal descending [V] vector. At the same time, assess the IK estimation average, [m], to obtain the estimation proportion of X classification for the study area.
- Multiply the uncertainty percentage region (under myGeoffice© this parameter may vary from 0% to 10%) by 900 in order to calculate the number of cells, [nc], to be included in the transition region. Then, it creates a new classification variable based on the following rule: TWO for the first IK probabilities whose estimations are greater than [m] and ZERO for the remaining ones. If the uncertainty percentage is different from zero, a new range of [nc] cells of [V] centered on [m] with ONEs must be re-assigned. Afterwards, the morphologic map should be laid out based on this new categorical variable (see Figure 55).

Figure 54. Under myGeoffice©, the thresholds presented to the user are based on the previous two cutoffs from the IK First/Second Cutoff variograph options. As a rhetorical example, the cutoff value to consider any location as A marble type or B marble type is 2.195

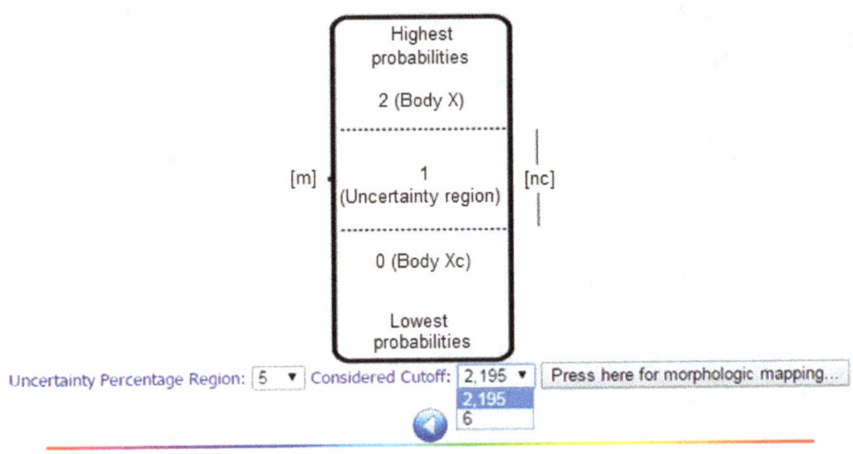

Figure 55. The BLUE pattern of the below map represents the first category of the morphologic region (marble type A, for instance), the GREEN stands for the uncertain or transition morphologic area while the RED design embodies the second classification (marble type B, for instance). The light blue and the yellow lines between the three categories do not hold any spatial interpretation (they correspond to some misspecification of the ActiveX® mapping technology)

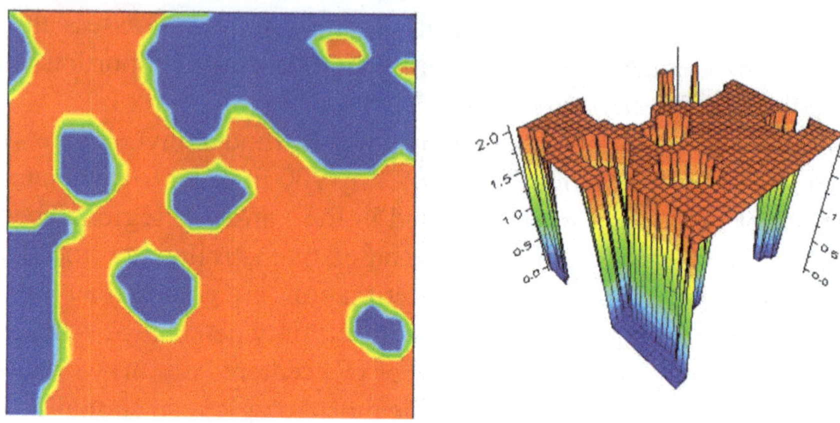

Figure 56. In this particular case, the t1 and t2 thresholds, respectively, equal 2.195 and 6

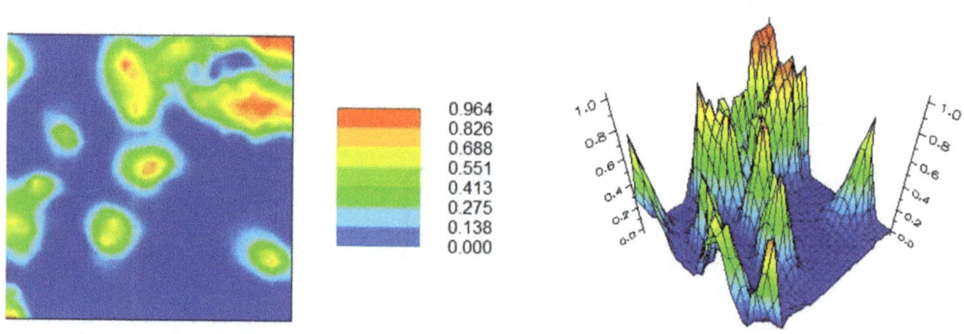

Probabilistic Interval

Like Shannon local entropy, both maps below can be interpreted as a local uncertainty measure. After all, uncertainty is a dimensionless parameter for which high values are bad and lower ones optimal (Negreiros, Painho, Aguilar & Aguilar, 2009). Specifically, if the probability difference between two generic IK thresholds (t1 and t2, where t1<t2) is small, for instance, the

spatial distribution tails become significant. Otherwise, if IK(t2)–IK(t1) is considerably larger, all major spatial interpolations will fall between those two cutoffs. That being said, a red pattern signifies that there is a high probability for that particular area to fall into the chosen [t1-t2] interpolation range. Conversely, blue color denotes a negligible probability of that region lying between the selected intervals (See Figure 56).

REFERENCES

Burrough, P., & McDonnell, R. (1998). *Principles of GIS*. Oxford University Press.

Clark, I. & Harper, W. (2000). *Practical geostatistics*. Ecosse North America.

Cressie, N. (1993, 2015). Statistics for spatial data. New York: John Wiley & Sons.

ESRI. (2005). *Using ArcGIS geostatistical analyst*. ESRI.

Goovaerts, P. (1997). *Geostatistics for natural resources evaluation*. Oxford University Press.

Isaaks, E., & Srivastava, M. (1989). *An introduction to applied geostatistics*. New York: Oxford University Press.

Negreiros, J., Costa, A. C., & Painho, M. (2010). Evaluation of stochastic geographical matters: Morphologic geostatistics, conditional sequential simulation and geographical weighted regression. *Trends in Applied Sciences Research, Science Alert–Asian Network for Scientific Information, 6*(3), 237-255.

Negreiros, J., Painho, M., Aguilar, M., & Aguilar, F. (2008). Spatial error and interpolation uncertainty appraisal within geographic information systems. *Research Journal of Applied Sciences, 3*(6), 471-479.

Negreiros, J., Painho, M., Aguilar, M., & Aguilar, F. (2009). *The local and global confidence uncertainty plumes of SAKWeb©*. Paper Presented at AGILE 29, Hannover, Germany.

Chapter 10
Sequential Simulation

ABSTRACT

In this chapter, the first main section analyzed the Conditional Cumulative Gaussian Simulation (CCGS) of myGeoffice©. Section 2 examines the Conditional Indicator Simulation (CIS).

INTRODUCTION

Unlike Kriging, stochastic simulation does not aim to minimize the local error variance but, instead, focuses on the reproduction of statistics such as the sample mean, variance, histogram or variogram model (Negreiros, 2015). In other words, it preserves the spatial variation of the attribute studied. In addition, it honors observation data values. For Deutsch and Journel (1998), in simulation, the reproduction of global features (texture) and statistics (histogram, covariance) take precedence over local accuracy. According to Goovaerts (1997), a simulated map looks more realistic than the map of the statistically best Kriging estimates because it reproduces the spatial variability modeled from the original data sample. Thus, stochastic simulation is increasingly preferred to Kriging for applications, where spatial variation of the measured field must be preserved and the extreme behavior of the variable under study should not be overlooked.

The trade-off cost for this better reproduction of spatial features by simulation maps is that the local property of the Kriging minimum error variance is lost. A practical consequence is the mean prediction error tends to be larger for simulated values than for Kriging estimates. Another difference

DOI: 10.4018/978-1-5225-3270-5.ch010

between stochastic simulation and Kriging is that many realizations, all of which match reasonably the same statistics (mean, variance, histogram, covariance structure), can be generated whereas there is only a single Kriging map that yields the minimum error variance at each location.

For these three authors, this set of equally-probable realizations is particularly useful for assessing the spatial uncertainty of the attribute under study and for investigating the performance of different scenarios (contamination of a water reservoir, for instance), where each realization may provide a different remediation/cleaning cost and, consequently, dissimilar decision-making. Hence, under this risk analysis perspective and based on 100 realizations, for instance, it is possible to quantify the common and extreme behaviors (minimum, average and maximum) for any spatial phenomenon or calculate the probability of how many of those realizations exceed a particular threshold.

A practical and remarkable example of stochastic geosimulation usage is given by Teixeira et al. (2012). Due the importance of sugarcane cultivation in the state of São Paulo, Brazil, and because of the uncertainty in the process of CO2 emission (FCO2) from agricultural soils, it is important to characterize this primary source of emissions for agriculture purposes, i.e., it plays a major role in global carbon and other cycling of nutrients and is also associated with positive feedback on global climate change.

The major aim of the study conducted by these researchers in Fazenda Santa Olga at Guariba was (1) to characterize the variability and spatial distribution of FCO2, (2) to compare the accuracy of the results of OK (Ordinary Kriging) and SGS (Sequential Gaussian Simulation) and (3) to determine the uncertainty in predicting the spatial variability of FCO2 (based on 141 data samples) using SGS (where the differences between the various outputs provide a measure of uncertainty for prediction purposes).

From the descriptive statistics perspective, the average CO2 soil emission ($1.57 \pm 0.07 \mu mol\ m^{-2} s^{-1}$) was lower than reported in other studies of Oxisols for sugarcane areas during this evaluation period, a difference that may be related to a lack of rainfall in the days before the experiment (according to the researchers).

Three hundred realizations were computed while the 30th, 68th, 176th and 214th were randomly selected (Figure 1). Further, minimum, medium (E-type), maximum and standard deviation maps were produced by counting the simulated points at each location of the 300 realizations. The standard deviation layout represents the point variation in all simulated realizations.

Figure 1. Gaussian simulation realization maps generated by Teixeira et al. (2012) of FCO2 emissions in Brazil

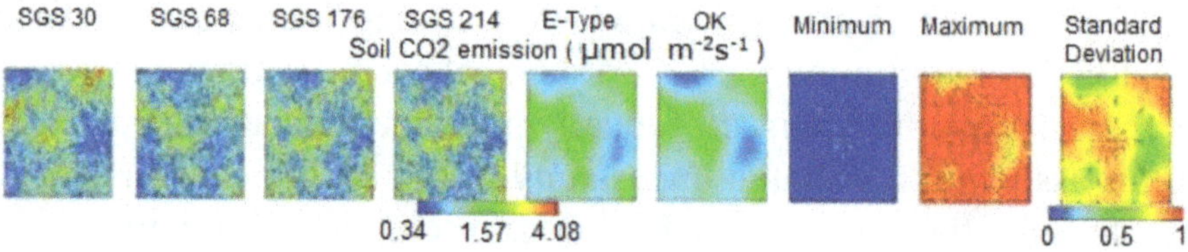

As expected, the deviations were zero at the points of the observed values due to the conditional characteristics of the Gaussian simulation. There is a predominance of high FCO2 values (>3.30μmol $m^{-2}s^{-1}$) in the upper-right and center of the map whereas low values (>1.80μmol $m^{-2}s^{-1}$) are distributed throughout the area but with greater consistency in the upper-left portion of the map.

This standard deviation map associated with the estimating FCO2 emissons reveals that the deviation was greater for areas with high estimated values (these areas displaying more uncertainty than the lowest sub-areas). On the other hand, description statistics displayed a high degree of similarity between the E-type estimates and those generated by Ordinary Kriging (OK). Again, the low performance of OK in the reproduction of the variogram highlights that OK should not be used when a replica of the original descriptive statistic and variability structure of the primary data is desired.

According to Grunwald, Reddy, Prenger and Fisher (2007), the average map represents the dominant property signal in the area while the range of possible results (characterized by the minimum and maximum values) represents the best and worst possible scenario for each variable. Over a short time period, the FCO2 showed a mean variation ranging from 0.36 to 3.85 μmol $m^{-2}s^{-1}$, resulting in a wide range of emissions (from 0.18 to 1.85 FCO2 ha^{-1}) under different scenarios.

In respect of the decision-making process, these outputs may lead to different agriculture policies (and profits/costs) regarding, *inter alia*, the type of fertilizers to be used, where to deploy them, the density of sugarcane cultivation, quality expectations of the sugarcane itself and frequency of irrigation. Therefore, this wide range of uncertainty (combined with the construction of different scenarios) provides useful information on possible patterns of soil respiration, something which is extremely erratic and difficult to predict.

In this chapter, the first main section analyzed the Conditional Cumulative Gaussian Simulation (CCGS) of myGeoffice©. Section 2 examines the Conditional Indicator Simulation (CIS).

CONDITIONAL GAUSSIAN

SGS Geocomputation

The Conditional Gaussian Sequential Simulation (CGSS) parametric technique is based on the assumption that any spatial distribution of a continuous random variable can be modelled by using a multivariate Gaussian model. For reference, sequential, p-field and annealing are other forms of simulation algorithms although only the former is covered by myGeoffice©. Furthermore, the ten pre-defined generated realizations of myGeoffice© take into account not only the original data but also all values simulated at previously visited locations, the conditional approach. Generically, the sequential Gaussian simulation of a continuous variable z(u) proceeds as follows:

1. Using the cumulative distribution function (cdf) of the variable under study, perform a Normal-score transform of the z original data into y-data, which follows a standard Normal cdf (the Gaussian restriction).
2. Define a random path that visits each node of the grid once. At each u node, retain a specified number of neighboring conditioning data, including both the initially transformed y-data and previously simulated grid node values (the conditional restriction).
3. Apply Simple Kriging (myGeoffice© also allows Ordinary Kriging) with the Normal score variogram (based on the y-values) to determine the estimation, $z^*(u)$, and the variance, $\sigma^*(u)$, at u location.
4. Draw a [0…1] value randomly from the Gaussian distribution $N(z^*(u),\sigma^*(u))$. The simulated value then equals $G^{-1}(z^*(u),\sigma^*(u))$. Afterwards, this simulated value is added to the present dataset.
5. Proceed to the next node and loop it until all nodes are simulated (the sequential restriction).
6. Back-transform of the y(u) simulated normal values into true simulated ones for each u location.

Figure 2. The second input parameter allows the user to setup the minimum and the maximum value of each realization to the lower and upper back-transformed tail implementation of SGSIM.exe (see (Deutsch & Journel, 1998) for further GSLib® details)

SGS Mapping I

See Figures 3 and 4.

SGS Mapping II

See Figures 5 and 6.

Figure 3. Under myGeoffice©, ten default realizations are generated according to the default input parameters. Due to their length, only two of them are presented here for illustrative purposes, including their descriptive statistics (as expected, similar but not equal outcomes can be found), 2D and 3D perspective. Analogous to all Ordinary Kriging interpolations in section "Ordinary Kriging", the present SGS are based on the same variograph achieved in section "Variogram Fitness."

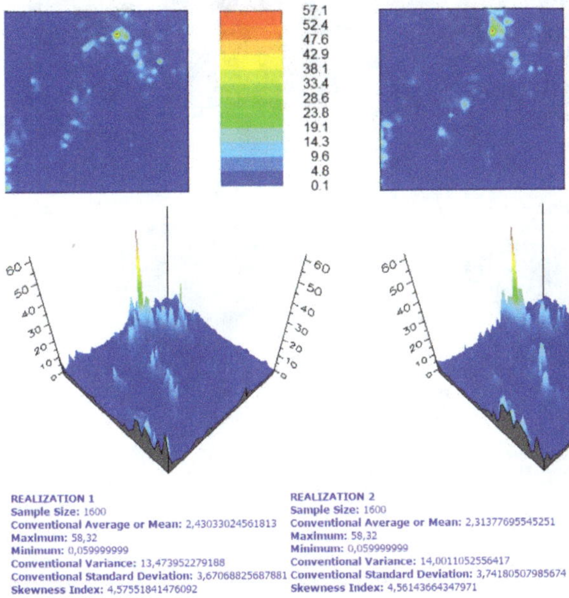

CONDITIONAL INDICATOR

Kriging is a deterministic method whose function has a unique solution and it does not attempt to represent the actual variability of the attribute studied. In effect, the smoothing property of any interpolation algorithm replaces local detail with a good average value. However, geologists are more interested in finer scaled details of reservoir heterogeneity than in a map of local estimates of the mean value (PetroWiki, 2013). Certainly, to capture heterogeneity and to quantify and assess uncertainty are two major reasons to perform stochastic simulation. As discussed earlier, stochastic simulation allows for quantification of the uncertainty and variability of the phenomenon under study. Sequential Gaussian Simulation (SGS) and Sequential Indicator Simulation (SIS) respect these principles.

Figure 4. Example of two realizations with no Gaussian data transformation of the original dataset (inevitably, diverse input parameter combinations can be set up, including different variogram models and thresholds for comparative simulation mapping purposes).

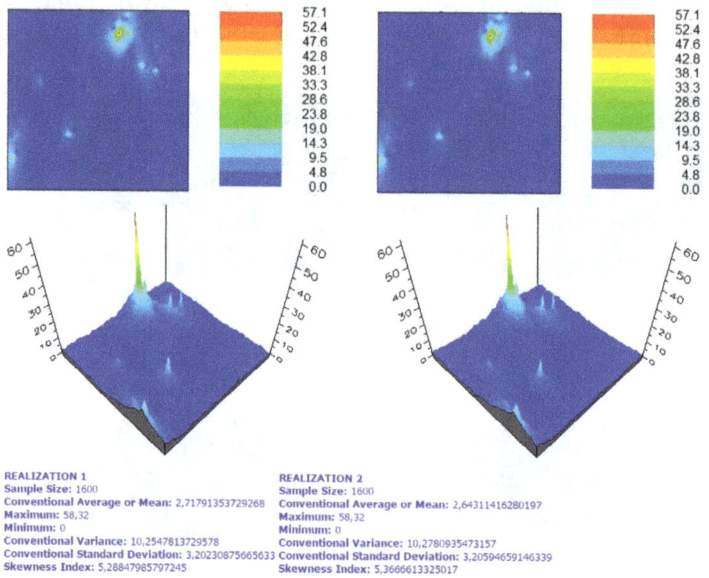

For Goovaerts (1997), the Gaussian model is inappropriate whenever the structural analysis indicates that extreme values could be better correlated in space than medium ones. Unlike SGS, this indicator approach permits the taking into account class-specific patterns of spatial continuity through different indicator variogram models. For instance, the second decile spatial continuity (and variogram) will be supposedly different from the ninth decile. Computationally, Sequential Indicator Simulation (SIS) proceeds as follows:

1. Discretize the range of variation at *u* location into (K+1) classes using K thresholds values, Z(K) (under myGeoffice©, K equals nine leading to ten non-overlapping classes). Then, it transforms each Z(u) datum into a vector of hard indicator data, defined as 1, if Z(u) <= Z(K), and 0, otherwise.
2. Define a random path visiting each node of the grid only once (similar to SGS).
3. At each *u* location:

Figure 5. The bottom-left map highlights the probability of the ten realizations being greater than 3. Computationally, for instance, if two or six out of the ten realizations for a particular location hold a value greater than the chosen threshold, the probability assigned is 0.2 and 0.6, respectively. As expected, the bottom-right layout represents the probability of the ten realizations being less or equal to 3.

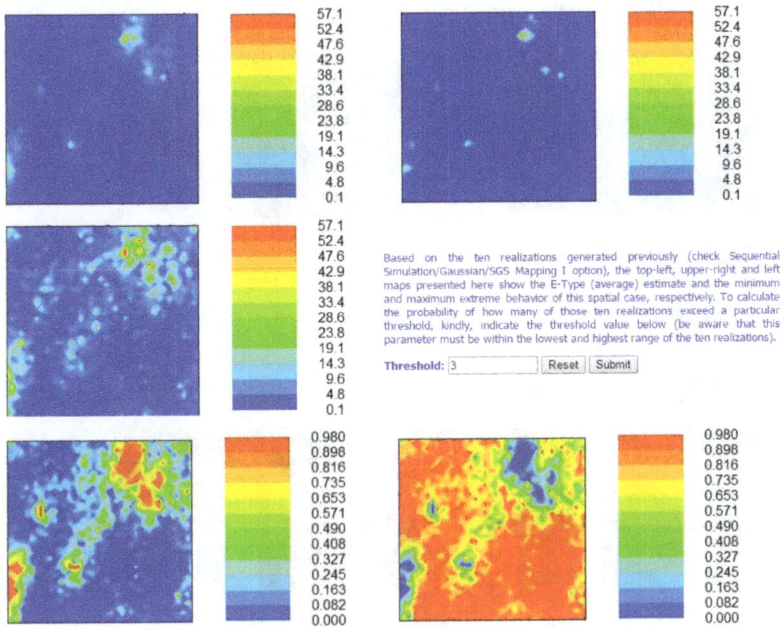

a. Determine the K ccdf values using Indicator Kriging (under myGeoffice© and for the sake of simplicity, MIK is used for the nine thresholds).
b. If necessary, correct the ccdf order relation using linear, power and hyperbolic interpolation/extrapolation techniques.
c. Draw a simulated value for u location from the [0…1] ccdf.
d. Add the simulated value to the conditional dataset.
e. Proceed to the next node along the random path by repeating the previous four steps.

4. To generate each realization, it is mandatory to repeat the entire procedure with a different random path. Like SGS, every time a new node is simulated, this new value is added to the next ccdf (the conditional assumption).

Figure 6. Analogous maps of the previous figures but based on no Gaussian data transformation of the original dataset. When comparing these results with the ones in Figure 4 and Figure 5, similar patterns can be found but no equal outputs. Confirmed by (Goovaerts, 1997), stochastic realizations do not match model statistics exactly, nor should they. Such discrepancies between realization and model statistics are referred to as ergodic fluctuations.

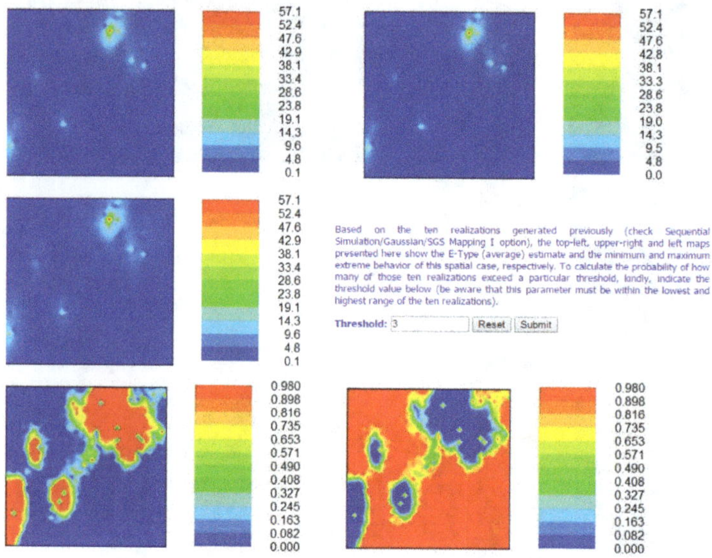

The major advantage of SIS over SGS is the flexibility of modeling with different variograms for each cutoff class (different spatial distributions) although from the user's perspective it can be a time consuming and tedious operation. This is the first reason why myGeoffice© pursues the Median Indicator Kriging (MIK) approach, where the variogram of the median value of the dataset is used for all cutoff classes. The second motivation is to avoid order relation issues among the assigned probabilities for the consecutive thresholds. Still, according to Deutsch and Journel (1998), the cost of using MIK is that all indicator variograms are proportional, i.e., standardized variograms are identical. Goovaerts (1997) confirms this perception: No matter how many thresholds are retained, only one (median) indicator variogram is considered and thus only a single IK system needs to be solved at each grid node. Consequently, all order relation deviations caused by the lack of data in some classes are eliminated. One loses, however, the flexibility to model class-specific patterns of spatial continuity.

As a last annotation regarding the choice of Sequential Gaussian Simulation (SGS) or Sequential Indicator Simulation (SIS) algorithms, no better methodology can be found because accuracy of any estimation can only be measured when the true value is known, leading otherwise to a deadlock situation. Statistically, to avoid a non-accurate predictor is a certainty. Increasing the spread of precision is a possible answer to avoid the non-accuracy issue (not desirable though). The ideal is to be simultaneously accurate and precise. According to the SGS and the SIS comparison done by

Figure 7. Simple Kriging should be the preferred algorithm within sequential simulation unless proven inappropriate (Deutsch & Journel, 1998). The researcher must also be aware that the indicator variogram for the median cutoff value of the dataset (2.195 in this particular case) has to already be setup as in section "Indicator Kriging" in order to run this option.

Conditional Sequential Indicator Simulation Geocomputation (Nine cutoffs, MIK)

Check if executable program (geocomputation of SIS realizations) were accomplished with success

```
c:\inetpub\wwwroot\executable\sisim.exe d:\data\232913710SIS9.par
SISIM Version: 2.907

variable type (1=continuous, 0=categorical)= 1
number of thresholds / categories = 9
thresholds / categories = 1.900000E-01 4.000000E-01 9.200000E-01
1.360000 2.170000 2.840000 4.290000
7.530000 9.420000
global cdf / pdf = 1.142857E-01 2.000000E-01 3.000000E-01
4.000000E-01 5.000000E-01 6.000000E-01 7.000000E-01
8.000000E-01 9.000000E-01
data file = d:\data\2329137109.SIS
input columns = 1 2 0 3
soft data file = d:\data\232913710direct9.ik
trimming limits -1.000000 1.000000E+20
data limits (tails) 6.000000E-02 58.320000
lower tail = 1 6.000000E-02
middle = 1 2.195000
upper tail = 1 58.320000
file for tab. quant. d:\data\2329137109.SIS
columns for vr wt = 3 0
debugging level = 0
debugging file = d:\data\232913710SIS9.dbg
output file = d:\data\232913710SIS9.out
number of simulations = 10
X grid specification = 40 5.000000E-01 1.200000
Y grid specification = 40 5.000000E-01 1.200000
Z grid specification = 1 1.000000 1.000000
random number seed = 69069
ndmax = 8
max prev sim nodes = 12
max soft indicator data = 0
search strategy = 1
multiple grid search flag = 0 3
max per octant = 0
search radii = 30.000000 30.000000 0.000000E+00
search anisotropy angles = 0.000000E+00 0.000000E+00 0.000000E+00
size of covariance lookup = 41 41 1
median IK switch = 0 2.195000
kriging type switch = 0
Reading input data

Data for SISIM: Variable number 3
Number of acceptable data = 140
Equal Weighted Average = 4.3504
Equal Weighted Variance = 44.9243

Setting up rotation matrices for variogram and search
```

Figure 8. Under myGeoffice©, ten default realizations were generated. Due to its length, only two of them are presented here for illustrative purposes, including their descriptive statistics (as expected, similar outcomes can be found), 2D and 3D perspectives. The alternative possibility of visualizing a set of realizations relies on the creation of an animated cartoon, where each successive frame should match the most similar realization of the previous one to allow the human eye to catch gradual changes (Goovaerts, 1997).

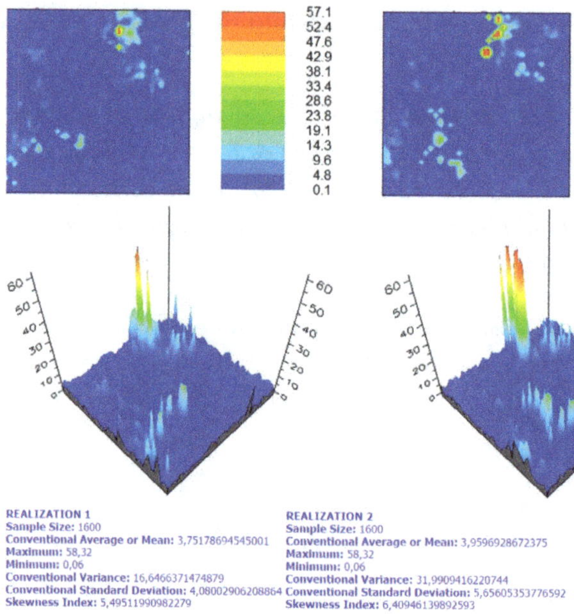

Jakab and Geiger (2014), the reproduction of the original distribution of high or low heterogeneous cases is best performed by SGS. On the other hand, SIS reproduces best uncertainties with high/low heterogeneities. It seems that for the same dataset, different algorithms can be considered best. As stated by Goovaerts (1997), no model and therefore no uncertainty measure can ever be objective - the point is to accept this limitation and clearly document all aspects of the model.

SIS Geocomputation

See Figure 7.

Figure 9. The bottom-left map highlights the probability of the ten realizations being greater than 3. Once more, the bottom-right layout represents the probability of the ten realizations being less than or equal to 3.

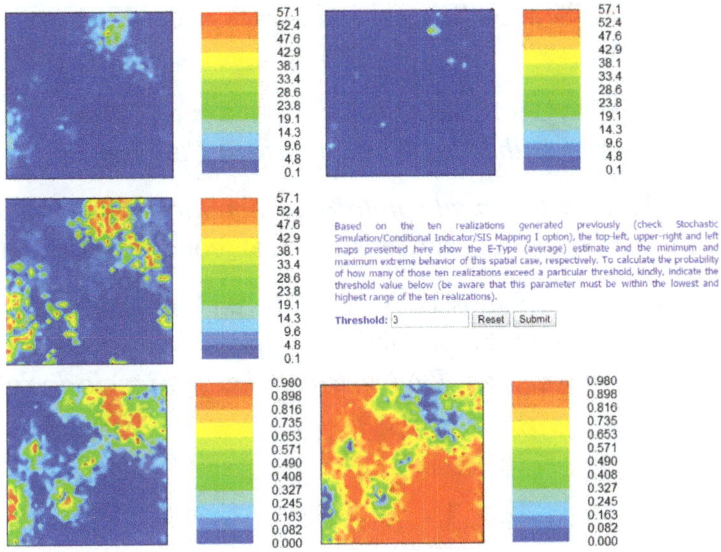

SIS Mapping I

See Figure 8.

SIS Mapping II

See Figure 9.

REFERENCES

Deutsch, C., & Journel, A. (1998). GSLIB. Oxford University Press.

Goovaerts, P. (1997). *Geostatistics for natural resources evaluation*. Oxford University Press.

Grunwald, S., Reddy, K., Prenger, J., & Fisher, M. (2007). Modeling of the spatial variability of biogeochemical soil properties in a freshwater ecosystem. *Ecol. Model*, 521-535.

Jakab, N., & Geiger, J. (2014). *Approaches to characterize the uncertainties related to stochastic simulations.* Retrieved in 2015 from SPE International (Hungarian Section): oslo.spe.org/HigherLogic/System/DownloadDocumentFile.ashx?DocumentFileKey=ede31381-965d-475d-912f-3a9ec3c4dcd5

Negreiros, J. (2015). myGeoffice©: Overview of a geographical information system. *Proceedings of 14th ISERD International Conference*, 12-16.

PetroWiki. (2013). *Geostatistical conditional simulation.* Retrieved in 2015 from petrowiki.org: petrowiki.org/Geostatistical_conditional_simulation

Teixeira, D., Bicalho, E., Panosso, A., Perillo, L., Iamagutti, J., & Pereira, G. (2012). Uncertainties in the prediction of spatial variability of soil CO2 emissions and related properties. *Revista Brasileira de Ciencia do Solo*, *26*(5).

Chapter 11 Geography: Epilogue

ABSTRACT

The primary goal of myGeoffice© is to empower Internet users with some geographical quantitative power in a direct and goal oriented way (particularly high-school and university students whose curriculum covers spatial topics). With the spread of smartphones, apps, laptops, tablets, e-learning, m-learning, 4G wireless connectivity, free wi-fi hot spots, apps, Web 2.0 tools and adaptive learning/progress tracking, it is hoped that myGeoffice© can be associated with a teaching strategy and incorporate these trends. This would assist with the basic understanding of spatial inferential and mathematical methods in the classroom and encourage geography as a career path.

INTRODUCTION

For a long time, Geography did not hold a specific mathematical approach for any interpretation of space and this was the major reason why geography bachelor degrees cover a wide variety of subjects (such as demography, geology, topography, hydrology, history, biology, morphology, climatology, statistics, sociology and economics). It seemed that geography did not have any conviction towards its own study object, space. This phenomenon of an absence of laws and techniques on how space rules the Third Rock from the Sun has finally shifted since the 1980s. Until then, geography was not a spatial science but a reflection on the universe, which suggested that it

DOI: 10.4018/978-1-5225-3270-5.ch011

"cooked" a global explanation from other sciences. Hence, the geographer became an intruder into the physical and social fields. Still, the Central Place Theory by Walter Christaller in 1933 is a noble positive outlying example on how geography and economics intersect, giving an insight into the nature of town development.

Finally, geography is creating its own research agenda to meet the needs of spatial location frameworks since what really matters for geographers is spatial (and time) attributes. From this viewpoint, geography becomes a widow of space (Santos, 1978) and space has a close brother called time: geography for space and history for time. Definitely, "Where is…?", "Is there a general spatial pattern?", "What are the anomalies?", and "Why do these phenomena pursue certain spatial distributions?" is the added value given by geographers to other sciences (Ferreira & Simões, 1994).

Again, it would be absurd to believe that quantitative methods could explain all Earth's phenomena. However, African decolonization or the Gulf wars should be studied in a spatial context within geography, a fragmented point of view. Mayhew (2015) criticizes his quantification for being unrealistic, bloodless and too deterministic, turning humans into automata and the importance of subjective experience.

However, Earth holds some spatial continuity, leading geography in a new effort to create its own general rules of space instead of begging methodologies from other sciences. Geography is the science of the diversity of space, based on its own laws (Longley, Goodchild, Maguire & Rhind, 2011) in detecting urban hierarchy and future industrial localizations, targeting military accessibility, identifying a hospital's regional influence or uncovering agriculture side-effects from the coexistence of different soil types, for instance. Certainly, this narrow view of geography emphasizes the major feature of its uniqueness to other research fields.

The red area in Figure 1 represents population losses between 1993 and 1997 from the Portuguese electoral roll of the town level (Lisbon and Setubal regions), while the lighter ones represent population gains for the same time period. As expected, a strong positive spatial autocorrelation can be found, a trouble-free detective job in this particular case, leading the researcher to ask questions such as what factors contributed to these shifts in population.

It is undoubtedly a geographer's job to interpolate the number of inhabitants of Kenya for a particular parish level, to extrapolate the number of socialist voters for Ireland at municipality level, to identify risk flood areas in São Tome and Principe, to set up new country' borders of the former USSR, to route the nearest available ambulance to an accident scene (including traffic

Figure 1. Population gains and losses on the Portuguese electoral roll at town level

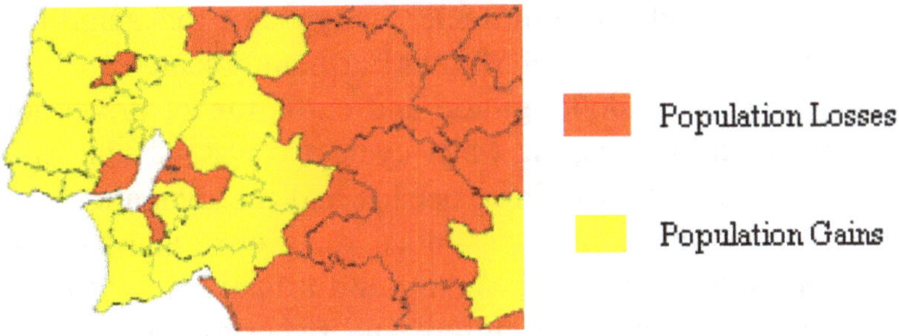

jam situations) in Beverly Hills, Los Angeles, to find how many customers live within the Kowloon area of Hong Kong, to locate the nearest drug store available at 3am in Singapore or to transform a territorial division dataset using a spatial overlay operation because of the recent postal code mapping.

Since the dynamic relationships between places reflect social, physical and economic processes, rejecting all traditional approaches would be as ridiculous as rejecting complementary conceptualizations. Therefore, by discussing Voronoi, spatial autocorrelation, Kriging and stochastic simulation, GWR, minimum spanning tree algorithms and point patterns become central to this book, where absolute and relative localization is fundamental in discovering the rules of space.

Spatial analysis must emphasize its program of searching for clues and evidence within spatial phenomena, which involves understanding, prediction and simulation and where location, time and geometry are significant. The difference between geography and other sciences is location, spatial structures and spatial processes because no other science field is concerned about the distribution of spatial phenomena. Further uses of geography include interpolating gold in South African mines, examining spatial relationships among religions in Israel, looking for local pockets of elephants in Bangladesh, extrapolating oil production in the Kamchatka and Alaskan peninsulas, finding an alternative solution for a gas pipeline leak in Mongolia, simulating the Egyptian Nile basin dam and the spread of infectious hepatitis C, calculating the shortest distance between Lima and Quito, South America, specifying the range area influence of a new police station in Canberra, Australia, or allocating voters in establishing new electoral boundaries in Warsaw, Poland. When astronomers or meteorologists, for instance, apply geographical methodologies to their own study objects, they become geographers and geography becomes highlighted.

Nowadays, typical geography and other closely related spatial subjects encompass a wide variety of issues such as types of rocks and territories, the flora and fauna biodiversity of Mother Earth, map scales and compasses, countries and flags, latitude/longitude and cartographic projections, time zones and economical spaces, the water cycle and watersheds, temperature and other weather factors (El Nino, oceans and the dynamics of streams, for instance), tectonic plates and volcanoes, the growth and composition of the world's population, cultural diversity and urbanization, migration and poverty, forests and deserts, agriculture and fisheries, tourism industries, environmental risks and protected natural regions, commercial routes and hurricanes, the Milky Way and rotation/translation of planet Earth… Without any doubt, all these topics are essential for the description of space, as we know it. Yet, in the current internet age, GIS is seldom included in this syllabus and curriculums used for true inferential analysis of spatial data (supplied from Landsat, for instance) are not very effective. Undoubtedly the state of contemporary education in this field leaves much to be desired.

The primary goal of myGeoffice© is to empower Internet users with some geographical quantitative power in a direct and goal oriented way (particularly high-school and university students whose curriculum covers spatial topics). With the spread of smartphones, apps, laptops, tablets, e-learning, m-learning, 4G wireless connectivity, free wi-fi hot spots, apps, Web 2.0 tools and adaptive learning/progress tracking, it is hoped that myGeoffice© can be associated with a teaching strategy and incorporate these trends. This would assist with the basic understanding of spatial inferential and mathematical methods in the classroom and encourage geography as a career path.

REFERENCES

Ferreira, C., & Simões, N. (1994). *Tratamento estatístico e gráfico em geografia*. Gradiva.

Longley, P., Goodchild, M., Maguire, D., & Rhind, D. (2011). *Geographical information systems and science*. Wiley.

Mayhew, S. (2015). *Dictionary of geography*. Oxford University Press.

Santos, M. (1978). *Por uma geografia nova*. São Paulo: Hucitec, Edusp.

Chapter 12
Education:
Contemplation

ABSTRACT

Given the present plug-in society of on-line services, today's youngsters became owners of digital handheld devices where Wikipedia, Twitter, Facebook, Google, Yahoo, Flickr, Viadeo, Amazon, Alibaba, WeChat, Line, Blogger, What'sApp, Instagram, Vine and Dropbox are regular daily services used as a common practice. Connectivity is oxygen nowadays. People spend hours engaging with social media, highlighting that this activity is playing a huge part of the growth and evolution of the online landscape (Kemp, 2014). Thirty years after the dawn of the Net, social media have become the first activity on the Web where everything is connected in real-time and are more personalized than ever in a universal cross-platform (the cloud concept). Definitely, real social life, based on activities such as going to a movie with your friends or children playing with toys is fading away.

INTRODUCTION

The number of selfies increased exponentially while the number of emails and spams followed the same trend. Being connected on a paradisiac beach or in the washroom has become normal. Ladies driving cars while simultaneously talking on their iPhones, putting on lipstick and looking at the GPS map is the rule rather than the exception.

DOI: 10.4018/978-1-5225-3270-5.ch012

The cyber-bullying effect, the online sexual predators, the fresh alphabet for children (A – Apple; B - Bluetooth, C- Chat, D – Download … W – Wifi; X – Xml; Y – Youtube; Z - Zerg), elderly being taught by their grandchildren, to fall asleep in front of a laptop due to physical exhaustion (the technostress awareness) and nomophobia (the fear of having no mobile network connection) are some of the daily life examples of behavior changes over the last ten years. Quoting Albert Einstein, "It has become appallingly clear that our technology has surpassed our humanity. The world will have a generation of idiots".

The e-businesses provided by this World Wide West have become countless, as well. If you need someone to travel with, just create a free account on triptogether.com. If you are looking for a date, just check on match.com. If you want to cheat on your soul mate, just click on ashleymadison.com. If you want to buy a machine gun, see gunbroker.com. If you need a university diploma, buy it on cheaper-than-tuition.com. If you have a kidney problem, have a look on organdonation.nhs.com. If you need a male escort, rent him on gentlemen4hire.com. Would you like to have a Russian fighter jet MIG 23? controller.com is the answer.

This empowerment or rather enslavement of this love/hate relationship of being online 24/7 continues on and on. When Facebook identifies what we write on its wall, Google's aim is to provide insights into users' interests and future actions for prediction purposes. This happens because if we do not pay for these services, our everyday life becomes the content to be explored for commercial purposes. There is no privacy (Leonhard, 2014). Period.

Governments contributed to this data cyber warfare situation too, particularly the Obama's NSA administration ("Yes, We Scan") over Brazil's Petrobras oil company and German spy scandal. On the other hand, China is alleged to have undertaken a widespread effort to purchase US military equipment. Such Chinese aggressiveness is well documented in multiple espionage cases, including Larry Wu-Tai Chin, Katrina Leung, Gwo-Bao Min, Chi Mak and Peter Lee (Wikipedia, 2015).

Gary Turk on his five-minute "Look Up" YouTube video captures in a perfect way the impact that the simple act of putting down the phone may have on our life. Here are some of his quotations to make us think and be ashamed of: "These media we call social are anything but, when we open our computers it's our doors we shut"; "four hundred and twenty two friends, yet I'm lonely; I speak to all of them every day, yet none of them really knows me". Is it so difficult to moderate technology in our life?

Finally, education is being overtaken by information technologies as a tsunami overtakes its shore. Learning has become a do-it-yourself, a just-in-time exploration of many sources to construct knowledge for a particular goal or interest (Leonhard, 2014).

Nowadays, students enjoy their wireless gadgets, allowing them to communicate with their teachers and classmates. On the other hand, schools and universities themselves are using social networks (SNs) to recruit their applicants. Besides, youth hold a rather close intimacy and familiarity with SNs. Yet, academic staff has been cautious with the use of SNs for distraction reasons. After all, Facebook quite frequently signifies gossip. According to Rosen (2012), students are only able to stay on task for an average of three to five minutes before losing their focus and, generally, their distractions are originated from technology, including (1) having more devices available in their studying environment such as iPods, laptops and smart phones; (2) texting; and (3) Facebook access. After all, SNs were not designed for formal learning purposes but rather for chatting and the sharing of pictures and videos.

Instead, Blackboard, Moodle, Schoology and other learning management systems (LMS) allow resources (files, labels, folders, URLs, pages) and activities (upload of assignments, forums, glossaries, journals, lessons, quizzes, questionnaires, wikis, chats, surveys) management for teaching purposes. A possibility could be the use of both systems in the classroom: discussions on Facebook (FB) and note taking on Moodle, for instance.

Many claim that social media in the classroom is needed now more than ever to nurture collaborative working practices. To host discussions on a specific topic on FB, to watch a lecture on YouTube, to deploy pictures of field trips in Instagram, to learn math from Khan academy, to bring TED talks into classroom debates, to use podcasts to enhance foreign language skills, to create questionnaires on Survey Monkey for research of PhD thesis, to use Prezi as an alternative to PPTs, to do class brainstorming with Word Cloud and Mind Map, to report plagiarism with smallseotools.com or to share PDFs within Slide Share are indubitably valuable tools that any instructor may and should use at present.

Does this mean that the traditional lecture is idle for the success of the learning experience? Under the formal academic path of knowledge (kindergarten, primary, secondary, bachelor, master, PhD, Post-doc), it is vital to stress that the professor is the ruler of the classroom and the glue between the content to be learnt and the learner. It is up to the professor to define the contents, goals, assessment tasks, tools to be used and the learning management process. According to the student's previous knowledge and subject's nature, his/her role can vary from a mere moderator to a heavy step-by-step broadcast teacher in a dynamic adjustment practice. For instance, math and urban geography courses will probably require two completely different approaches. If the former may follow the traditional whiteboard/handwriting approach to cope with the hundreds of possibilities of exercises (the teacher as a teacher), the latter requires web platforms (such as mygeoffice.org) and social media under the concept of the teacher as a moderator.

Furthermore, within this electronic classroom, the teacher's focal role is to provide assistance, guidance and information regarding the benefits and drawbacks of the use of technologies. As schools become more mobile and internet-connected, it is significant to consider the responsibility of giving students guidance in becoming productive, self-aware and self-disciplined citizens of the web. Yet, how can the society, in general, and the educator, in particular, moderate and set up self-discipline and responsibility within students when facing WWW inside of the teaching room? How to draw a line between being professional and being entertained? Certainly, this is the first aim for any educator in this modern era and a huge problem to be solved for the next coming years.

Concerning assessment, should teaching courses follow a quiz or a paper/project-based assessment? This answer belongs to the educator. If the teacher covers the six steps of Bloom's revised taxonomy (remember, understand, apply, analyze, evaluate, create), both types of assessment will be most likely used. Once again, social media in education are just a set of tools available to any lecturer for a possible better education support and are to be used in a complementary mode (inside and outside) of the classroom. Data and information belong to a computer digitalization process but the use and interpretation of knowledge and wisdom belong to human beings.

Another mandatory concern to be addressed in education is closely related to the learning progression from the learners' point of view: Does the student aspire to study by himself in a self-tutoring way (using electronic books or SNs, for instance) or does s/he want to enroll in courses and rely on the teacher's traditional approach? What do Bill Gates, Michael Dell, Larry Page or Mark Zuckerberg have in common? All of them dropped out of university to become self-made, earned authorities in their fields of expertise. Following Rohn (2014), "formal education will make you a living; self-education will make you a fortune".

Under this informal path of knowledge, Web 2.0 tools (go2web20.net) and learning objects (learnactivity.com/lo) play even a more significant key role because informal learning is a very personal project. In order to succeed, these specific individuals hold an internal strong incentive to obtain new information, allowing them to reach the last level of the Bloom scale (in the quickest way) with tangible outcomes. It is financially cheaper, too. As expected, the price to pay for this option is sweat, hard work and a high level of uncertainty about the final results. Furthermore, no formal diploma or certificate is issued to the learner by any official institution for future society recognition, which represents another high-risk factor.

This dispute over formal (intentional) and informal (incidental) education implementation has a long lasting history and it has been subject of an ongoing debate about its effects within society. Complexity, self-paced rhythm, fulfillment of immediate needs, personalized empowerment and flexibility are characteristics of informal learning (Cora, Machon, Milton, Crystal, & Connie, 2015). ICT courses fit this approach quite well (by creating the famous computer geeks community). On the other hand, formal learning promotes a paced learning, conventional in the sense that it favors organizations for fitting their schedules, an equal opportunity to receive education and "one program fitting all learners". Quite often, this leads to a superficial learning, that is, memorization of content to recite in a quiz/exam and to forget afterwards. As expected, a top-down instruction is expected to be enforced. Frequently, the teaching of major subjects follows this trend. As a complement, additional private tutoring nurtures children's academic supervision for modern families with a high divorce rate, full-time working parents and single mothers struggling between office and household responsibilities.

Another key difference between both strategies is discipline. Discipline is a tailor for formal education to keep the class moving ahead and smoothly sustainable (Cora, Machon, Milton, Crystal & Connie, 2015). If formal education requires discipline in every school corner during fifteen years in an official schedule setting, informal learning comes from life experiences, curiosity and personal capability. Informal education does not respect discipline but looks for innovation and freedom in flexible, fun and social environments such as blogs and wikis, visiting museums, attending fairs and exhibitions or participating in scientific conferences. Having the freedom to choose according to their interests creates a greater desire to learn among self-motivated learners.

This is particularly true for those who excel in drama, music, sports, photography, painting, film making, fashion and industry design, for instance. To realize their true creative potential, it is certainly easier to endeavor future projects in an informal setting than it is in a formal one. This context can be especially beneficial to toddlers and teenagers with special needs (Bryson, Brewer, Sibieta & Butt, 2012). Another common adversity in fields such as humanities, for instance, is the low market demand for jobs.

It is also crucial to stress that formal and informal learning should not be thought of as completely distinct entities, but rather as being part of a continuum (Clark, 2013). Whatever option is chosen by each individual, both choices should be available in a complementary manner. The concept of blended learning (face-to-face & digital media) is a consequence of this idea. This contemplation can be confirmed with the use of inquiry-oriented sessions. In a geography class, for example, the conventional introduction lecture to earthquakes can be followed with WebQuest.org, where the lecturer supplies the questions to be answered, materials to be researched and how the evaluation will be undertaken. Another possibility is taking the formal learners to a specific environment for teaching purposes. The English educator may cover verbal expressions and vocabulary in a farm where students can experiment harvesting vegetables and fruits in a motivating setting.

Another remarkable viewpoint is given by Robert Kiyosaki, the author of Rich Dad, Poor Dad: "My rich dad wanted me to be an entrepreneur. He believed college was for those who would be employees, not entrepreneurs, because it didn't teach you anything about money, investing or running a business. Instead, it taught you how to do as you're told to advance up the corporate ladder." It seems that formal education shapes anyone to be a contributing citizen for society but it does not train you to become a boss or start up your own business (Andy, Anissa, Eda & Tiffany, 2015). According to these authors, there are only 5% of individuals that fall into this group because many people in this world could not take that much amount of risk in life. The majority of human beings tends to stay in their comfort zone and do not enjoy facing challenges and difficulties. It seems that formal teaching is for followers whereas leaders prefer informal learning.

Is it possible to implement a school system that could balance both learning perspectives with enjoyable interactive education? Is it possible to set up an official learning framework where teenagers have more fun and freedom to choose from a wide range of topics according to their interests (the customization concept)? From the student perspective, this kind of learning would definitely be more meaningful and effective. From the government perspective, it would be more costly. From the teacher's perspective, many of them would be unemployed due to the non-popularity of their teaching subjects (probabilities, philosophy, econometrics, finance, ledger accounting). From the perspective of the enterprise and the scientific community, applicants without the mandatory technical background may not meet the job needs.

Formal learning will always be an integral component of schooling to better prepare and provide learners with the certifications needed for future employment. Informal learning, on the other hand, offers a great deal of educational processes in a motivating environment. Again, is it possible to merge them? There are plenty of factors to be considered within education. Taking into account how learners get motivated and involved could be already a giant challenge. Motivation is an essential ingredient for learning where formal lessons should be presented in such a way that students are able to understand and experience what is being taught (learning by doing), feel and understand its relevance. Still, how can any teacher handle and respond effectively when a class has more than 20 students?

Education is more than schooling. Lessons should not be confined to classrooms. Creating informal learning situations as part of the curriculum becomes indispensable to provide a diversified learning environment, including technological advancements and life-long learning (Saldanha, Inacio, Ampon, Sobhani & Fellini, 2015). If educators aim at a better education system, governments should continue to develop a system wherein there is a combination of formal and informal learning. Informal learning can reinforce the knowledge learned formally with the aim of developing students into autonomous learners.

For sure, the educational interaction depends on three agents: learners, educators and intangible digital knowledge. Moreover, the dramatic rise in use of digital media has changed the way learning and teaching is taking place and has led us to rethink the way we teach with digital technologies (Thota & Negreiros, 2014). Social media in education are here to stay. Keeping shared blogs of each class, creating a social classroom on Edmodo or using Google hangouts to bring people together is already on the way.

At last, several evidences towards informal and formal learning and the need to reshape 21st century hybrid teaching in knowledge economies and open societies are a requirement. For instance, several subjects are embraced (more than 10, quite repeatedly) by F1-F6 students. Does this duck feeding approach make sense? A possible answer is to consider a set of fundamental courses for everyone such as math, languages, physical education, drawing and technology and leave the remaining courses (geography, physics, chemistry, history, music, philosophy…) as optional, according to the preference of each individual. From the educational perspective, less is more and, from the perspective of the employment market, specialization is better.

But the most eminent setback of our school system nowadays is in regard to its aims and goals. At present, the dominant grade culture based on quizzes such as SAT, GMAT, TOEFL, GRE, school and national exams taken every semester, ISAT, IELTS, PISA, ETS, among others, undermine the education sector because of professional competitive advantage reasons where students are expected to perform well in order to achieve the desired university degree and to attain a reasonable CV for the current job market. Somehow, the final GPA became the first reason (and probably the only one) to be in school, an evil requisite from society for ranking, standardization and recognition purposes. Meanwhile, learners stop enjoying the process of learning where teachers have to labor (Robinson, 2006). As a result, major dropouts have become a reality; resilient students endure while the creative process and the joy for learning is relegated to the background.

REFERENCES

Andy, Eda, & Tiffany. (2015). *Formal VS Informal Education*. Unpublished.

Bryson, C., Brewer, M., Sibieta, L., & Butt, S. (2012). *The role of informal childcare: a synthesis and critical review of the evidence*. Institute for Social and Economic Research.

Clark, D. (2013). *Informal and Formal Learning*. Retrieved from www.knowledgejump.com/learning/informal.html

Cora, L. J., Machon, J. C., Milton, W. C., Crystal, L. L., & Connie, C. M. (2015). *Formal VS Informal Learning*. Unpublished.

Kemp, S. (2014). *We are social*. Retrieved from wearesocial.sg/blog/2014/01/social-digital-mobile-apac-2014/.

Leonhard, G. (2014). *Gerd Home Page*. Retrieved from www.futuristgerd.com/

Rohn, J. (2014). *BrainyQuote*. Retrieved from www.brainyquote.com/quotes/authors /j/jim_rohn.html

Rosen, L. (2012). *Driven to distraction: How to help wired students learn to focus*. Retrieved from www.eschoolnews.com/2012/11/13/driven-to-distraction-how-to-help-wired-students-learn-to-focus/

Saldanha, A., Inacio, A. R., Ampon, K. A., Sobhani, N., & Fellini, R. (2015). *Formal VS Informal Learning*. Unpublished.

Thota, N., & Negreiros, J. (2015). Teachers: Experience Report. *Journal of University Teaching & Learning Practice, 12*(1).

Appendix 1

STATISTICAL TABLES

Examples on How to Use This Standard Normal Table

1. Use the right-tailed z-test, if the given test statistic is 2.34: the p-value of P(Z>2.34) equals 1-P(Z<2.34) = 1-0.9904 = 0.0096. Since the p-value is less than .05, the right-tailed z-test is significant (rejection of H0 with a level of confidence of 95%).
2. Use a two-tailed z-test, if the test statistic is 1.06: the p-value P(Z<-1.06 or Z>1.06) equals 2×P(Z>1.06) = 2×[1-P(Z<1.06)] = 2×(1-0.8554) = 0.2892. Since this p-value>.05, the two-tailed z-test is not significant at the .05 level (acceptance of H0 with a confidence level of 95%).
3. The 97.5th quantile of the standard Normal distribution is 1.96.

Under this T distribution table, the column header states the probabilities to the left of the critical value, for example, t(19, 0.95)=1.729.

As usual, the areas given across the top are the areas to the right of the critical value. For instance, the critical Chi-square value for an alpha (type I) error of 5% (df=4) equals 9.488. From a practical point of view, if your Chi-square test is greater than 9.488, you are able to reject your null hypothesis (H0) with a 95% level of confidence.

This table presents the critical values of the F distribution for an alpha (type I error) of 0.05 (level of significance), for example, F(2,19;0.95)=3.522.

For practical purposes, if the number of available samples (n) is 32 and the number of explanatory variables excluding the constant term (k) equals 5 then dL=1.11 and dU=1.82.

Appendix 1

Table 1. Standard Gaussian table

z	.00	.01	.02	.03	.04	.05	.06	.07	.08	.09
.0	.5000	.5040	.5080	.5120	.5160	.5199	.5239	.5279	.5319	.5359
.1	.5398	.5438	.5478	.5517	.5557	.5596	.5636	.5675	.5714	.5753
.2	.5793	.5832	.5871	.5910	.5948	.5987	.6026	.6064	.6103	.6141
.3	.6179	.6217	.6255	.6293	.6331	.6368	.6406	.6443	.6480	.6517
.4	.6554	.6591	.6628	.6664	.6700	.6736	.6772	.6808	.6844	.6879
.5	.6915	.6950	.6985	.7019	.7054	.7088	.7123	.7157	.7190	.7224
.6	.7257	.7291	.7324	.7357	.7389	.7422	.7454	.7486	.7517	.7549
.7	.7580	.7611	.7642	.7673	.7704	.7734	.7764	.7794	.7823	.7852
.8	.7881	.7910	.7939	.7967	.7995	.8023	.8051	.8078	.8106	.8133
.9	.8159	.8186	.8212	.8238	.8264	.8289	.8315	.8340	.8365	.8389
1.0	.8413	.8438	.8461	.8485	.8508	.8531	.8554	.8577	.8599	.8621
1.1	.8643	.8665	.8686	.8708	.8729	.8749	.8770	.8790	.8810	.8830
1.2	.8849	.8869	.8888	.8907	.8925	.8944	.8962	.8980	.8997	.9015
1.3	.9032	.9049	.9066	.9082	.9099	.9115	.9131	.9147	.9162	.9177
1.4	.9192	.9207	.9222	.9236	.9251	.9265	.9279	.9292	.9306	.9319
1.5	.9332	.9345	.9357	.9370	.9382	.9394	.9406	.9418	.9429	.9441
1.6	.9452	.9463	.9474	.9484	.9495	.9505	.9515	.9525	.9535	.9545
1.7	.9554	.9564	.9573	.9582	.9591	.9599	.9608	.9616	.9625	.9633
1.8	.9641	.9649	.9656	.9664	.9671	.9678	.9686	.9693	.9699	.9706
1.9	.9713	.9719	.9726	.9732	.9738	.9744	.9750	.9756	.9761	.9767
2.0	.9772	.9778	.9783	.9788	.9793	.9798	.9803	.9808	.9812	.9817
2.1	.9821	.9826	.9830	.9834	.9838	.9842	.9846	.9850	.9854	.9857
2.2	.9861	.9864	.9868	.9871	.9875	.9878	.9881	.9884	.9887	.9890
2.3	.9893	.9896	.9898	.9901	.9904	.9906	.9909	.9911	.9913	.9916
2.4	.9918	.9920	.9922	.9925	.9927	.9929	.9931	.9932	.9934	.9936
2.5	.9938	.9940	.9941	.9943	.9945	.9946	.9948	.9949	.9951	.9952
2.6	.9953	.9955	.9956	.9957	.9959	.9960	.9961	.9962	.9963	.9964
2.7	.9965	.9966	.9967	.9968	.9969	.9970	.9971	.9972	.9973	.9974
2.8	.9974	.9975	.9976	.9977	.9977	.9978	.9979	.9979	.9980	.9981
2.9	.9981	.9982	.9982	.9983	.9984	.9984	.9985	.9985	.9986	.9986
3.0	.9987	.9987	.9987	.9988	.9988	.9989	.9989	.9989	.9990	.9990
3.1	.9990	.9991	.9991	.9991	.9992	.9992	.9992	.9992	.9993	.9993
3.2	.9993	.9993	.9994	.9994	.9994	.9994	.9994	.9995	.9995	.9995
3.3	.9995	.9995	.9995	.9996	.9996	.9996	.9996	.9996	.9996	.9997
3.4	.9997	.9997	.9997	.9997	.9997	.9997	.9997	.9997	.9997	.9998

(Source: www.normaltable.com)

Table 2. Standard t-table

df	$t_{.80}$	$t_{.90}$	$t_{.95}$	$t_{.975}$	$t_{.99}$	$t_{.995}$
1	1.376	3.078	6.314	12.71	31.82	63.66
2	1.061	1.886	2.920	4.303	6.965	9.925
3	.978	1.638	2.353	3.182	4.541	5.841
4	.941	1.533	2.132	2.776	3.747	4.604
5	.920	1.476	2.015	2.571	3.365	4.032
6	.906	1.440	1.943	2.447	3.143	3.707
7	.896	1.415	1.895	2.365	2.998	3.499
8	.889	1.397	1.860	2.306	2.896	3.355
9	.883	1.383	1.833	2.262	2.821	3.250
10	.879	1.372	1.812	2.228	2.764	3.169
11	.876	1.363	1.796	2.201	2.718	3.106
12	.873	1.356	1.782	2.179	2.681	3.055
13	.870	1.350	1.771	2.160	2.650	3.012
14	.868	1.345	1.761	2.145	2.624	2.977
15	.866	1.341	1.753	2.131	2.602	2.947
16	.865	1.337	1.746	2.120	2.583	2.921
17	.863	1.333	1.740	2.110	2.567	2.898
18	.862	1.330	1.734	2.101	2.552	2.878
19	.861	1.328	1.729	2.093	2.539	2.861
20	.860	1.325	1.725	2.086	2.528	2.845
21	.859	1.323	1.721	2.080	2.518	2.831
22	.858	1.321	1.717	2.074	2.508	2.819
23	.858	1.319	1.714	2.069	2.500	2.807
24	.857	1.318	1.711	2.064	2.492	2.797
25	.856	1.316	1.708	2.060	2.485	2.787
26	.856	1.315	1.706	2.056	2.479	2.779
27	.855	1.314	1.703	2.052	2.473	2.771
28	.855	1.313	1.701	2.048	2.467	2.763
29	.854	1.311	1.699	2.045	2.462	2.756
30	.854	1.310	1.697	2.042	2.457	2.750
40	.851	1.303	1.684	2.021	2.423	2.704
60	.848	1.296	1.671	2.000	2.390	2.660
80	.846	1.292	1.664	1.990	2.374	2.639
100	.845	1.290	1.660	1.984	2.364	2.626
120	.845	1.289	1.658	1.980	2.358	2.617
Inf	.842	1.282	1.645	1.960	2.326	2.576

(source: www.normaltable.com)

Appendix 1

Table 3. Standard chi-square table

df	0.99	0.975	0.95	0.90	0.10	0.05	0.025	0.01	0.005
1		0.001	0.004	0.016	2.706	3.841	5.024	6.635	7.879
2	0.020	0.051	0.103	0.211	4.605	5.991	7.378	9.210	10.597
3	0.115	0.216	0.352	0.584	6.251	7.815	9.348	11.345	12.838
4	0.297	0.484	0.711	1.064	7.779	9.488	11.143	13.277	14.860
5	0.554	0.831	1.145	1.610	9.236	11.070	12.833	15.086	16.750
6	0.872	1.237	1.635	2.204	10.645	12.592	14.449	16.812	18.548
7	1.239	1.690	2.167	2.833	12.017	14.067	16.013	18.475	20.278
8	1.646	2.180	2.733	3.490	13.362	15.507	17.535	20.090	21.955
9	2.088	2.700	3.325	4.168	14.684	16.919	19.023	21.666	23.589
10	2.558	3.247	3.940	4.865	15.987	18.307	20.483	23.209	25.188
11	3.053	3.816	4.575	5.578	17.275	19.675	21.920	24.725	26.757
12	3.571	4.404	5.226	6.304	18.549	21.026	23.337	26.217	28.300
13	4.107	5.009	5.892	7.042	19.812	22.362	24.736	27.688	29.819
14	4.660	5.629	6.571	7.790	21.064	23.685	26.119	29.141	31.319
15	5.229	6.262	7.261	8.547	22.307	24.996	27.488	30.578	32.801
16	5.812	6.908	7.962	9.312	23.542	26.296	28.845	32.000	34.267
17	6.408	7.564	8.672	10.085	24.769	27.587	30.191	33.409	35.718
18	7.015	8.231	9.390	10.865	25.989	28.869	31.526	34.805	37.156
19	7.633	8.907	10.117	11.651	27.204	30.144	32.852	36.191	38.582
20	8.260	9.591	10.851	12.443	28.412	31.410	34.170	37.566	39.997
21	8.897	10.283	11.591	13.240	29.615	32.671	35.479	38.932	41.401
22	9.542	10.982	12.338	14.041	30.813	33.924	36.781	40.289	42.796
23	10.196	11.689	13.091	14.848	32.007	35.172	38.076	41.638	44.181
24	10.856	12.401	13.848	15.659	33.196	36.415	39.364	42.980	45.559
25	11.524	13.120	14.611	16.473	34.382	37.652	40.646	44.314	46.928
26	12.198	13.844	15.379	17.292	35.563	38.885	41.923	45.642	48.290
27	12.879	14.573	16.151	18.114	36.741	40.113	43.195	46.963	49.645
28	13.565	15.308	16.928	18.939	37.916	41.337	44.461	48.278	50.993
29	14.256	16.047	17.708	19.768	39.087	42.557	45.722	49.588	52.336
30	14.953	16.791	18.493	20.599	40.256	43.773	46.979	50.892	53.672
40	22.164	24.433	26.509	29.051	51.805	55.758	59.342	63.691	66.766
50	29.707	32.357	34.764	37.689	63.167	67.505	71.420	76.154	79.490
60	37.485	40.482	43.188	46.459	74.397	79.082	83.298	88.379	91.952
70	45.442	48.758	51.739	55.329	85.527	90.531	95.023	100.425	104.215
80	53.540	57.153	60.391	64.278	96.578	101.879	106.629	112.329	116.321
90	61.754	65.647	69.126	73.291	107.565	113.145	118.136	124.116	128.299
100	70.065	74.222	77.929	82.358	118.498	124.342	129.561	135.807	140.169

(Source: https://people.richland.edu/james/lecture/m170/tbl-chi.html)

Table 4. Standard f-table for an alpha error of 0.05

df1=	1	2	3	4	5	6	7	8	9	10	12	15	20	30	40	60	120	Inf
df_2=1	161.4	199.5	215.7	224.6	230.2	234.0	236.8	238.9	240.5	241.9	243.9	245.9	248.0	250.1	251.1	252.2	253.3	254.3
2	18.51	19.00	19.16	19.25	19.30	19.33	19.35	19.37	19.38	19.40	19.41	19.43	19.45	19.46	19.47	19.48	19.49	19.50
3	10.13	9.552	9.277	9.117	9.014	8.941	8.887	8.845	8.812	8.786	8.745	8.703	8.660	8.617	8.594	8.572	8.549	8.526
4	7.709	6.944	6.591	6.388	6.256	6.163	6.094	6.041	5.999	5.964	5.912	5.858	5.803	5.746	5.717	5.688	5.658	5.628
5	6.608	5.786	5.409	5.192	5.050	4.950	4.876	4.818	4.772	4.735	4.678	4.619	4.558	4.496	4.464	4.431	4.398	4.365
6	5.987	5.143	4.757	4.534	4.387	4.284	4.207	4.147	4.099	4.060	4.000	3.938	3.874	3.808	3.774	3.740	3.705	3.669
7	5.591	4.737	4.347	4.120	3.972	3.866	3.787	3.726	3.677	3.637	3.575	3.511	3.445	3.376	3.340	3.304	3.267	3.230
8	5.318	4.459	4.066	3.838	3.688	3.581	3.500	3.438	3.388	3.347	3.284	3.218	3.150	3.079	3.043	3.005	2.967	2.928
9	5.117	4.256	3.863	3.633	3.482	3.374	3.293	3.230	3.179	3.137	3.073	3.006	2.936	2.864	2.826	2.787	2.748	2.707
10	4.965	4.103	3.708	3.478	3.326	3.217	3.135	3.072	3.020	2.978	2.913	2.845	2.774	2.700	2.661	2.621	2.580	2.538
11	4.844	3.982	3.587	3.357	3.204	3.095	3.012	2.948	2.896	2.854	2.788	2.719	2.646	2.570	2.531	2.490	2.448	2.404
12	4.747	3.885	3.490	3.259	3.106	2.996	2.913	2.849	2.796	2.753	2.687	2.617	2.544	2.466	2.426	2.384	2.341	2.296
13	4.667	3.806	3.411	3.179	3.025	2.915	2.832	2.767	2.714	2.671	2.604	2.533	2.459	2.380	2.339	2.297	2.252	2.206
14	4.600	3.739	3.344	3.112	2.958	2.848	2.764	2.699	2.646	2.602	2.534	2.463	2.388	2.308	2.266	2.223	2.178	2.131
15	4.543	3.682	3.287	3.056	2.901	2.790	2.707	2.641	2.588	2.544	2.475	2.403	2.328	2.247	2.204	2.160	2.114	2.066
16	4.494	3.634	3.239	3.007	2.852	2.741	2.657	2.591	2.538	2.494	2.425	2.352	2.276	2.194	2.151	2.106	2.059	2.010
17	4.451	3.592	3.197	2.965	2.810	2.699	2.614	2.548	2.494	2.450	2.381	2.308	2.230	2.148	2.104	2.058	2.011	1.960
18	4.414	3.555	3.160	2.928	2.773	2.661	2.577	2.510	2.456	2.412	2.342	2.269	2.191	2.107	2.063	2.017	1.968	1.917
19	4.381	3.522	3.127	2.895	2.740	2.628	2.544	2.477	2.423	2.378	2.308	2.234	2.156	2.071	2.026	1.980	1.930	1.878
20	4.351	3.493	3.098	2.866	2.711	2.599	2.514	2.447	2.393	2.348	2.278	2.203	2.124	2.039	1.994	1.946	1.896	1.843
21	4.325	3.467	3.072	2.840	2.685	2.573	2.488	2.420	2.366	2.321	2.250	2.176	2.096	2.010	1.965	1.916	1.866	1.812

continued on following page

Table 4. Continued

df1=	1	2	3	4	5	6	7	8	9	10	12	15	20	30	40	60	120	Inf
22	4.301	3.443	3.049	2.817	2.661	2.549	2.464	2.397	2.342	2.297	2.226	2.151	2.071	1.984	1.938	1.889	1.838	1.783
23	4.279	3.422	3.028	2.796	2.640	2.528	2.442	2.375	2.320	2.275	2.204	2.128	2.048	1.961	1.914	1.865	1.813	1.757
24	4.260	3.403	3.009	2.776	2.621	2.508	2.423	2.355	2.300	2.255	2.183	2.108	2.027	1.939	1.892	1.842	1.790	1.733
25	4.242	3.385	2.991	2.759	2.603	2.490	2.405	2.337	2.282	2.236	2.165	2.089	2.007	1.919	1.872	1.822	1.768	1.711
26	4.225	3.369	2.975	2.743	2.587	2.474	2.388	2.321	2.265	2.220	2.148	2.072	1.990	1.901	1.853	1.803	1.749	1.691
27	4.210	3.354	2.960	2.728	2.572	2.459	2.373	2.305	2.250	2.204	2.132	2.056	1.974	1.884	1.836	1.785	1.731	1.672
28	4.196	3.340	2.947	2.714	2.558	2.445	2.359	2.291	2.236	2.190	2.118	2.041	1.959	1.869	1.820	1.769	1.714	1.654
29	4.183	3.328	2.934	2.701	2.545	2.432	2.346	2.278	2.223	2.177	2.104	2.027	1.945	1.854	1.806	1.754	1.698	1.638
30	4.171	3.316	2.922	2.690	2.534	2.421	2.334	2.266	2.211	2.165	2.092	2.015	1.932	1.841	1.792	1.740	1.683	1.622
40	4.085	3.232	2.839	2.606	2.449	2.336	2.249	2.180	2.124	2.077	2.003	1.924	1.839	1.744	1.693	1.637	1.577	1.509
60	4.001	3.150	2.758	2.525	2.368	2.254	2.167	2.097	2.040	1.993	1.917	1.836	1.748	1.649	1.594	1.534	1.467	1.389
120	3.920	3.072	2.680	2.447	2.290	2.175	2.087	2.016	1.959	1.910	1.834	1.751	1.659	1.554	1.495	1.429	1.352	1.254
inf	3.842	2.996	2.605	2.372	2.214	2.099	2.010	1.938	1.880	1.831	1.752	1.666	1.571	1.459	1.394	1.318	1.221	1.000

(Source: www.normaltable.com)

Table 5. The Durbin-Watson d statistic: significance points of dL and dU at 0.05 level of significance

	k=1		k=2		k=3		k=4		k=5		k=6		k=7		k=8		k=9		k=10	
n	dL	dU	dL	dU	dL	dU	dL	dU	dL	dU	dL	dU	dL	dU	dL	dU	dL	dU	dL	dU
6	0.61	1.40	—	—	—	—	—	—	—	—	—	—	—	—	—	—	—	—	—	—
7	0.70	1.36	0.47	1.90	—	—	—	—	—	—	—	—	—	—	—	—	—	—	—	—
8	0.76	1.33	0.56	1.78	0.37	2.29	—	—	—	—	—	—	—	—	—	—	—	—	—	—
9	0.82	1.32	0.63	1.70	0.46	2.13	0.30	2.59	—	—	—	—	—	—	—	—	—	—	—	—
10	0.88	1.32	0.70	1.64	0.53	2.02	0.38	2.41	0.24	2.82	—	—	—	—	—	—	—	—	—	—
11	0.93	1.32	0.66	1.60	0.60	1.93	0.44	2.28	0.32	2.65	0.20	3.01	—	—	—	—	—	—	—	—
12	0.97	1.33	0.81	1.58	0.66	1.86	0.51	2.18	0.38	2.51	0.27	2.83	0.17	3.15	—	—	—	—	—	—
13	1.01	1.34	0.86	1.56	0.72	1.82	0.57	2.09	0.45	2.39	0.33	2.69	0.23	2.99	0.15	3.27	—	—	—	—
14	1.05	1.35	0.91	1.55	0.77	1.78	0.63	2.03	0.51	2.30	0.39	2.57	0.29	2.85	0.20	3.11	0.13	3.36	—	—
15	1.08	1.36	0.95	1.54	0.81	1.75	0.69	1.98	0.56	2.22	0.45	2.47	0.34	2.73	0.25	2.98	0.18	3.22	0.11	3.44
16	1.11	1.37	0.98	1.54	0.86	1.73	0.73	1.94	0.62	2.16	0.50	2.39	0.40	2.62	0.30	2.86	0.22	3.09	0.16	3.30
17	1.13	1.38	1.02	1.54	0.90	1.71	0.78	1.90	0.66	2.10	0.55	2.32	0.45	2.54	0.36	2.76	0.27	2.98	0.20	3.18
18	1.16	1.39	1.05	1.54	0.93	1.70	0.82	1.87	0.71	2.06	0.60	2.26	0.50	2.46	0.41	2.67	0.32	2.87	0.24	3.07
19	1.18	1.40	1.07	1.54	0.97	1.69	0.86	1.85	0.75	2.02	0.65	2.21	0.55	2.40	0.46	2.59	0.37	2.78	0.29	2.97
20	1.20	1.41	1.10	1.54	1.	1.68	0.89	1.83	0.79	1.99	0.69	2.16	0.60	2.34	0.50	2.52	0.42	2.70	0.34	2.89
21	1.22	1.42	1.13	1.54	1.03	1.67	0.93	1.81	0.83	1.96	0.73	2.12	0.64	2.29	0.55	2.46	0.46	2.63	0.38	2.81
22	1.24	1.43	1.15	1.54	1.05	1.66	0.96	1.80	0.86	1.94	0.77	2.09	0.68	2.25	0.59	2.41	0.50	2.57	0.42	2.73
23	1.26	1.44	1.17	1.54	1.08	1.66	0.99	1.79	0.90	1.92	0.80	2.06	0.72	2.21	0.63	2.36	0.55	2.51	0.47	2.67
24	1.27	1.45	1.19	1.55	1.10	1.66	1.01	1.78	0.93	1.90	0.84	2.04	0.75	2.17	0.67	2.32	0.58	2.46	0.51	2.61
25	1.29	1.45	1.21	1.55	1.12	1.65	1.04	1.77	0.95	1.89	0.87	2.01	0.78	2.14	0.70	2.28	0.62	2.42	0.54	2.56

continued on following page

Table 5. Continued

	k=1		k=2		k=3		k=4		k=5		k=6		k=7		k=8		k=9		k=10	
n	**dL**	**dU**	**dL**	**dU**	**dL**	**dU**	**dL**	**dU**	**dL**	**dU**	**dL**	**dU**	**dL**	**dU**	**dL**	**dU**	**dL**	**dU**	**dL**	**dU**
26	1.30	1.46	1.22	1.55	1.14	1.65	1.06	1.76	0.98	1.87	0.90	1.99	0.82	2.12	0.74	2.25	0.66	2.38	0.58	2.51
27	1.32	1.47	1.24	1.56	1.16	1.65	1.08	1.75	1.	1.86	0.93	1.97	0.85	2.09	0.77	2.22	0.69	2.34	0.62	2.47
28	1.33	1.48	1.26	1.56	1.18	1.65	1.10	1.75	1.03	1.85	0.95	1.96	0.87	2.07	0.80	2.19	0.72	2.31	0.65	2.43
29	1.34	1.48	1.27	1.56	1.20	1.65	1.12	1.74	1.05	1.84	0.98	1.94	0.90	2.05	0.83	2.16	0.75	2.28	0.68	2.40
30	1.35	1.49	1.28	1.57	1.21	1.65	1.14	1.74	1.07	1.83	1.	1.93	0.93	2.03	0.85	2.14	0.78	2.25	0.71	2.36
31	1.36	1.50	1.30	1.57	1.23	1.65	1.16	1.74	1.09	1.83	1.02	1.92	0.95	2.02	0.88	2.12	0.81	2.23	0.74	2.33
32	1.37	1.50	1.31	1.57	1.24	1.65	1.18	1.73	1.11	1.82	1.04	1.91	0.97	2.	0.90	2.10	0.84	2.20	0.77	2.31
33	1.38	1.51	1.32	1.58	1.26	1.65	1.19	1.73	1.13	1.81	1.06	1.90	0.99	1.99	0.93	2.09	0.86	2.18	0.80	2.28
34	1.39	1.51	1.33	1.58	1.27	1.65	1.21	1.73	1.14	1.81	1.08	1.89	1.02	1.98	0.95	2.07	0.89	2.16	0.82	2.26
35	1.40	1.52	1.34	1.58	1.28	1.65	1.22	1.73	1.16	1.80	1.10	1.88	1.03	1.97	0.97	2.05	0.91	2.14	0.85	2.24
36	1.41	1.53	1.35	1.59	1.30	1.65	1.24	1.72	1.18	1.80	1.11	1.88	1.05	1.96	0.99	2.04	0.93	2.13	0.87	2.22
37	1.42	1.53	1.36	1.59	1.31	1.66	1.25	1.72	1.19	1.80	1.13	1.87	1.07	1.95	1.01	2.03	0.95	2.11	0.89	2.20
38	1.43	1.54	1.37	1.59	1.32	1.66	1.26	1.72	1.20	1.79	1.15	1.86	1.09	1.94	1.03	2.02	0.97	2.10	0.91	2.18
39	1.44	1.54	1.38	1.60	1.33	1.66	1.27	1.72	1.22	1.79	1.16	1.86	1.10	1.93	1.05	2.01	0.99	2.09	0.93	2.16
40	1.44	1.54	1.39	1.60	1.34	1.66	1.29	1.72	1.23	1.79	1.18	1.85	1.12	1.92	1.06	2.	1.01	2.07	0.95	2.15
45	1.48	1.57	1.43	1.62	1.38	1.67	1.34	1.72	1.29	1.78	1.24	1.84	1.19	1.90	1.14	1.96	1.09	2.02	1.04	2.09
50	1.50	1.59	1.46	1.63	1.42	1.67	1.38	1.72	1.34	1.77	1.29	1.82	1.25	1.88	1.20	1.93	1.16	1.99	1.11	2.04
55	1.53	1.60	1.49	1.64	1.45	1.68	1.41	1.72	1.37	1.77	1.33	1.81	1.29	1.86	1.25	1.91	1.21	1.96	1.17	2.01
60	1.55	1.62	1.51	1.65	1.48	1.69	1.44	1.73	1.41	1.77	1.37	1.81	1.34	1.85	1.30	1.89	1.26	1.94	1.22	1.98
65	1.57	1.63	1.54	1.66	1.50	1.70	1.47	1.73	1.44	1.77	1.40	1.81	1.37	1.84	1.34	1.88	1.30	1.92	1.27	1.96

continued on following page

Table 5. Continued

	k=1		k=2		k=3		k=4		k=5		k=6		k=7		k=8		k=9		k=10	
n	**dL**	**dU**	**dL**	**dU**	**dL**	**dU**	**dL**	**dU**	**dL**	**dU**	**dL**	**dU**	**dL**	**dU**	**dL**	**dU**	**dL**	**dU**	**dL**	**dU**
70	1.58	1.64	1.55	1.67	1.53	1.70	1.49	1.74	1.46	1.77	1.43	1.80	1.40	1.84	1.37	1.87	1.34	1.91	1.31	1.95
75	1.60	1.65	1.57	1.68	1.54	1.71	1.52	1.74	1.49	1.77	1.46	1.80	1.43	1.83	1.40	1.87	1.37	1.90	1.34	1.94
80	1.61	1.66	1.59	1.69	1.56	1.72	1.53	1.74	1.51	1.77	1.48	1.80	1.45	1.83	1.43	1.86	1.40	1.89	1.37	1.93
85	1.62	1.67	1.60	1.70	1.58	1.72	1.55	1.75	1.53	1.77	1.50	1.80	1.47	1.83	1.45	1.86	1.42	1.89	1.40	1.92
90	1.64	1.68	1.61	1.70	1.59	1.73	1.57	1.75	1.54	1.78	1.52	1.80	1.49	1.83	1.47	1.85	1.45	1.88	1.42	1.91
95	1.65	1.69	1.62	1.71	1.60	1.73	1.58	1.76	1.56	1.78	1.54	1.80	1.51	1.83	1.49	1.85	1.47	1.88	1.44	1.90
100	1.65	1.69	1.63	1.72	1.61	1.74	1.59	1.76	1.57	1.78	1.55	1.80	1.53	1.83	1.51	1.85	1.48	1.87	1.46	1.90
150	1.72	1.75	1.71	1.76	1.69	1.77	1.68	1.79	1.67	1.80	1.65	1.82	1.64	1.83	1.62	1.85	1.61	1.86	1.59	1.88
	k=11		k=12		k=13		k=14		k=15		k=16		k=17		k=18		k=19		k=20	
n	dL	dU	dL	dU	dL	dU	dL	dU	dL	dU	dL	dU	dL	dU	dL	dU	dL	dU	dL	dU
16	0.10	3.50	—	—	—	—	—	—	—	—	—	—	—	—	—	—	—	—	—	—
17	0.14	3.38	0.09	3.56	—	—	—	—	—	—	—	—	—	—	—	—	—	—	—	—
18	0.18	3.27	0.12	3.44	0.08	3.60	—	—	—	—	—	—	—	—	—	—	—	—	—	—
19	0.22	3.16	0.16	3.34	0.11	3.50	0.07	3.64	—	—	—	—	—	—	—	—	—	—	—	—
20	0.26	3.06	0.20	3.23	0.15	3.40	0.10	3.54	0.06	3.68	—	—	—	—	—	—	—	—	—	—
21	0.31	2.98	0.24	3.14	0.18	3.30	0.13	3.45	0.09	3.58	0.06	3.71	—	—	—	—	—	—	—	—
22	0.35	2.90	0.28	3.06	0.22	3.21	0.17	3.36	0.12	3.50	0.08	3.62	0.05	3.73	—	—	—	—	—	—
23	0.39	2.83	0.32	2.98	0.26	3.13	0.20	3.27	0.15	3.41	0.11	3.54	0.08	3.65	0.05	3.75	—	—	—	—
24	0.43	2.76	0.36	2.91	0.30	3.05	0.24	3.19	0.19	3.33	0.14	3.45	0.10	3.57	0.07	3.68	0.04	3.77	—	—
25	0.47	2.70	0.40	2.84	0.34	2.98	0.28	3.12	0.22	3.25	0.17	3.38	0.13	3.49	0.09	3.60	0.07	3.70	0.04	3.79

continued on following page

Table 5. Continued

	k=1		k=2		k=3		k=4		k=5		k=6		k=7		k=8		k=9		k=10	
n	**dL**	**dU**	**dL**	**dU**	**dL**	**dU**	**dL**	**dU**	**dL**	**dU**	**dL**	**dU**	**dL**	**dU**	**dL**	**dU**	**dL**	**dU**	**dL**	**dU**
26	0.51	2.65	0.44	2.78	0.37	2.92	0.31	3.05	0.26	3.18	0.21	3.30	0.16	3.42	0.12	3.53	0.09	3.63	0.06	3.72
27	0.54	2.60	0.48	2.73	0.41	2.86	0.35	2.99	0.29	3.11	0.24	3.23	0.19	3.35	0.15	3.46	0.11	3.56	0.08	3.66
28	0.58	2.56	0.51	2.68	0.45	2.81	0.38	2.93	0.33	3.05	0.27	3.17	0.22	3.28	0.18	3.39	0.14	3.50	0.10	3.59
29	0.61	2.52	0.54	2.63	0.48	2.76	0.42	2.87	0.36	2.99	0.31	3.11	0.25	3.22	0.21	3.33	0.17	3.43	0.13	3.53
30	0.64	2.48	0.58	2.59	0.51	2.71	0.45	2.82	0.39	2.94	0.34	3.05	0.29	3.16	0.24	3.27	0.20	3.37	0.16	3.47
31	0.67	2.44	0.61	2.55	0.55	2.67	0.48	2.78	0.43	2.89	0.37	3.	0.32	3.10	0.27	3.21	0.22	3.31	0.18	3.41
32	0.70	2.41	0.64	2.52	0.58	2.63	0.52	2.73	0.46	2.84	0.40	2.95	0.35	3.05	0.30	3.15	0.25	3.25	0.21	3.35
33	0.73	2.38	0.67	2.48	0.61	2.59	0.55	2.69	0.49	2.80	0.43	2.90	0.38	3.	0.33	3.10	0.28	3.20	0.24	3.29
34	0.76	2.36	0.70	2.45	0.63	2.55	0.58	2.65	0.52	2.75	0.46	2.85	0.41	2.95	0.36	3.05	0.31	3.15	0.27	3.24
35	0.78	2.33	0.72	2.43	0.66	2.52	0.60	2.62	0.55	2.72	0.49	2.81	0.44	2.91	0.39	3.01	0.34	3.10	0.30	3.19
36	0.81	2.31	0.75	2.40	0.69	2.49	0.63	2.59	0.58	2.68	0.52	2.77	0.47	2.87	0.42	2.96	0.37	3.05	0.32	3.14
37	0.83	2.29	0.77	2.37	0.71	2.46	0.66	2.56	0.60	2.65	0.55	2.74	0.50	2.83	0.45	2.92	0.40	3.01	0.35	3.10
38	0.85	2.27	0.80	2.35	0.74	2.44	0.68	2.53	0.63	2.61	0.58	2.70	0.52	2.79	0.47	2.88	0.42	2.97	0.38	3.05
39	0.88	2.25	0.82	2.33	0.76	2.41	0.71	2.50	0.65	2.59	0.60	2.67	0.55	2.76	0.50	2.84	0.45	2.93	0.40	3.01
40	0.90	2.23	0.84	2.31	0.79	2.39	0.73	2.47	0.68	2.56	0.63	2.64	0.58	2.72	0.53	2.81	0.48	2.89	0.43	2.97
45	0.99	2.16	0.94	2.23	0.89	2.30	0.84	2.37	0.79	2.44	0.74	2.51	0.69	2.59	0.64	2.66	0.60	2.73	0.55	2.81
50	1.06	2.10	1.02	2.16	0.97	2.23	0.93	2.29	0.88	2.35	0.84	2.41	0.79	2.48	0.75	2.54	0.70	2.61	0.66	2.68
55	1.13	2.06	1.09	2.12	1.05	2.17	1.	2.23	0.96	2.28	0.92	2.34	0.88	2.40	0.84	2.45	0.80	2.51	0.75	2.57
60	1.18	2.03	1.15	2.08	1.11	2.13	1.07	2.18	1.03	2.23	0.99	2.28	0.95	2.33	0.91	2.38	0.87	2.43	0.84	2.49
65	1.23	2.01	1.20	2.05	1.16	2.09	1.12	2.14	1.09	2.18	1.05	2.23	1.02	2.28	0.98	2.32	0.94	2.37	0.91	2.42

continued on following page

Table 5. Continued

	k=1		k=2		k=3		k=4		k=5		k=6		k=7		k=8		k=9		k=10	
n	dL	dU	dL	dU	dL	dU	dL	dU	dL	dU	dL	dU	dL	dU	dL	dU	dL	dU	dL	dU
70	1.27	1.99	1.24	2.03	1.21	2.07	1.17	2.11	1.14	2.15	1.11	2.19	1.07	2.23	1.04	2.28	1.01	2.32	0.97	2.36
75	1.31	1.97	1.28	2.01	1.25	2.04	1.22	2.08	1.18	2.12	1.15	2.16	1.12	2.20	1.09	2.24	1.06	2.28	1.03	2.32
80	1.34	1.96	1.31	1.99	1.28	2.02	1.25	2.06	1.22	2.09	1.20	2.13	1.17	2.17	1.14	2.20	1.11	2.24	1.08	2.28
85	1.37	1.95	1.34	1.98	1.32	2.01	1.29	2.04	1.26	2.07	1.23	2.11	1.21	2.14	1.18	2.17	1.15	2.21	1.12	2.24
90	1.40	1.94	1.37	1.97	1.34	2.	1.32	2.03	1.29	2.06	1.27	2.09	1.24	2.12	1.21	2.15	1.19	2.18	1.16	2.21
95	1.42	1.93	1.39	1.96	1.37	1.98	1.35	2.01	1.32	2.04	1.30	2.07	1.27	2.10	1.25	2.13	1.22	2.16	1.20	2.19
100	1.44	1.92	1.42	1.95	1.39	1.97	1.37	2.	1.35	2.03	1.32	2.05	1.30	2.08	1.28	2.11	1.25	2.14	1.23	2.16
150	1.58	1.89	1.56	1.91	1.55	1.92	1.54	1.94	1.52	1.96	1.50	1.97	1.49	1.99	1.47	2.01	1.46	2.02	1.44	2.04

(Source: Gujarati [2004] and available at www.fcmfmpep.org.br/disciplinas/turma1/MB-751/Aula02/Statistical%20tables.pdf)

Appendix 1

Figure 1.

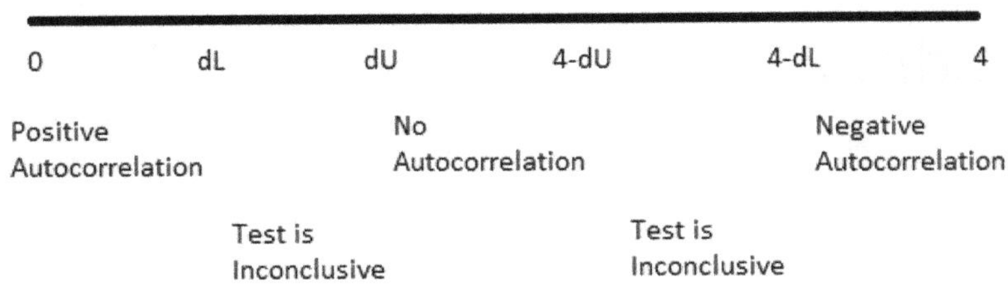

Appendix 2

C LANGUAGE HIGHLIGHTS

INTRODUCTION

This appendix aims to review some of the C computer language features (scattered by small programs for demonstration purposes) frequently used with myGeoffice© construction, particularly when a user wants to implement his/her version of C code (presented in Appendix C, Geographically Weighted Regression – GWR). For an extensive and full understand of this powerful programming language, kindly read specialized literature such as *The C Programming Language (ANSI C Version)* by Dennis Richie and Ken Thompson (ISBN 978-0131103627, Prentice Hall Software Series, 2nd version, 272 pages, 1988).

The C language was developed in 1970 for the construction of the Unix® OS. Despite its age, it is still one of the most popular programming languages due to its effectiveness and efficiency in terms of CPU speed (see *bloodshed.net/dev/devcpp.html* for a C free Windows® compiler). Like the Pascal computer language, it may contain multiple functions inside of a program although the *main()* function is considered mandatory because this is where everything starts. Another important aspect concerns the libraries of functions supplied by C such as *#include <stdio.h>*, *#include <ctype.h>*, *#include <stdlib.h>*, *#include <graphics.h>* or *#include <math.h>*.

VARIABLES

A variable is a memory location to which programmers assign a name. There are two major types of variables: Local (only valid within the function) and global (defined outside of any function and to be valid everywhere). A register variable advises the compiler that the variable in question will be heavily used (*register int z:*). Table 1 displays the definition and scope of the basic default data types that are associated with variables.

On the other hand, a constant is a fixed value that can never be changed throughout the program such as *const pi 3.14159.* An extremely useful function supplied by C libraries is the *printf()*, allowing the user to display on-screen (by default) any variable content according to its data type, including the following: *%c* (print a character), *%d* (print an integer), *%s* (print a string), *%f* (print a float) and *%lf* (print a double).

Reading Keyboard Characters, Arithmetic Functions, and Comments

The simplest way to read keyboard characters is via the *scanf()* function (see Table 2). For instance, the input of a real number implies two instructions: *fflush(stdin);scanf("%lf",&var);*. Moreover, there are five basic arithmetic functions: +, -, /, * and % (the latter returns the remainder of an integer division). If the user enjoys setting up comments inside the program, the following instruction should be respected: /* *This is a comment* */.

Table 1.

Type	Memory allocation (bits)	Range
char	8	-128 - 127
unsigned char	8	0 - 255
signed char	8	-128 - 127
int	16	-32.768 - 32.767
unsigned int	16	0 - 65.535
short int	16	-32.768 - 32.767
long int	32	-2.147.483.648 - 2.147.483.647
unsigned long int	32	0 - 4.294.967.295
float	32	3.4E-38 - 3.4E+38
double	64	1.7E-308 - 1.7E+308
long double	80	3.4E-4932 - 1.1E+4932

Table 2.

```
#include <stdio.h>
int main(void){
char c1, c2;
int n1, n2;
printf( "Write two characters: ");
fflush(stdin);scanf( "%c %c",&c1,&c2);
printf( "%c%c \n",c1,c2);
printf( "Write two numbers: ");
fflush(stdin);scanf( "%d %d",&n1,&n2);
printf( "Quotient: %d; Remainder: %d\n",n1/n2,n1%n2);
printf( "Write a character:\n");
fflush(stdin);scanf( "%c",&c1);
printf( "The third ASCII character after %c is' %c\n",c1,(char)((int)c1+3));
printf( "Write an integer:\n");
fflush(stdin);scanf( "d",&n1);
if (n1%2==0) printf( "Even number\n");
else printf( "Odd number\n");}
```

```
Write two characters: g r
gr
Write two numbers: 34 77
Quotient: 0; Remainder: 34
Write a character:
e
The third ASCII character after e is' h
Write an integer:
222
Even number
```

The *if (boolean test) {...} else {...}* statement in the previous program is a selection command. Its operation is defined by the result of the conditional test that is carried out which is either true (any value other than zero) or False (zero). An extra selection alternative is the *switch()/break* statements.

Random Values and Iterations

The code of Table 3 illustrates the C language's capability to generate a random value (for gaming purposes, for instance). The increment or decrement in variables (*i=i+1;* or *i=i-1;*) can be substituted with *i++;* or *i--;*. In terms of relational operators, the usual ones are found: > (greater than); >= (greater than or equal to); < (less than); <= (less than or equal to); == (equal to); *!=* (different from). The usual logical elementary functions are also available: *&&* (and), || (or) and *!* (not).

Further, the C language possesses three types of iterations or block cycles:

Table 3.

```
#include <stdio.h>
#include <math.h>
int main(void){
long t;
time(&t);
srand((unsigned)t);
printf("Random value: %d",rand());}
```

```
C:\>11.exe
Random generated value: 19802
C:\>_
```

- *for (initialization;Boolean expression;increment) {statement block};*
- *while (Boolean expression) {statement block};*
- *do {statement block} while (Boolean expression);*

The following code of Table 5 highlights some other capabilities previously pointed out.

Arrays, Constant Insertion, and String Management

An array is a list of related variables with a common name and defined by the following syntax: *type-name variable-variable [size]*. Hereafter, an array

Table 4.

```
#include <stdio.h>
int factorial(int n) {
int f = 1, t;
t=n;
while (t > 1) {
f = f * t;
t--;}
return f;}
int combin(int n, int k) {
return factorial(n)/(factorial(k)*factorial(n-k));}
int main(void) {
int n, k;
printf("n, k? ");fflush(stdin);scanf("%d %d",&n,&k);
printf("C(n,k)=%d\n",combin(n,k));}
```

```
D:\>11.exe
n, k? 10 3
C(n,k) = 120
```

Table 5.

```
#include <stdio.h>
float x, y, z;
char op;
int OpXY(){
switch (op){
 case '+': z=x+y; break;
 case '-': z=x-y; break;
 case '*': z=x*y; break;
 case '/': z=x/y; break;
 default: return 0;}
return 1;}
int main(void) {
printf("Write two numbers:\n");
fflush(stdin);scanf("%f %f", &x, &y);
printf("Which operation (+,-,*,/)?\n");
fflush(stdin);scanf("%c", &op);
while (OpXY()) {
 printf("Result: %f\n\n",z);
 printf("Which operation (+,-,*,/)?\n");
 fflush(stdin);scanf("%c", &op);}
printf("Invalid operation\n");}
```

```
D:\>11.exe
Write two numbers:
23 4
Which operation (+,-,*,/)?
/
Result: 5.750000

Which operation (+,-,*,/)?
q
Invalid operation

D:\>
```

element is accessed by its index (in C, zero is the first one). Another point of interest concerns content insertion for a given position. For instance, the following statement, *scanf("%d", &account[9])*, allows one to read an integer from the keyboard (by default) for the tenth position of the *account* array. In addition, a string is defined as a character array ending with the null character (*'\0'*). For string management operations, the C library also provides various functions defined under the *string.h* library:

- *strcpy(destination, source)*;
- *strcat(destination, source)*;
- *strcmp(string1, string2)*;
- *strlen(string1)*;

Table 6.

#include <stdio.h> *int main(void){* *char str[20];* *int i;* *for (i=1; i<=20; i++) putchar('*');* *putchar('\n');* *printf("String input (less than 20 characters):\n");* *gets(str);* *for(i=0;str[i]!='\0';i++) printf("%c",str[i]);}*
D:\>11.exe ******************** String input (less than 20 characters): Like a rolling stone Like a rolling stone D:\>_
#include <string.h> *int main(void) {* *char str1[80], str2[80];* *int i;* *printf("First string: ");fflush(stdin);gets(str1);* *printf("Second string: ");fflush(stdin);gets(str2);* *printf("The length of string %s is %d characters\n",str1,strlen(str1));* *printf("The length of string %s is %d characters\n",str2,strlen(str2));* *i=strcmp(str1,str2);* *if (i==0) printf("Both strings are equal. \n");* *else* *if(i<0) printf("Alphabetically, %s ranks greater than %s \n",str1,str2);* *else printf("Alphabetically, %s ranks less than %s \n",str1,str2);* *if (strlen(str1)+strlen(str2)<80) {* *strcat(str1,str2);* *printf("%s\n", str1); }* *strcpy(str1,str2);* *printf("%s %s\n",str1,str2);}*
D:\>11.exe First string: roma Second string: amor The length of string roma is 4 characters The length of string amor is 4 characters Alphabetically, roma ranks less than amor romaamor amor amor

Pointers

A pointer is a variable that contains the memory address of an object such as a float or a structure. For example, a pointer to an integer is *int *p;* Briefly, the C language contains two special functions associated with pointers, * and *&*. The latter returns the address of a variable while the former yields the value stored at the particular address specified by the pointer. It is also possible to add or subtract values from a pointer, i.e., the statement **p++;* involves an increment of the address pointed out by p. Yet, *(*p)++;* increments the internal value pointed out by p. The following two small programs stress some of the pointer's capabilities (see Table 7).

Transfer of Parameters and Arguments

The transfer of parameters to any function can be achieved in two ways: By value or by reference. The first method copies the value of the argument to the formal parameter of the function. If the transfer is by reference, the variable address will be supplied instead and thus any changed value of the variable inside the function will cause a change that will be reflected throughout the whole program (see Table 8).

On the other hand, the transfer of arguments from the command line to a C program is accomplished through the optional parameters *argc* and *argv*. If the first one represents the total number of arguments, the second one denotes the string that holds all the arguments of the command line. Thus, *argv[0]* contains the name of the program, *argv[1]* saves the second parameter and so on. Two closely related functions are the *atoi()* and the *atof()* which convert text to integers and text to floats, respectively (see Table 9).

Streams

Within C code, a stream is an interface to other devices such as a printer, disk or DVD. In its simplest form, a stream is a logical interface with a file, allowing the dual flow of data regardless of the device. There are basically two types of streams, text and binary. A text stream is composed of ASCII characters whereas a binary stream can be used with any kind of data, including binary. Briefly, the basic functions that allow file manipulation using streams are as follows:

- *FILE *fopen(char *filename, char *mode);*
- *int fclose(FILE *fp);*
- *int fgetc(FILE *fp);*
- *int fputc(int ch, FILE *fp);*
- *int feof(FILE *fp);*
- *int ferror(FILE *fp);*

Appendix 2

Table 7.

```
#include <stdio.h>
void Calc(float *v int num,float *xmin, float *xmax){
 int i;
 *xmin=*xmax=*v;
 for (i=0;i<num;i++){
 if (v[i]<*xmin) *xmin=v[i];
 if (v[i]>*xmax) *xmax=v[i];}
}
int main(void){
float Vector[]={10,20,30,40,50,11,12,5,-33}, Maior, Menor;
Calc(Vector,9,&Menor, &Maior);
printf("Maximum: %f\n Minimum %f\n",Maior,Menor);}
```

```
D:\>11.exe
Maximum: 50.000000
 Minimum -33.000000
```

```
#include <stdio.h>
int main(void){
  char s[1];
  char *ptr = s;
  printf("\tWrite a string: ");
  fflush(stdin);gets(s);
  if (*ptr=='\0') return 0;
  while (*ptr!='\0') putchar(*ptr++);
  ptr--;
while (ptr>=s) putchar(*ptr--);}
```

```
D:\>11.exe
        Write a string: myGeoffice: Live Free or Die
myGeoffice: Live Free or DieeiD ro eerF eviL :eciffoeGym
D:\>
```

Table 8.

```
#include <stdio.h>
void AddOne(int x, int*y){
 x=x+1;
*y=(*y)+1;
printf("AddOne function: %d %d\n",x,*y);}
int main(void){
 int a=0,b=0;
 AddOne(a,&b);
 printf("Main function: %d %d\n",a,b);}
```

```
D:\>11.exe
AddOne function: 1 1
Main function: 0 1

D:\>_
```

Table 9.

```
#include <stdio.h>
#include <stdlib.h>
int main(int argc, char *argv[]) {
int i,total;
for (i=1,total=0;i<argc;i++)
 total+=atoi(argv[i]);
printf("%d\n",total);}
```

```
D:\>11.exe 12 24 36 48 60
180

D:\>_
```

- *int fputs(char *str, FILE *fp);*
- *char *fgets(char *str, int a, FILE *fp);*
- *int fprintf(FILE *fp, char *string control, ...);*
- *int fscanf(FILE *fp, char *string control, ...);*
- *int fseek (FILE *fp, long offset, int origin);*
- *long ftell (FILE *fp);*
- *int remove (char *filename);*
- *void rewind (FILE *fp);*

Structures

A structure is a cluster of related fields through a common name where field access is achieved using the dot operator. To understand the following code of Table 11, three comments should be made previously:

- Two basic structures are defined: date(day,month,year) and name(first,last)
- Based on a previously defined structure, the employee one has two singular fields (age and salary) and two structure fields (id and hired)..
- The local variable f is an employee structure.

Appendix 2

Table 10.

#include <stdio.h> *#include <stdlib.h>* *#define MAX_STR 120* *int main(int argc, char *argv[]) {* *FILE * fp;* *char s[MAX_STR+1];* *int i=0;* *if (argc!=2) exit(1);* *if ((fp=fopen(argv[1],"r"))==NULL){* *fprintf(stderr,"Impossible to open file %s\n\n",argv[1]);* *exit(1);}* *while (fgets(s,MAX_STR+1,fp)!=NULL) printf("%2d: %s",++i,s);* *fclose(fp);* *exit(0);}*
D:\>11.exe file1.txt 1: Pink Floyd 2: Doors 3: Cat Stevens 4: U2 D:\>11.exe music.txt Impossible to open file music.txt D:\>
#include <stdio.h> *int main(void){* *int i, k;* *char* city[4]={"Brasilia","Salvador","Manaus","Recife"};* *float temp[4]={22.5, 21.8, 2.2, 14},tmp;* *FILE* f;* *f=fopen("exp.txt","w");* *for (i=0; i<4; i++)* *k=fprintf(f,"%s %f\n",cidade[i],temp[i]);* *fclose(f);* *printf("New file created\n");* *f=fopen("exp.txt","r");* *while (fscanf(f,"%s %f",city,&tmp) != EOF)* *printf("%s %f\n",city,tmp);* *fclose(f);}*
D:\>11.exe New file created Brasilia 22.500000 Salvador 21.799999 Manaus 2.200000 Recife 14.000000 D:\>

Table 11.

```
#include <stdio.h>
struct date {
int day, month, year;};
struct name{
char first[20];
char last[3][15];};
struct employee{
struct name id;
int age;
struct date hired;
float salary;};
int main(void) {
struct employee f={{"Pedro","Domingos","Paulo","Jorge"},45,{14,3,23},270};
printf("%s %s %s %s\n",f.id.first,f.id.last[0],f.id.last[1],f.id.last[2]);
printf("Idade: %2d\n",f.age);
printf("%2d/%2d/%2d\n",f.hired.day,f.hired.month,f.hired.year);
printf("Salary: %6.1f\n",f.salary); }
```

```
D:\>11.exe
 Pedro Domingos Paulo Jorge
Idade: 45
 14/ 3/2003
Salary: 27000.0

D:\>
```

Unions

A union is a unique memory location used by several different types of variables. According to the following program, the *ascii* variable can be interpreted both as an integer value (*ascii.ch*) or as a byte structure (allowing bit access). As expected, this is only possible as the memory location is the same (see Table 12).

Enumerations

Another further possibility is the defining of enumerations, a way of allocating names to integer values. A classic example of an enumeration type is the definition of the weekdays where the value for each day begins with 0 for Monday, 1 for Tuesday and so on through 6 for Sunday (see Table 13).

Appendix 2

Table 12.

```
#include <stdio.h>
struct byte{
int a:1;
int b:1;
int c:1;
int d:1;
int e:1;
int f:1;
int g:1;
int h:1;};
union bits{
int ch;
struct byte bit;
} ascii;
void decode(union bits b){
if(b.bit.h) printf("1");
else printf("0");
if(b.bit.g) printf("1");
else printf("0");
if(b.bit.f) printf("1");
else printf("0");
if(b.bit.e) printf("1");
else printf("0");
if(b.bit.d) printf("1");
else printf("0");
if(b.bit.c) printf("1");
else printf("0");
if(b.bit.b) printf("1");
else printf("0");
if(b.bit.a) printf("1");
else printf("0");}
int main(void){
printf("\n Press q to end this program\n");
do {
printf("\n Click on a character of your keyboard: ");
fflush(stdin);ascii.ch = getchar();
printf(" %c -> ",ascii.ch);
decode(ascii);
} while (ascii.ch!='q');}
```

```
D:\>11.exe

Press q to end this program

Click on a character of your keyboard: a
a -> 01100001
Click on a character of your keyboard: b
b -> 01100010
Click on a character of your keyboard: y
y -> 01111001
Click on a character of your keyboard: z
z -> 01111010
Click on a character of your keyboard: 3
3 -> 00110011
Click on a character of your keyboard: ;
; -> 00111011
Click on a character of your keyboard:
  -> 00100000
Click on a character of your keyboard: q
q -> 01110001
D:\>_
```

Table 13.

```
#include <stdio.h>
enum days {Monday, Tuesday, Wednesday, Thursday, Friday, Saturday, Sunday};
int main(void){
register int a,b;
a = Monday;
b = Tuesday;
b++;
if (b==Wednesday)
printf("Today is %d",b);}
```

```
D:\>11.exe
Today is 2
D:\>
```

?, Typedef, and Bitwise Operators

The existence of the *?* ternary operator allows the replacement of the conditional expression, *if (condition) var=exp1; else var=exp2;* with the more simple *var=(condition)? exp1:exp2;* The *typedef* keyword allows the creation of new data type names. For instance, the declaration *typedef float abc;* makes the name *abc* a synonym of *float*. Another useful feature of C language is its ability to test or assign values at bit mode by using *&* (to mask off some set of bits) and | (to turn bits on) operators, respectively, as the following code illustrates (see Table 14).

The C language uses the stack concept, a linear list in which all insertions and deletions are made at the end of the list (LIFO - Last Input First Output). This means that the last element to be inserted (push) will be the first element to be removed (pop). Typically, this structure is used to implement recursive functions and explains some of the considerations involved with pointers and arrays of structures (see Table 15).

Table 14.

```
#include <stdio.h>
int main(void){
unsigned char ch;
puts("Press a character to test if the fourth bit is 0 or 1:");
ch = getchar();
if (ch & 8) puts("\nFourth bit: ON");
else puts("\nFourth bit: OFF");}
```

```
D:\>11.exe
 Press a character to test if the fourth bit is 0 or 1:
a

Fourth bit: OFF

D:\>11.exe
 Press a character to test if the fourth bit is 0 or 1:
z

Fourth bit: ON

D:\>
```

Table 15.

```
#include <stdio.h>
#include <string.h>
#include <stdlib.h>
struct student {
char name[50];
int grade;
struct student* next;
} *first;
int pop(void) {
struct student *p;
if (first != NULL){
 p = first;
 first = first->next;
 printf("\n Withdrawn element: %s %d",p->name,p->grade);
 free(p);
}
else
 printf("\n Empty stack!!!"); }
int push(void) {
struct student *new1;
char s[50];
new1 = (struct student *) malloc(sizeof(struct student));
(*new1).next=NULL;
printf("\nStudent name: ");fflush(stdin);gets(new1->name);
printf("\nStudent grade: ");fflush(stdin);scanf("%d",&new1->grade);
if (first == NULL)
first = new1;
else {
(*new1).next = first;
first = new1;}
}
int list(void) {
struct student* p;
p = first;
while (p != NULL){
printf("\n %s %d",p->name,p->grade);
p=(*p).next; }
}
int main(void) {
char op;
while (1) {
printf("\n\tStack management: i(nsert), d(elete), l(ist), e(xit)\n");
printf("\n\t>> Which operation? ");
fflush(stdin);scanf(" %c", &op);
switch (op){
 case 'i': push(); break;
 case 'd': pop(); break;
 case 'l': list(); break;
 case 'e': exit(0);
 default: printf("Operation not defined!\n");}
}
}
```

continued on following page

Table 15 Continued

```
D:\>11.exe
        Stack management: i(nsert), d(elete), l(ist), e(xit)

        >> Which operation? l

        Stack management: i(nsert), d(elete), l(ist), e(xit)

        >> Which operation? i

Student name: Neena

Student grade: 87

        Stack management: i(nsert), d(elete), l(ist), e(xit)

        >> Which operation? i

Student name: Tjay

Student grade: 77

        Stack management: i(nsert), d(elete), l(ist), e(xit)

        >> Which operation? l
Tjay 77
Neena 87
        Stack management: i(nsert), d(elete), l(ist), e(xit)

        >> Which operation? d

Withdrawn element: Tjay 77
        Stack management: i(nsert), d(elete), l(ist), e(xit)

        >> Which operation? l

Neena 87
        Stack management: i(nsert), d(elete), l(ist), e(xit)

        >> Which operation? e

D:\>
```

Appendix 3

Figure 1.

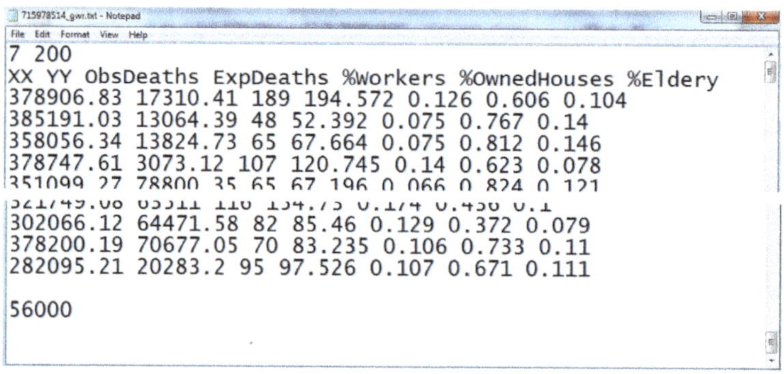
7 200
XX YY ObsDeaths ExpDeaths %Workers %OwnedHouses %Eldery
378906.83 17310.41 189 194.572 0.126 0.606 0.104
385191.03 13064.39 48 52.392 0.075 0.767 0.14
358056.34 13824.73 65 67.664 0.075 0.812 0.146
378747.61 3073.12 107 120.745 0.14 0.623 0.078
351099.27 78800 35 65 67.196 0.066 0.824 0.121
321749.08 65311 110 134.73 0.174 0.456 0.1
302066.12 64471.58 82 85.46 0.129 0.372 0.079
378200.19 70677.05 70 83.235 0.106 0.733 0.11
282095.21 20283.2 95 97.526 0.107 0.671 0.111

56000

GEOGRAPHICALLY WEIGHTED REGRESSION C CODE

```
#include <stdio.h>
#include <stdlib.h>
#include <string.h>
#include <math.h>
/* Pre-definition of functions */
double determinant(double[][],int);
void cofactor(double[][],int);
void transpose(double[][],double[][],int);
/* Global variable */
double minverse[5][5];
int main(int argc, char *argv[]) {
FILE *input_file;
FILE *output_file;
int i,j,k,mvariables,nrecords,numsamples,par;
double z[2],x[2][7],y[2][5],yt[5][2],w[2]
[2],distance,kernel,beta[5];
double aux, aux1,result[5][2], result1[5]
```

```
[5],d,result2[5],result3[5], estimation;
double RSS, RSSponto, AICc;
char str1[2], str2[2], str3[2], str4[2], str5[2], str6[2],
str7[2], strfile[2];
if (argc!=2) {
 printf("\n There are no two arguments to be processed. Bear in
mind that this executable (gwr.exe) must be run with a single
text input file as an argument. \n");
 exit(1);}
if ((input_file = fopen(argv[1],"r"))==NULL){
 printf("\n\n It is not possible to read the input file (gwr.
exe). \n");
 exit(2);}
if ((input_file = fopen(argv[1],"r"))==NULL){
 printf("\n\n It is not possible to read the input file (gwr.
exe). \n");
 exit(3);}
fscanf(input_file,"%d %d\n",&mvariables,&nrecords);
if (mvariables<4 || mvariables>7 || nrecords<10 || nrecords>2)
{
 printf("\n Bad value for the number of variables (below 4 or
above 7, for instance) or number of records (below 10 or above
2, for instance) of the text input file (gwr.exe).\n");
 exit(4);}
str1[0]='\0';str2[0]='\0';str3[0]='\0';str4[0]='\0';str5[0]='\0
';str6[0]='\0';str7[0]='\0';strfile[0]='\0';
if (mvariables==4){
 fscanf(input_file,"%s %s %s %s",str1,str2,str3,str4);
 printf("\n Dependent variable: %s; First Independent variable:
%s\n",str3, str4);}
 else if (mvariables==5){
fscanf(input_file,"%s %s %s %s %s",str1,str2,str3,str4,str5);
printf("\n Dependent variable: %s; First independent variable:
%s; Second independent variable: %s\n", str3, str4,str5);}
 else if (mvariables==6){
fscanf(input_file,"%s %s %s %s %s %s",str1,str2,str3,str4,str5,
str6);
printf("\n Dependent variable: %s; First independent variable:
%s; Second independent variable: %s; Third independent
variable: %s\n",str3, str4,str5,str6);}
 else {
fscanf(input_file,"%s %s %s %s %s %s %s",str1,str2,str3,str4,st
r5,str6,str7);
printf("\n Dependent variable: %s; First independent variable:
%s; Second independent variable: %s; Third independent
variable: %s; Fourth independent variable: %s
\n",str3,str4,str5,str6,str7);}
/* Variables Initialization */
for (i=0;i<2;i++){
```

```
 x[i][0]=0;x[i][1]=0;x[i][2]=0;x[i][3]=0;x[i][4]=0;x[i]
[5]=0;x[i][6]=0;
 z[i]=0;y[i][0]=1;y[i][1]=0;y[i][2]=0;y[i][3]=0;y[i][4]=0;yt[0]
[i]=1;yt[1][i]=0;yt[2][i]=0;yt[3][i]=0; yt[4][i]=0;}
for (i=0;i<nrecords;i++){
 if (mvariables==4){
 par=2;
 fscanf(input_file,"%lf %lf %lf %lf\n",&x[i][0],&x[i][1],&x[i]
[2],&x[i][3]);}
 else if (mvariables==5){
 par=3;
 fscanf(input_file,"%lf %lf %lf %lf %lf\n",&x[i][0],&x[i]
[1],&x[i][2],&x[i][3],&x[i][4]);}
 else if (mvariables==6){
 par=4;
 fscanf(input_file,"%lf %lf %lf %lf %lf %lf\n",&x[i][0],&x[i]
[1],&x[i][2],&x[i][3],&x[i][4],&x[i][5]);}
 else {
 par=5;
 fscanf(input_file,"%lf %lf %lf %lf %lf %lf %lf\n",&x[i]
[0],&x[i][1],&x[i][2],&x[i][3],&x[i][4],&x[i][5], &x[i][6]);}
 z[i]=x[i][2];y[i][1]=x[i][3];y[i][2]=x[i][4];y[i][3]=x[i]
[5];y[i][4]=x[i][6];yt[1][i]=x[i][3];
 yt[2][i]=x[i][4];yt[3][i]=x[i][5];yt[4][i]=x[i][6];}
fscanf(input_file,"%lf\n",&kernel);
fclose(input_file);
printf("The Kernel Bandwidth (KB) of the GWR Gaussian function,
W(x)=EXP(-0.5(Distance/KB)^2, equals %lf\n",kernel);
RSS=0;
strcpy(strfile,"gwroutput.txt");
output_file=fopen(strfile,"w");
for (numsamples=0;numsamples<nrecords;numsamples++){
 /* Initialization of W[2][2] matrix */
 for (i=0;i<2;i++)
 for (j=0;j<2;j++) w[i][j]=0;
 /* Initialization of minverse[5][5] matrix */
 for (i=0;i<5;i++)
 for (j=0;j<5;j++) minverse[i][j]=0;
 /* Initialization of beta[5] matrix */
 for (i=0;i<5;i++) beta[i]=0;
 /* Diagonal W[2][2] matrix */
 for (j=0;j<nrecords;j++){
 distance=sqrt((x[j][0]-x[numsamples][0])*(x[j]
[0]-x[numsamples][0])+(x[j][1]-x[numsamples][1])* (x[j]
[1]-x[numsamples][1]));
 w[j][j]=exp(-0.5*(distance/kernel)*(distance/kernel));}
 /* yt[5][2] multiply by W[2][2] giving Result[5][2] matrix
including initialization */
 for (i=0;i<5;i++)
```

```
 for (j=0;j<2;j++) result[i][j]=0;
 for (i=0;i<5;i++)
 for (j=0;j<2;j++){
 aux=0;
 for(k=0;k<2;k++) aux=aux+yt[i][k]*w[k][j];
 result[i][j]=aux;}
 /* Result[5][2] multiply by y[2][5] giving Result1[5][5]
matrix including initialization */
 for (i=0;i<5;i++)
 for (j=0;j<5;j++) result1[i][j]=0;
 for (i=0;i<5;i++){
 for (j=0;j<5;j++){
 aux=0;
 for(k=0;k<2;k++) aux=aux+result[i][k]*y[k][j];
 result1[i][j]=aux;}
 }
 /* The result1[i][i]=1 if i > mvariables-3 in order to assess
its inverse */
 for (i=0;i<5;i++)
 if (i>mvariables-3) result1[i][i]=1;
 /* Assessment of the inverse of Result1[5][5] matrix by
generating minverse[5][5]*/
 d=determinant(result1,5);
 if (d==0){
 printf("\n The inverse of the present Result1[5][5] matrix is
not possible since it determinant is zero. Quite often, the
kernel bandwidth setup is quite small leading to a situation
where no neighbors samples can be found for the internal
computation of the W matrix (gwr.exe).\n");
 fclose(output_file);
 exit(5);}
 else
 cofactor(result1,5);
 /* Result[5][2] multiply by z[2] giving Result2[5] matrix
including initialization */
 for (i=0;i<5;i++) result2[i]=0;
 for (i=0;i<5;i++){
 aux=0;
 for(k=0;k<2;k++) aux=aux+result[i][k]*z[k];
 result2[i]=aux;}
 /* minverse[5][5] multiply by result2[5] giving Result3[5]
matrix including initialization */
 for (i=0;i<5;i++) result3[i]=0;
 for (i=0;i<5;i++)
 result3[i]=minverse[i][0]*result2[0]+minverse[i]
[1]*result2[1]+minverse[i][2]*result2[2]+minverse[i][3]*
result2[3]+minverse[i][4]*result2[4];
 estimation=0;
 printf("\n GEOGRAPHICALLY WEIGHTED REGRESSION [Sample %d at
```

```
(%lf,%lf) location with a %lf
value]\n",numsamples+1,x[numsamples][0], x[numsamples]
[1],x[numsamples][2]);
 if (mvariables==4) printf("%s: %lf\n",str4,x[numsamples][3]);
 else if (mvariables==5)
 printf("%s: %lf ; %s: %lf\n",str4,x[numsamples]
[3],str5,x[numsamples][4]);
 else if (mvariables==6)
 printf("%s: %lf; %s: %lf; %s: %lf\n",str4,x[numsamples]
[3],str5,x[numsamples][4],str6, x[numsamples][5]);
 else
 printf("%s: %lf; %s: %lf; %s: %lf; %s:
%lf\n",str4,x[numsamples][3],str5,x[numsamples][4],str6,
x[numsamples][5],str7,x[numsamples][6]);
 for (i=0;i<5;i++){
 printf("BETA%d: %lf",i,result3[i]);
 if (i==4) printf("\n");}
estimation=result3[0]+result3[1]*x[numsamples]
[3]+result3[2]*x[numsamples][4]+result3[3]*x[numsamples]
[5]+result3[4]*x[numsamples][6];
 RSSponto=0;
 for (j=0;j<nrecords;j++)
 RSSponto=RSSponto+w[j][j]*(x[numsamples]
[2]-estimation)*(x[numsamples][2]-estimation);
 RSS=RSS+RSSponto;
 printf("GWR estimation: %lf; Residual (real-estimation): %lf;
Local weighted RSS: %lf\n",estimation, x[numsamples]
[2]-estimation,RSSponto);
 printf("\n");
 if (numsamples+1!=nrecords)
 fprintf(output_file,"%d %lf %lf %lf %lf %lf %lf %lf %lf %lf
%lf\n",numsamples+1,x[numsamples][0], x[numsamples]
[1],x[numsamples][2],estimation,x[numsamples][2]-estimation,res
ult3[0],result3[1],result3[2], result3[3],result3[4]);
 else
 fprintf(output_file, "%d %lf %lf %lf %lf %lf %lf %lf %lf %lf
%lf",numsamples+1,x[numsamples][0], x[numsamples]
[1],x[numsamples][2],estimation,x[numsamples][2]-estimation,res
ult3[0],result3[1],result3[2], result3[3],result3[4]);}
if ((nrecords-par-1)!=0) {
 AICc=nrecords*log(RSS)+2*par+(2*par*(par+1))/(nrecords-par-1);
 printf("\n Total Residual (Weighted) Sum of Squares (RSS):
%lf",RSS);
 printf("\n Corrected Akaike Information Criteria,
AICc=N*LOG(RSS)+2*K+(2*K*(K+1))/(N-K-1), where N and K equals
%d and %d respectively: %lf",nrecords,par,AICc);}
else{
 AICc=nrecords*log(RSS)+2*par;
 printf("\n Total Residual (Weighted) Sum of Squares (RSS):
```

```
%lf",RSS);
 printf("\n Akaike Information Criteria, AIC=N*LOG(RSS)+2*K,
where N and K equals %d and %d respectively:
%lf",nrecords,par,AICc);}
fclose(output_file);
exit(0);}
double determinant(double a[5][5], int k){
double s=1,det=0,b[5][5];
int i,j,m,n,c;
if (k==1){return (a[0][0]);}
else{
 det=0;
 for (c=0;c<k;c++){
 m=0; n=0;
 for (i=0;i<k;i++){
 for (j=0;j<k;j++){
 b[i][j]=0;
 if (i != 0 && j != c){
 b[m][n]=a[i][j];
 if (n<(k-2)) n++;
 else{n=0; m++;}
 }
 }
 }
 det=det+s*(a[0][c]*determinant(b,k-1));
 s=-1*s;}
 }
return (det);}
void cofactor(double num[5][5],int f){
double b[5][5],fac[5][5];
int p,q,m,n,i,j;
for (q=0;q<f;q++){
 for (p=0;p<f;p++){
 m=0; n=0;
 for (i=0;i<f;i++){
 for (j=0;j<f;j++){
 if (i != q && j != p){
 b[m][n]=num[i][j];
 if (n<(f-2)) n++;
 else{n=0;m++;}
 }
 }
 }
 fac[q][p]=pow(-1,q+p)*determinant(b,f-1);
 }
}
transpose(num,fac,f);}
/*Finding transpose of matrix*/
void transpose(double num[5][5],double fac[5][5],int r){
```

```
int i,j;
double b[5][5],d;
for (i=0;i<r;i++){
 for (j=0;j<r;j++){b[i][j]=fac[j][i];}
}
d=determinant(num,r);
for (i=0;i<r;i++){
 for (j=0;j<r;j++){minverse[i][j]=b[i][j]/d;}
}
}
```

Appendix 4

JAVA PLATFORM HIGHLIGHTS

Introduction

In 1990, Sun Microsystems® commenced a project called Green in order to develop electronic software. At the same time, James Gosling began to write software in C++ to be added to VCRs and PDAs. However, he soon realized that C++ was the wrong tool for the job. Although C++ is flexible enough to create embedded applications, it is very prone to bugs, which may block the system. Gosling's solution was a new language called Oak. By preserving the familiar syntax of C++, the potentially dangerous references to operating system resources was omitted. For instance, Oak incorporated the memory manager directly into the language, enabling the programmer to concentrate only on the application to be developed. As Oak matured, the World Wide Web had a dramatic period of growth and the Sun® team realized that Oak was the perfect tool for the Internet. Finally, in 1995 Oak was renamed as Java for commercial reasons and was officially announced at SunWorld 95 in San Francisco, USA.

For this appendix, the free Java development tool chosen was Bluej® (www.bluej.org) but others can be downloaded in the marketplace such as Eclipse® or NetBeans®. Regardless of the user's level of knowledge and compiler, one should distinguish the following concepts:

- **JavaScript:** Programming language to be interpreted at the client-side (at the expense of the *<script></script>* tag of HTML files).
- **Java Applet:** Java applications compiled for the client and referenced by the *<applet></applet>* tag. For security reasons, Applets cannot read or write files on the client or run any remote program. Furthermore,

this technology is not currently supported by Chrome®, Edge® and Opera®.

- **Java Application:** Object oriented programming (OOP) language like C#.
- **JavaServer Pages:** Like ASP.Net® and PHP® (server-side technologies), this is a technique to create documents that combine static HTML and Java expressions embedded in tags so as to generate dynamic content.
- **JDBC:** Java DataBase Connectivity is a set of classes that can be used to develop client-server database applications. For reference, these drivers can be found at www.oracle.com/technetwork/java/javase/jdbc/index.html.

As an OOP language, classes (abstractions of the real world), objects (class materialization or realization created with the *new* operator), local variables of objects, methods and static global variables, constants, polymorphism, inheritance, sub-classes, overhead (methods with the same name but with different arguments), garbage collector (automatic management of RAM), messages (data exchanged between objects), overriding of methods, exceptions (error handling), interfaces, packages (set of classes and interfaces to be imported into new programs), encapsulation (process that prevents class variables to be read or modified by other classes) and class libraries (*java.applet*, *java.awt*, *java.beans*, *java.io*, *java.lang*, *java.math*, *java.rmi*, *java.sql*, *java.security*, *java.swing*, *java.util*, *java.text*...) are intrinsic concepts to this programming platform. Java Applets, for example, organize classes in a hierarchical manner, i.e., one can construct new classes from an existing one. In fact, an Applet created by someone is a sub-class of the *java.applet.Applet* class.

Applets

Another central feature is the inclusion of the Applets code in the HTML 4 version. The tag that allows this insertion is *<applet>...</applet>*. For HTML 5 files, *<object>...</object>* should be used instead, as the next illustration shows (see section 2.3 of this book on how to run Applets in your browser to avoid security issues).

When one Applet is executed, several internal events take place in a well-defined sequence. Java uses the optional *init()* method to perform the initialization operation, followed by the *start()* method. The *stop()* method is invoked whenever the user leaves the Web page. As one would expect, many

Figure 1.

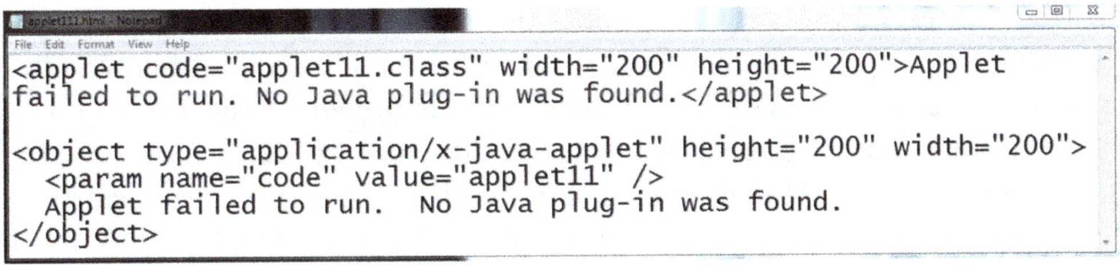

```
<applet code="applet11.class" width="200" height="200">Applet
failed to run. No Java plug-in was found.</applet>

<object type="application/x-java-applet" height="200" width="200">
  <param name="code" value="applet11" />
  Applet failed to run.  No Java plug-in was found.
</object>
```

Figure 2.

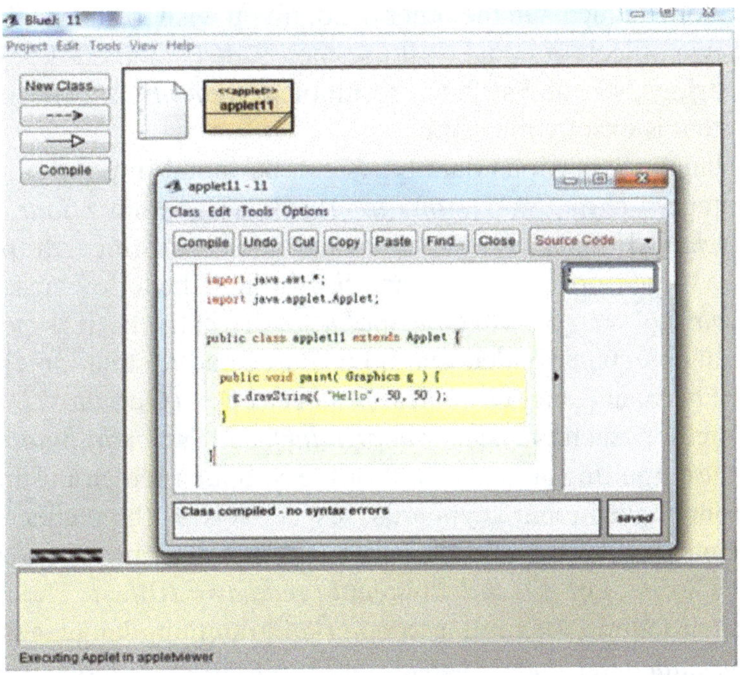

interactions occur such as size change of the window browser or overlapping it with another window. These events are controlled by the *paint()* and *repaint()* methods. At the end, when the user closes the Web page, the *destroy()* method is responsible to release all the resources used by the Applet itself.

Programming Rules and Basics

Some good programming rules are as follows: (A) Each variable must hold a unique name; (B) Choose names that identify the meaning of the variables; (C) All statements end with a semicolon; (D) To create a string, *String message=new String();* is a possibility; (E) Another flexible solution to handle an array of characters is the *StringBuffer* class because it presents appealing methods for its management such as *append()* and *insert()*; (F) The declaration of an integer array with a dimension of ten would be *int[] numbers=new int[10]* (analogous to C#, the first index of the array starts at zero); (G) The generation of a random number between 0 and 1 is achieved by means of the statement *java.lang.Math.random();* (H) The *public static void main(String[] args)* method is always public in order to be called by the Java interpreter. (I) Concatenation of two strings can be carried out with the + operator, for instance. On the other hand, if you wish to check whether the content of two objects is equal, just use the *equals()* method. For example, given two strings, str1 and str2, the result of *str1.equals(str2)* becomes true if their content is exactly the same.

Like C language, various data types can be found in a program (class) such as integers (*byte*, *short*, *int*, *long*), floating point (*float*, *double*), characters (*char*), Boolean, arrays, tables and classes. Casting is the process of converting a value from one type to another through the following statement: *new_variable type=(type) variable_value;*. Alternatively, it is possible to cast between objects, provided that both source and destination classes are related by inheritance, e.g. *Graphics2D screen2D = (Graphics2D) screen;*

Global methods such as *System.out.println();* and *System.in.readln();* allow by default the outputting of final results on the computer screen and the reading of input from conventional keyboards, respectively. If the reader wishes to detect the Enter key, type the following condition statement: *if((char)System.in.read()=='\n'){...}* or *if ((char)System.in.read()==10){...}*. The following conditional structures are similar to other programming languages, namely *if(condition)...then...else* and *switch(condition){case:...}*, loop structures as *while(condition){...}*, *for(initial value;stop condition;increment){...}* and *do{...} while(condition)*. The same rules apply to the *break* and *continue* statements.

Java ranks the variables according to two types, primitive or built-in (*float*, *Boolean*, *int*, *char*...) and non-primitive (objects and arrays). Java endorses the first type to a function on a passage by value (the original variable always remains unchangeable, a situation not followed by the second option). Another remarkable structure is the *try{...}catch{...}finally{...}*. This allows detection and control of program errors and exception handling procedures. The throws option makes it possible to specify the type of exception (*NumberFormatException*, *EOFException*, *InterruptedException*,

Table 1.

Operator	Examples	Operator meaning
=	x = y+z;	Simple assignment
+=	x += y;	x=x+y Addition
-=	x -= y;	x=x-y Subtraction
/=	x /= y;	x=x/y Division
*=	x *= y;	x=x*y Multiplication
%	x %/ y;	x=x%y Division remainder
^=	x ^= y;	x=x^y XOR
&=	x &= y;	x=x&y AND
|=	x |= y;	x=x|y OR
<<=	x <<= y;	x=x<<y Shift on the left
>>=	x >>=y;	x=x>>y Shift on the right
>>>=	x >>>=y;	x=x>>>y Fill in with a zero the shift on the right.

IOException...). Let us imagine that the reader wants to open a file that does not exist or tries to perform a division by zero. It is up to the programmer to decide what should be done when the program detects either situation.

Another qualifier that can be applied to methods is access restriction by following three reserved words: *public*, *private* and *protected. Public* allows the method in question to be available to all other classes of the program. The *private* keyword makes a method accessible only within the class where it is defined while *protected* makes it accessible only to its class and subclasses. Further, Java allows two types of comments. The first type is to precede each line with two slashes (//). If you want a contiguous block, you should begin the text with /* and end it with */. Note that all character literals are delimited by single quotes whereas strings are enclosed by double quotes.

The layout of any Applet may be carried out by *drawString("Text", nline, ncolumn)*. Similar to other programming languages, the coordinate's origin for the Cartesian axes is on the upper left corner. The keyboard event control is another possibility through the AWT (Abstract Window Toolkit) class. Another common background color option is the use of *setBackground(Color. green)*, for example.

Any interactivity necessarily involves control over the mouse. Java presents, among others, the following methods to achieve this goal: *mouseEntered(Event e)*, *mouseExited(Event e)*, *mousePressed (Event e)*, *mouseReleased(Event e)*, *mouseClicked(Event e)*, *mouseMoved(Event e)* and *mouse Dragged(Event e)*. Each one is activated by the respective mouse action that captures the coordinates of the current cursor position. To receive the coordinates, the programmer may employ the *e.getX()* and *e.getY()* methods.

The construction of a standardized GUI can be achieved through the following components: *Button(string)*, *TextArea(Nlin,ncol)*, *Scrollbar()*,

Figure 3.

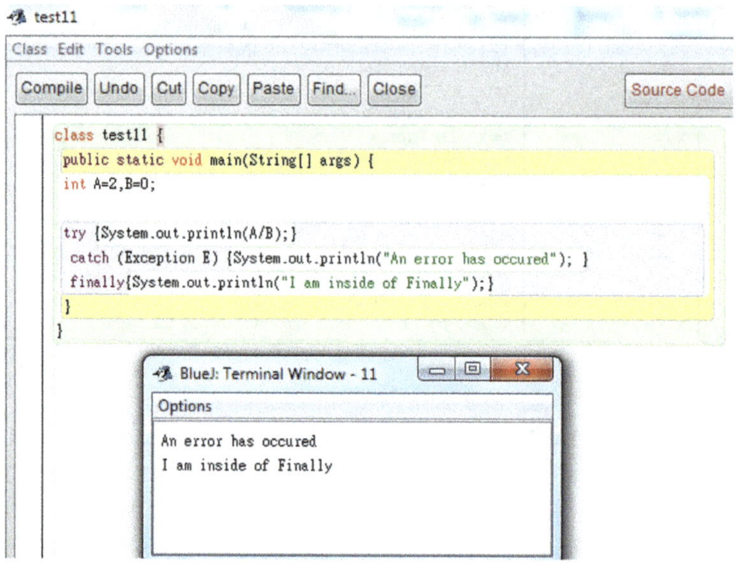

Figure 4.

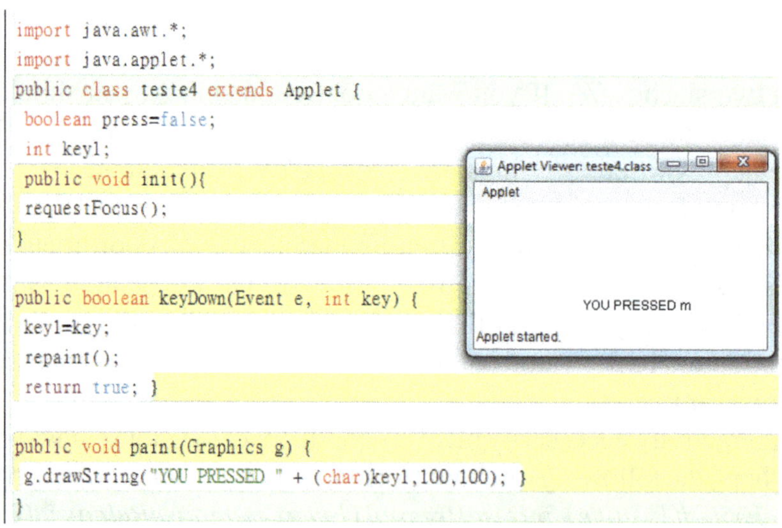

TextField(width), *Layout()* and *Label(string, alignment)*, where alignment can take values as *Label.LEFT*, *Label.RIGHT* or *Label.CENTER*.

The drawing of a line involves the use of the *drawLine(int x1,int y1, int x2, int y2)* method. As an extension, a few more graphics methods available

Appendix 4

Table 2.

```
// Peter McBride (1997), Java Introduction, Editorial Presença (Portuguese)
import java.applet.*;
import java.awt.*;
public class applet11 extends Applet{
Checkbox fontBold=new Checkbox ("Bold",false);
Checkbox fontItalic=new Checkbox ("Italic",false);
int negro=0, italico=0;
CheckboxGroup fontNome =new CheckboxGroup();
Checkbox sans= new Checkbox("Sans Serif", fontNome,true);
Checkbox serif= new Checkbox("Serif", fontNome,false);
Checkbox mono= new Checkbox("Monospaced", fontNome,false);
String fName=new String("Sans Serif");
public void init(){
add(fontBold); add(fontItalic); add(sans); add(serif); add(mono); }
public boolean action(Event evt,Object arg){
if (fontBold.getState()==true) negro=1; else negro=0;
if (fontItalic.getState()==true) italico=2; else italico =0;
if (fontNome.getSelectedCheckbox()==sans) fName="Sans Serif";
else if (fontNome.getSelectedCheckbox()==serif) fName="Serif";
else fName="Monospaced";
repaint();
return true;}
public void paint(Graphics g){
g.setFont(new Font(fName,negro+italico,18));
g.drawString("This type of letter is "+fName,30,50); }
}
```

Figure 5.

```
import java.applet.*;
import java.awt.*;
public class rgb extends Applet {
 public void paint(Graphics g){
  for(int i=1;i<=8;i++){
    Color color = new Color(i*50-1,0,0);
    g.setColor(color);
    g.fillRect(i*40+2,2,28,28); }
 }
}
```

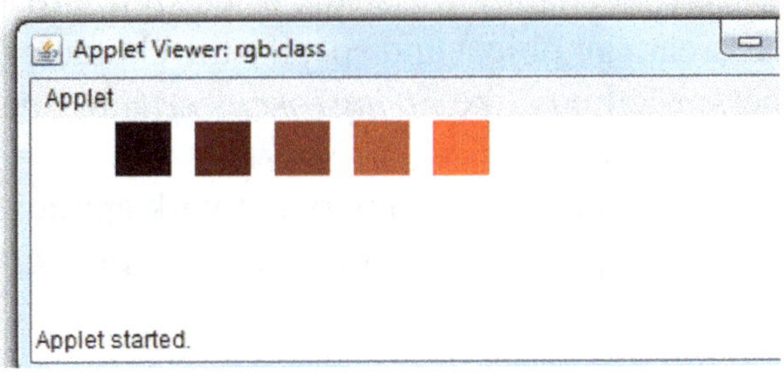

on this platform are presented next: *drawRoundRect(int x, int y, int width, int height, int arclar, int arcalt); drawOval(int x, int y, int width, int height); drawRect(int x, int y, int width, int height); drawArc(int x, int y, int width, int height, int startangle, int arcangle); drawPolygon(int xpoints[], int ypoints[], int npoints); fillRect(int x, int y, int width, int height); fillRoundRect(int x, int y, int width, int height, int arclar, int arcalt); fillOval(int x, int y, int width, int height); fillArc(int x, int y, int width, int height, int startangle, int arcangle); fillPolygon(int xpoints[], int ypoints[], int npoints); fill3DRect(int x, int y, int x1, int y1, true);*

The days of Web pages including only text have come to an end. Currently, any page has GIFs and JPEG images. Java provides the *Image getImage(URL url)* option where the URL parameter points to any image file. The *drawImage()* is responsible for presenting it on the canvas. Inserting sounds on a Web page is also feasible through the *AudioClip* object and the sound management is controlled by the following: *play(); loop();* and *stop();*

The use of *threads* (objects that implement the Runnable interface, allowing concurrent execution with other parts of the class) is a central concept with animation. By definition, a process is an independent entity that is identified by its unique PID (Process Identification) and runs simultaneously with other programs, i.e., both thread and process share a similar purpose in performing a task without causing an excessive allocation of resources. The simplest way to include a thread is to activate the Runnable interface in the developed class by using the *start()* and *run()* methods..

Text Data

In Java, all text data is written using file streams via the *RandomAccessFile()* class. However, other possibilities exist: (A) *FileInputStream* and *FileOutputStream* (B) *DataInputStream* and *DataOutputStream* (C) *BufferedInputStream* and *BufferedOutputStream* (D) *FileReader* and *FileWriter*. Regardless of the option, the library *java.util* defines the *StringTokenizer* class that enables the string to be split into tokens. In order to do so, one must create an object and specify the character that to be used to split its members. Methods like *StringTokenizer()* and *nextToken()* have this functionality.

For the reader's reference, the running of network applications is another option for this platform (http, telnet and finger, for example) through the Socket

Appendix 4

Figure 6.

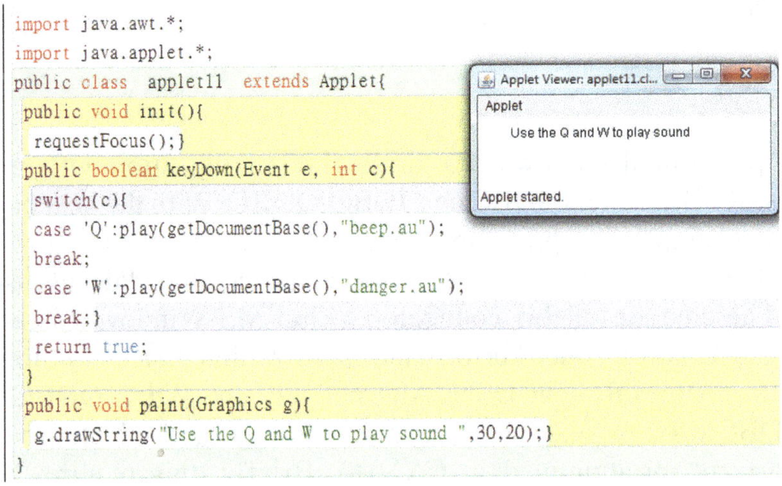

Table 3.

```
import java.applet.*;
import java.io.*;
import java.util.*;
import java.awt.*;
public class applet11 extends Applet{
  String data; String stock_nome; float qtd,qtd_depois,preco; String linhaTexto;
  public void paint(Graphics g){
  try{
  int nlinha=0;
  long filesize1;
  RandomAccessFile file;
  file= new RandomAccessFile("quote.dat","r");
  filesize1 = file.length();
  resize(3,3);
  while((linhaTexto=file.readLine())!=null){
   StringTokenizer st;
   st=new StringTokenizer(linhaTexto);
   g.drawString(linhaTexto,5,25+nlinha);
   nlinha+=15;
   stock_nome = st.nextToken();
   data=st.nextToken();
   qtd = Float.valueOf(st.nextToken()).floatValue();
   qtd_depois=Float.valueOf(st.nextToken()).floatValue();
   preco=Float.valueOf(st.nextToken()).floatValue();
   System.out.println("Line Content: "+linhaTexto);
   System.out.println("Stock name :"+stock_nome);
   System.out.println("Date : "+data);
   System.out.println("Quantity : "+qtd);
   System.out.println("Quantity after: "+qtd_depois);
   System.out.println("Price : "+preco); }
   file.close(); }
  catch(IOException e){ System.out.println("File error: "+ e); }
  }
}
```

class (similar to the Unix standard) - *Socket conn=new Socket(HostName, PortNumber)*.

Databases

The Java platform also allows two forms of communication with relational databases (RDB), namely, JDBC and JDBC-ODBC. With the next particular Windows® example, the presented code will allow access to MS-Access® RDB via JDBC-ODBC (www.microsoft.com/enus/download/details.aspx?id=13255). The creation of this connector within MS-Windows® is achieved through the Control Panel/Performance and Maintenance/Administrative Tools/Data Sources/System DSN. Alternatively, the JDBC-ODBC Bridge allows JDBC drivers (made with Java technology) and which are used as ODBC drivers (programmed in C/C++). Briefly, this requires requires three procedures: (A) Inclusion of the driver *sun.jdbc.odbc.JdbcOdbcDriver* (requires prior integration with Java 2); (B) Existence of an ODBC driver for the respective RBD; (C) Setting up of the DSN system of the ODBC connection to the RBD in use.

Following the above program, the first task is to load the driver (*package java.lang*). Naturally, the *ClassNotFoundException* can be added to test exceptions and system errors. The connection establishment is accomplished

Table 4.

```
import java.sql.Connection;
import java.sql.DriverManager;
import java.sql.ResultSet;
import java.sql.SQLException;
import java.sql.Statement;
public class MDBAccess {
public static void main(String[] arguments) {
String data = "jdbc:odbc:CandiesODBC";
try {
 Class.forName("sun.jdbc.odbc.JdbcOdbcDriver");
 Connection conn=DriverManager.getConnection(data, "","");
 Statement st = conn.createStatement();
 ResultSet rec = st.executeQuery("SELECT * FROM TCandies");
 System.out.println("CandiesID\t\tName\t\tPrice");
 while(rec.next()) {
 System.out.println(rec.getInt(1)+"\t\t"+rec.getString(2) + "\t\t" + rec.getDouble(3));}
 st.close();
}catch (SQLException s) {
 System.out.println("SQL Error: " + s.toString() + " " + s.getErrorCode() + " " + s.getSQLState());}
 catch (Exception e) {
 System.out.println("Error: " + e.toString() + e.getMessage()); }
}
```

by the *DriveManager* (*java.sql* package) and the *getConnection(string1,string 2,string3)* methods. The SQL query is represented by the *executeQuery(string)* whose outcome is returned to the *ResultSet*. To obtain the content of the record fields, the programmer can use one of the following methods: *getDate(field)*, *getDouble(field)*, *getFloat(field)*, *getString(field)* or *getInt(field)*. Obviously, the choice depends on the original data type.

It should be stressed again, that the purpose of this Appendix is not to present a deep insight into the Java platform, an impossibility given the size constraints of this appendix. Rather, the purpose is to outline only several important topics to understand the code created under chapter 3 of this book.

Appendix 5

BASIC STATISTIC FORMULAS

Table 1.

Population Parameters
Population mean = $\mu = \Sigma X_i/N$
Population standard deviation = σ = sqrt[$\Sigma(X_i-\mu)^2/N$]
Population variance = $\sigma^2 = \Sigma(X_i-\mu)^2/N$
Standardized score = $Z = (X-\mu)/\sigma$
Sample Statistics (assuming random sampling)
Sample mean = $\overline{x} = \Sigma x_i/n$
Sample standard deviation = s = sqrt[$\Sigma(x_i-\overline{x})^2/(n-1)$]
Sample variance = $s^2 = \Sigma(x_i-\overline{x})^2/(n-1)$
Correlation Between Two Variables
Pearson product-moment correlation = r = $\Sigma(xy)$/sqrt[$(\Sigma x^2)(\Sigma y^2)$]
Simple Linear Regression
Simple linear regression line: $\hat{y} = b_0+b_1x$
Regression coefficient = $b_1 = \Sigma[(x_i-\overline{x})(y_i-\overline{y})]/\Sigma[(x_i-\overline{x})^2]$
Regression slope intercept = $b_0 = \overline{y} - b_1\overline{x}$
Standard error of regression slope = s_{b1} = sqrt[$\Sigma(y_i-\hat{y}_i)^2/(n-2)$]/sqrt[$\Sigma(x_i-\overline{x})^2$]
Counting
n factorial: n! = n×(n-1)×(n-2)×...×3×2×1 (by convention, 0! = 1).
Permutations of n things, taken r at a time: ${}_nP_r = n!/(n-r)!$
Combinations of n things, taken r at a time: ${}_nC_r = n!/r!(n-r)! = {}_nP_r/r!$
Probability
Rule of addition: $P(A\cup B) = P(A)+P(B)-P(A\cap B)$
Rule of multiplication: $P(A\cap B) = P(A)\times P(B\vert A)$
Rule of subtraction: $P(A') = 1-P(A)$
Random Variables
Expected value of X = E(X) = $\mu(x) = \Sigma[x_iP(x_i)]$
Variance of X = Var(X) = $\sigma^2 = \Sigma[x_i-E(x)]^2P(x_i) = \Sigma[x_i-\mu_x]^2P(x_i)$
Normal random variable = z-score = $z = (X-\mu)/\sigma$
Chi-square statistic = $X^2 = [(n-1)s^2]/\sigma^2$
f statistic = $f = [s_1^2/\sigma_1^2]/[s_2^2/\sigma_2^2]$
Standard Error
Standard error of the mean = SE $\overline{x}$ = s $\overline{x}$ = s/sqrt(n)

continued on following page

Appendix 5

Table 1. Continued

Discrete Probability Distributions
Binomial formula: $P(X=x) = b(x;n,P) = {}_{n}C_{x} \times P^{x} \times (1-P)^{n-x}$
Mean of binomial distribution $= \mu_x = n \times P$
Variance of binomial distribution $= \sigma_x^2 = n \times P \times (1-P)$
Poisson formula: $P(x;\mu) = (e^{-\mu})(\mu^{x})/x!$
Mean of Poisson distribution $= \mu_x = \mu$
Variance of Poisson distribution $= \sigma_x^2 = \mu$
Confidance Interval Estimation
Confidence interval: Sample statistic$\pm$Critical value$\times$Standard error of statistic
Confidence interval for $\mu = \bar{x} \pm z \times \sigma/\text{sqrt}(n)$
Confidence interval for $\bar{x} = \bar{x} \pm t \times s_x/\text{sqrt}(n)$
Confidence interval for two-sample t-statistic $= (\bar{x}_1 - \bar{x}_2) \pm t \times \text{sqrt}(s_1^2/n_1 + s_2^2/n_2)$
Confidence interval for linear regression slope $= b1 \pm t \times S_{b1}$
Hypothesis Testing
One sample z-test for means: z-score $= (\bar{x} - \mu)/[\sigma/\text{sqrt}(n)]$
One-sample t-test for means: t-score $= t = (\bar{x} - \mu)/[s_x/\text{sqrt}(n)]$
Two-sample t-test for means: t-score $= t = (\bar{x}_1 - \bar{x}_2)/\text{sqrt}(s_1^2/n_1 + s_2^2/n_2)$
Chi-square test statistic $= X^2 = \Sigma[(\text{Observed-Expected})^2/\text{Expected}]$
Degrees of Freedom
One-sample t-test: DF = n-1
Simple linear regression, test slope: DF = n-2
Chi-square goodness of fit test: DF = k-1
Chi-square test for independence and homogeneity: DF = (r-1)×(c-1)

Appendix 6

Figure 1.

IMAGE ENHANCEMENT AND DESCRIPTIVE STATISTICS WITH A LINEAR TRANSFORMATION HISTOGRAM

```
import java.io.*;
public class LinearEnhancement {
public static void main (String[] args) throws IOException {
String nome_arq_ent="nemo5-270nc.raw";
String nome_arq_sai="nemo5-270realce.raw";
String nome_arq_res="Statistics.txt";
FileInputStream arq_ent;
FileOutputStream arq_sai;
FileWriter arq_res;
int nlin=270, ncol =5;
arq_ent = new FileInputStream(nome_arq_ent);
arq_sai = new FileOutputStream(nome_arq_sai);
arq_res = new FileWriter(nome_arq_res);
double media = (double)0;
for(int l=0; l<nlin; l++)
 for(int c=0; c<ncol; c++)
 media+= (double)arq_ent.read();
media/=(double)(nlin*ncol);
```

```
arq_ent.close();

arq_ent = new FileInputStream(nome_arq_ent);
double variancia = (double)0;
for(int l=0; l<nlin; l++)
 for(int c=0; c<ncol; c++)
 variancia+= (double)((arq_ent.read()-media)*(arq_ent.read()-
media));
variancia/=(double)(nlin*ncol);
arq_ent.close();

arq_ent = new FileInputStream(nome_arq_ent);
int histo[]= new int[256];
for(int i=0;i<256;i++) histo[i]=0;
 for(int l=0; l<nlin; l++)
 for(int c=0; c<ncol; c++)
 histo[arq_ent.read()]++;
arq_res.write("Statistics of the image RAW file: "+nome_arq_
ent);
arq_res.write("\n\nNumber of columns = "+ncol);
arq_res.write("\nNumber of lines = "+nlin);
arq_res.write("\nTotal of pixels = "+nlin*ncol);
arq_res.write("\nAverage value = "+media+" ~ "+Math.
round(media));
arq_res.write("\nVariance = "+variancia);
arq_res.write("\nStandard deviation = "+Math.sqrt(variancia));
arq_res.write("\nHistogram ");
int max = 0;
for(int i=0;i<256;i++){
 arq_res.write("\nFreq["+i+"] = "+histo[i]);
 if(histo[i]>histo[max])max = i; }
arq_res.write("\n\nMode = "+max);
arq_res.write("\n\nEnd of report");
arq_ent.close();

arq_ent = new FileInputStream(nome_arq_ent);
int xmin = 255; int xmax = 0; int aux, freq_min=50;
for(int l=0; l<nlin; l++)
 for(int c=0; c<ncol; c++)
 if(histo[aux = arq_ent.read()]> freq_min){
 if(aux < xmin) xmin = aux;
 if(aux > xmax) xmax = aux; }
System.out.println("Xmin = "+xmin+" Xmax = "+xmax);
arq_ent.close();
arq_ent = new FileInputStream(nome_arq_ent);
double a = 255.0/(double)(xmax-xmin);
double b = -a*(double)xmin;
System.out.println("a = "+a+" b = "+b);
int y;
```

```
for(int l=0; l<nlin; l++)
 for(int c=0; c<ncol; c++){
 y = (int)(a*(double)arq_ent.read()+b);
 if (y<0) y=0; if(y>255)y=255;
 arq_sai.write((byte)y); }
arq_ent.close();
arq_sai.close();
arq_res.close();
}
}
```

Partial Output Saved on Statistics.txt File

Statistics of the image RAW file: nemo5-270nc.raw
Number of columns = 5
Number of lines = 270
Total of pixels = 1350
Average value = 104.09327407407407 ~ 104
Variance = 6063.305651859719
Standard deviation = 77.8672309245662
Histogram
Freq[0] = 0
Freq[1] = 0
Freq[2] = 0
Freq[3] = 0
Freq[4] = 0
Freq[5] = 0
Freq[6] = 0
Freq[7] = 0
Freq[8] = 0
Freq[9] = 2
Freq[10] = 0
Freq[11] = 2
Freq[12] = 5
Freq[13] = 12
Freq[14] = 19
...
Freq[237] = 18
Freq[238] = 13
Freq[239] = 14

Freq[240] = 10
Freq[241] = 1
Freq[242] = 5
Freq[243] = 2
Freq[244] = 2
Freq[245] = 0
Freq[246] = 1
Freq[247] = 0
Freq[248] = 0
Freq[249] = 0
Freq[250] = 0
Freq[251] = 0
Freq[252] = 0
Freq[253] = 0
Freq[254] = 0
Freq[255] = 0
Mode = 101
End of report

Figure 2.

Appendix 7

THE RUN LENGTH ENCODING IMAGE COMPRESSION

Compression is a computational technique applied to a digital image (and other files) to reduce the size of the original file. Compression allows increased data transmission speed as the amount of information transmitted is reduced. Two factors are essential when analyzing compression - the compression rate and the speed of this processing. The compression rate equals the ratio between the final size of the compressed image and the original one. Clearly, the smaller the compressed image becomes, the greater the compression rate. The processing speed is vital since it adds "overhead" time when opening the image.

In a generic multimedia database which stores and searches for digital pictures, one must install compression and decompression procedures of the original images. If these actions are time-consuming, these procedures may not be justified in applications that require fast processing. Simultaneously, compression allows increased data transmission speed since the amount of information to be transmitted decreases.

Image compression procedures can be classified into two types, lossless (procedures without loss and thus reversible) and lossy (irreversible). The former enables the recovery of the original image using a decoder applied to the compressed image. This type of technique can be classified into two categories: dictionary (RLE (Run-length encoding) and LZW (Lempel–Ziv–Welch)) and statistics (Shannon Huffman & Gilbert, 1952). Although, it is beyond the scope of this appendix to review all compression procedures, the Run Length Encoding (RLE) and Huffman methods will be briefly examined.

Figure 1 illustrates the use of the lossless RLE algorithm. As one can quickly see, the Y information contains the frequency of appearance (from top to bottom and from left to right) of the digital X image (1 times the 15 digit, 2 times the 12 digit, 1 times the 10 digit … 7 times the 10 digit). Assuming

Figure 1.

```
      15 12 12 10 12 12 15        01 15 02 12 01 10 02 12 01 15
X=    12 12 10 10 10 12 12     Y= 02 12 03 10 03 12
      12 10 10 10 10 10 12        05 10 01 12
      10 10 10 10 10 10 10        07 10
```

that each digital picture level is encoded with one byte (8 bits), the X image is represented with 28 bytes. After RLE encoding, it is possible to obtain the same information with 22 bytes and without any loss of information. Thus, the input image was reduced by 6 bytes. Unsurprisingly, this type of encoding generates a higher compression index for homogeneous images and can create larger output files than the input image file. Can the reader explain this last statement?

Lossy picture compression procedures enable the original image to be recovered, though with a certain loss of definition and exactness, i.e., it is not possible to restore the original picture. However, the great advantage of this procedure is the ability to achieve higher compression rates.

Consider a X image and assume for instance that the threshold difference between its digital values equals 4, i.e., any digital value can be taken as the base if the threshold difference is less than 4. The following example shows the use of lossy RLE coding by considering a threshold of 2 for X image. The Y output is as follows: 1 times the 15 digit (note that the difference between fifteen and twelve is greater than two, i.e., these numbers cannot be aggregated into the same subset), 5 times the 12 digit (the difference between twelve and ten is two, allowing the algorithm to consider the same value for compression purposes) ... 21 times the 12 digit (see Figure 2).

When comparing this Y vector content to the previous one, a much greater compression ratio is noticeable, i.e., a reduction of 28 bytes to 8 bytes. But, as mentioned previously, this lossy encoding does not allow for reconstruction

Figure 2.

```
      15 12 12 10 12 12 15        01 15 05 12 01 15
X=    12 12 10 10 10 12 12     Y= 21 12
      12 10 10 10 10 10 12
      10 10 10 10 10 10 10
```

of the original X image with the same level of detail, unlike lossless RLE compression.

The following Java code illustrates lossless RLE compression by using the fish722-453.raw input image (Figure 3) and the compressed and uncompressed files (fish722-453.rwc and fish722-453n.raw, respectively). Logically, the size of the compressed file is less than the original uncompressed file (203Kb against 320Kb).

```
import java.io.*;
public class compressionRLE{
public static void main (String[] args) throws IOException {
 String nome_arq_in="fish722-453.raw";
 String nome_arq_out="fish722-453.rwc";
 FileInputStream arq_in;
 FileOutputStream arq_out;
 int nlin=453, ncol =722;
 arq_in = new FileInputStream(nome_arq_in);
 arq_out = new FileOutputStream(nome_arq_out);
 int cont = 1, valorc =-1, valor, limiar=0, dif;
 valorc = arq_in.read();
 for(int l=0; l<nlin; l++)
for(int c=0; c<ncol; c++){
valor=(int)arq_in.read();
dif = valor-valorc; if(dif<0)dif*=-1;
if((dif > limiar)||(cont==120)){
 arq_out.write(cont);
 arq_out.write(valor);
 cont = 0;}
cont++;
valorc = valor; }
 arq_in.close();
 arq_out.close();
 arq_in = new FileInputStream(nome_arq_out);
 arq_out = new FileOutputStream("fish722-453n.raw");
 for(int l=0; l<nlin; l++)
for(int c=0; c<ncol; c++){
cont = arq_in.read();
valor = arq_in.read();
for(int k =0; k<cont; k++)
 arq_out.write(valor);}
 arq_in.close();
 arq_out.close();
 System.out.println("\nTermino Normal");}
}
```

Appendix 7

Figure 3.

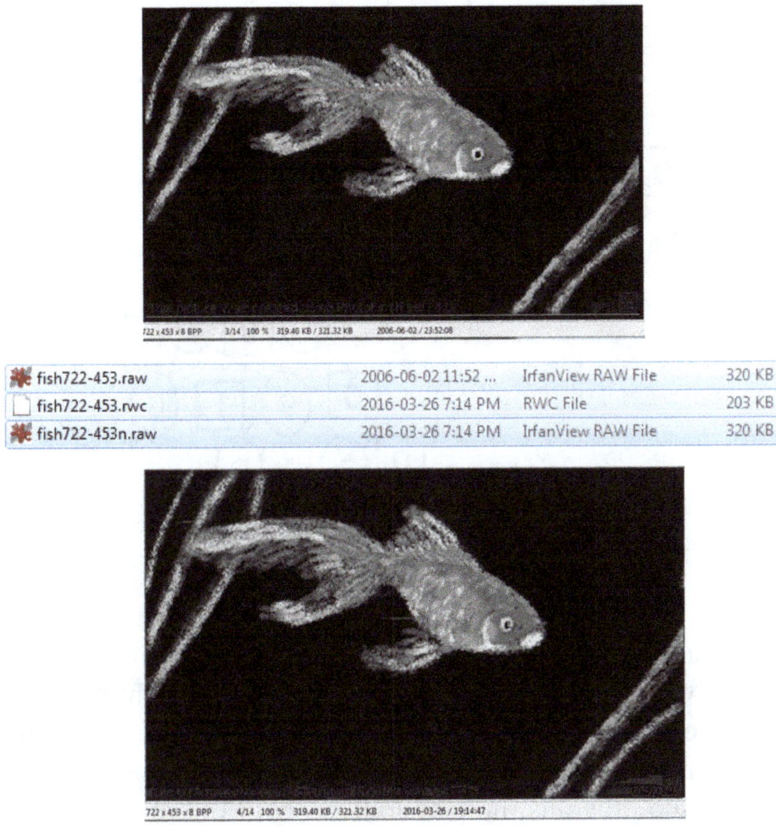

As a complementary perspective for the reader, the Huffman tree is addressed next as an alternative compression method. As a greedy algorithm, Huffman's methodology is one of the most elegant and oldest forms of data compression, particularly for text files. Based on the concept of minimum redundancy, the idea is to encode the available symbols according to their frequency order (highest to lowest frequency). Imagine you have a text with the following one hundred characters whose frequencies are as follows: A –> 20; B –> 27; C –> 11; D –> 7; E –> 20; F –> 13; G –> 1; H -> 1. The construction of the Huffman tree is based on the next six steps:

1. According to their probability of occurrence, sort all existing characters in ascending order.

Figure 4.

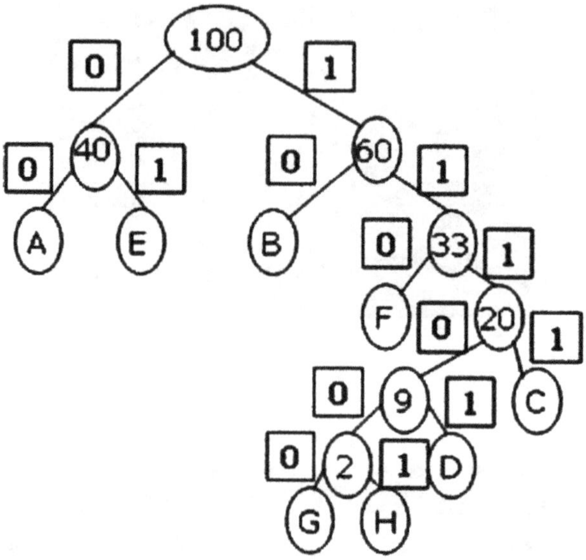

2. In constructing the inverted tree, observe the following rule: place the character with the lowest probability and, in the right branch, put the character or string of characters with the next lowest probability.
3. Add both probabilities and create a super junction node.
4. Carry out the previous three operations until the root of the Huffman tree reaches a probability of 1.
5. Encode it into binary terms according to the following rule: 0 to the left; 1 otherwise (or vice versa).
6. Read each combination from top to bottom and encode each character accordingly.

The above image shows the Huffman binary tree, consistent with the defined probabilities. Its construction deserves the following comments:

1. **G(1) and H(1):** Since both characters hold equal probabilities of occurrence, it would be possible to reverse their codification in the binary tree.
2. **GH(2) and D(7):** Elements with the lowest occurrence are put together.
3. **GHD(9) and C(11):** As above.
4. **F(13) and GHDC(20):** As above.

5. **B(27) and FGHDC(33):** Note that the B character has a lower probability than the FGHDC string and thus remains on the left.
6. **A(20) and E(20):** Again, both characters have a lower probability than the BFGHDC (60) string, leading this sub-tree to be set up on the left.
7. **AE(40) and BFGHDC(60):** Binding of the preceding sub-trees.
8. **AEBFGHDC(1):** End of the binary Huffman tree.

The next step is to assign 1s and 0s to the edges that connect the tree's nodes, following this rule: 0 if the parent node connects to a child node on the left side, otherwise 1. The encoding of each character consists of reading all 0s and 1s along the path of the root tree. The coding for the above example is as follows:

- A – 00
- B – 10
- C – 1111
- D - 11101
- E – 01
- F – 110
- G – 1110
- H – 1111

Consider the following text with nine characters, ACDBEHFCG. Since each ASCII character size is one byte (8 bits), the expected length of this message would be 72 bits (9×8). With this compression algorithm, the contents become 1111 11101 10 01 1111 110 111 1110. The new length equals 33 bits, meaning a 45.83% (33/72) compression rate.

Although the use of the Huffman tree algorithm has its advantages, its most negative aspect lies in the compression process itself since it is necessary to read the data twice, i.e., first to build the table of frequencies and, secondly, to compress the original data. Note that the reading of a compressed message can only be performed without ambiguity due to the prefix free property (any sequence of bits is unique to any specific character) of this methodology.

List of Acronyms

AIC: Akaike Information Criterion
AMC: Arithmetic Mean Center
API: Application Program Interface
BLUE: Best Linear Unbiased Estimator
CCDF: Conditional Cumulative Distribution Function
CIS: Contact Image Sensor
CLT: Central Limit Theorem
CYM: Cyan, Yellow, Magenta
DF: Degrees of Freedom
DIP: Digital Image Processing
DTM: Digital Terrain Model
EDA: Exploratory Data Analysis
ESDA: Exploratory Spatial Data Analysis
E-Type: Conditional Expectation Estimate
IIS: Internet Information Server
IK: Indicator Kriging
GLS: Generalized Least Squares
GS-LIB: Geostatistical Software Library
GUI: Graphical User Interface
GWR: Geographically Weighted Regression
IDW: Inverse Distance Weighted
JRE: Java Runtime Environment
KT: Kriging with Trend
LADS: Localized Autocorrelation Diagnostic Statistic
LK: Lognormal Kriging
MA: Moving Average
MIK (Median IK): Median Indicator Kriging
NN: Nearest Neighbor
NST: Normal Score Transformation

OK: Ordinary Kriging
OLS: Ordinary Least Square
OSPF: Open Shortest Path First
PMT: Photo Multiplier Tube
RBF: Radial Basis Function
RF: Random Function
RGB: Red, Green, Blue
RLE: Run Length Encoding
SGS: Sequential Gaussian Simulation
SK: Simple Kriging
SIS: Sequential Indicator Simulation
SSI: Short Sampling Interval
SQRT: Square Root
UK: Universal Kriging
WWW: World Wide Web

Index

H

I

J

L

M

N

O

P

R

S

T

U

V

W

www.ingramcontent.com/pod-product-compliance
Lightning Source LLC
LaVergne TN
LVHW081531300926
849161LV00006B/1387
9781522532705